T0383212

Quantum Optomechanics

Quantum Optomechanics

Warwick P. Bowen
Gerard J. Milburn

CRC Press
Taylor & Francis Group
Boca Raton   London   New York

CRC Press is an imprint of the
Taylor & Francis Group, an **informa** business

CRC Press
Taylor & Francis Group
6000 Broken Sound Parkway NW, Suite 300
Boca Raton, FL 33487-2742

© 2016 by Taylor & Francis Group, LLC
CRC Press is an imprint of Taylor & Francis Group, an Informa business

No claim to original U.S. Government works

Printed on acid-free paper
Version Date: 20151014

International Standard Book Number-13: 978-1-4822-5915-5 (Hardback)

Visit the Taylor & Francis Web site at
http://www.taylorandfrancis.com

and the CRC Press Web site at
http://www.crcpress.com

To E.K.K. and C.A.H.

Contents

Preface   xi

Author biographies   xv

Acknowledgements   xvii

CHAPTER   1 ▪ Quantum harmonic oscillators   1

1.1   QUANTISING THE HARMONIC OSCILLATOR   2

1.2   FLUCTUATIONS AND DISSIPATION IN A QUANTUM
      HARMONIC OSCILLATOR   8

1.3   MODELLING OPEN SYSTEM DYNAMICS VIA QUANTUM
      LANGEVIN EQUATIONS   18

1.4   QUANTUM LANGEVIN EQUATION WITHIN THE
      ROTATING WAVE APPROXIMATION   28

CHAPTER   2 ▪ Radiation pressure interaction   37

2.1   BASIC RADIATION PRESSURE INTERACTION   37

2.2   EFFECTIVE QUANTISATION   39

2.3   DISPERSIVE OPTOMECHANICS   40

2.4   ELECTRO- AND OPTO-MECHANICAL SYSTEMS   43

2.5   MECHANICAL AND OPTICAL DECOHERENCE RATES   45

2.6   DYNAMICS OF DISPERSIVE OPTOMECHANICAL
      SYSTEMS   47

2.7   LINEARISATION OF THE OPTOMECHANICAL
      HAMILTONIAN   51

2.8   DISSIPATIVE OPTOMECHANICS   55

CHAPTER   3 ▪ Linear quantum measurement of mechanical
              motion   57

3.1   FREE-MASS STANDARD QUANTUM LIMIT   58

3.2	RADIATION PRESSURE SHOT NOISE	60
3.3	MEASUREMENT OF MECHANICAL MOTION	64
3.4	STANDARD QUANTUM LIMIT ON MECHANICAL POSITION MEASUREMENT	76
3.5	STANDARD QUANTUM LIMIT FOR GRAVITATIONAL WAVE INTERFEROMETRY	86
3.6	STANDARD QUANTUM LIMIT FOR FORCE MEASUREMENT	88

CHAPTER 4 ■ Coherent interaction between light and mechanics   93

4.1	STRONG COUPLING	94
4.2	OPTICAL COOLING OF MECHANICAL MOTION	96
4.3	OPTOMECHANICALLY INDUCED TRANSPARENCY	114
4.4	OPTOMECHANICAL ENTANGLEMENT	122
4.5	MECHANICAL SQUEEZING OF LIGHT	140

CHAPTER 5 ■ Linear quantum control of mechanical motion   151

5.1	STOCHASTIC MASTER EQUATION INCLUDING DISSIPATION	152
5.2	FEEDBACK COOLING	156
5.3	BACK-ACTION EVADING MEASUREMENT	166
5.4	SURPASSING THE STANDARD QUANTUM LIMIT USING SQUEEZED LIGHT	179

CHAPTER 6 ■ Single-photon optomechanics   193

6.1	OPTOMECHANICAL PHOTON BLOCKADE	194
6.2	SINGLE-PHOTON STATES	196
6.3	SINGLE-PHOTON PULSE INCIDENT ON A SINGLE-SIDED OPTOMECHANICAL CAVITY	202
6.4	DOUBLE-CAVITY OPTOMECHANICAL SYSTEM	206
6.5	MACROSCOPIC SUPERPOSITION STATES USING SINGLE-PHOTON OPTOMECHANICS	219
6.6	SINGLE-SIDEBAND-PHOTON OPTOMECHANICS	230

CHAPTER 7 ■ Nonlinear optomechanics   231

7.1	DUFFING NONLINEARITY	231
7.2	QUANTUM DUFFING OSCILLATOR	233

7.3	NONLINEAR DAMPING	240
7.4	SELF-PULSING AND LIMIT CYCLES	241
7.5	NONLINEAR MEASUREMENT OF A MECHANICAL RESONATOR	249

CHAPTER 8 ▪ Hybrid optomechanical systems — 255

8.1	ELECTROMECHANICAL SYSTEMS	256
8.2	COUPLING A MECHANICAL RESONATOR AND A TWO-LEVEL SYSTEM	261
8.3	MICROWAVE TO OPTICAL INTERFACE.	270

CHAPTER 9 ▪ Arrays of optomechanical systems — 277

9.1	SYNCHRONISATION IN OPTOMECHANICAL ARRAYS	277
9.2	IRREVERSIBLY COUPLED ARRAYS OF OPTOMECHANICAL SYSTEMS	293

CHAPTER 10 ▪ Gravitational quantum physics and optomechanics — 299

10.1	WHAT IS GRAVITATIONAL DECOHERENCE?	300
10.2	OPTOMECHANICAL TESTS OF GRAVITATIONAL DECOHERENCE	308
10.3	TESTS OF NONSTANDARD GRAVITATIONAL EFFECTS USING GEOMETRIC PHASE	315

APPENDIX A ▪ Linear detection of optical fields — 319

A.1	EFFECT OF INEFFICIENCIES	319
A.2	LINEAR DETECTION OF OPTICAL FIELDS	322
A.3	POWER SPECTRAL DENSITY OBTAINED BY HETERODYNE DETECTION	323
A.4	CHARACTERISING THE OPTOMECHANICAL COOPERATIVITY	324
A.5	CHARACTERISING THE TEMPERATURE OF A MECHANICAL OSCILLATOR	326

References	329
Index	355

Preface

Quantum optomechanics has its roots in the study of the mechanical action of light that goes back to the 19th century. More relevant to this century, it is an example of an emerging capability to engineer quantum systems de novo. Engineered quantum systems use modern fabrication techniques to enable the construction of macroscopic systems designed to exhibit novel quantum behaviour in collective degrees of freedom.

The aims of this book are to introduce to the reader some of the modern developments in quantum optomechanics, from both an experimental and a theoretical standpoint. We would hope that the book will prove valuable to physicists and engineers both actively involved in the field and wishing to familiarise themselves with it. We do not seek to provide a full review of the research in the field. The book does, however, cover a range of important topics, including optomechanical cooling and entanglement, quantum limits on measurement precision and how to overcome them via back-action evading measurements, feedback control, single photon and nonlinear optomechanics, optomechanical synchronisation, coupling of optomechanical systems to microwave circuits and to two-level systems such as atoms and superconducting qubits, and optomechanical tests of gravitational decoherence.

The relationship of light to mechanical motion of matter is a long and convoluted story in the Western philosophical tradition. The Greek atomists regarded light as particulate and all atoms could participate in motion. In 55 BCE the Roman epicurean poet Lucretius captured this most elegantly in *De Rerum Natura*: "*The light and heat of the sun; these are composed of minute atoms which, when they are shoved off, lose no time in shooting right across the interspace of air in the direction imparted by the shove*". In opposition to the corporealists, Aristotle, Plato, and the neoplatonists argued that light was incorporeal. How light could interact with matter, for example, in refraction, then required some explanation.

It is often claimed that Kepler gave the first statement of the mechanical action of light in his explanation of why comet tails point away from the sun.

> "*The head is like a conglobulate nebula and somewhat transparent; the train or beard is an effluvium from the head, expelled through the rays of the sun into the opposed zone and in its continued effusion the head is finally exhausted and consumed so that the tail represents the death of the head*".

Indeed, in discussing refraction Kepler viewed the action of light as mechanical, likening it to the action of a missile striking a panel. However, we must be cautious with this interpretation, for we are missing a key point in Kepler's views, which, for modern scientists, is very hard to understand. Kepler was a neoplatonist and denied the corporealist view of light [181]. As a neoplatonist he regarded light as a universal principle of animation and to argue that the light of the sun influenced the tails of comets was of a piece with a view that light influenced all material motion.

In the 19th century the idea that light could have a mechanical action was widely held. Kelvin, in "On the Mechanical Action of Radiant Heat or Light" in 1852, pointed out that heating of material indices by light was "*mechanical effect of the dynamical kind ... merely the excitation or the augmentation of certain motions among its particles*".

The discovery of the laser in the mid-20th century provided a new tool to study and exploit forces from light. Bell Labs scientist, Arthur Ashkin, reported the action of optical gradient forces from a tightly focussed laser field on micron-sized particles in 1970 [19]. This discovery later informed the development of optical tweezers, which has had a big impact in biology. Subsequently the study of light forces on atomic motion got under way, culminating in a series of breakthroughs, including laser cooling and atomic gas Bose–Einstein condensation, and was recognised in three Nobel prize awards to Chu, Phillips, and Cohen-Tannoudji in 1997; to Ketterle, Cornell, and Wieman in 2001 and to Wineland and Haroche in 2012. An emerging quantum technology based on laser cooled atoms enables new kinds of gravity gradiometry [264] and inertial sensing [132].

Ion trapping is perhaps the best example of a quantum optomechanical system [176]. Using laser pulses to control the coupling between internal degrees of freedom and the mechanical motion, ions may be cooled to the ground state of collective motion from which highly nonclassical quantum states can be produced. Ion traps are the most advanced technology for implementing quantum information processing. However, ion trapping differs from the optomechanical systems discussed in this book, which are concerned with the quantum control of bulk elastic degrees of freedom involving supra-atomic scales.

Engineered quantum systems represent a new paradigm for the study of quantum physics with significant outcomes for applications. The quantum theory originated in the study of natural systems: atoms, molecules, solids and light. Despite the difficulty in reconciling quantum theory with our classical intuitions, it is a remarkably successful theory. It is often claimed that one should not be alarmed by this, as it is unreasonable to expect our classical intuitions to apply in such an unfamiliar domain as atomic physics. However, quantum theory does not contain within it any law that forbids us from applying it to bigger and bigger things, even the entire universe in the case of quantum cosmology. The quantum–classical border is not co-located with the microscopic–macroscopic border.

Until recently the possibility of quantum phenomena manifesting themselves in the ordinary everyday world of macroscopic physics could be safely confined to gedanken experiments. The last decade or so has changed this. We now see a rapidly developing capability to engineer supra-atomic systems designed to exhibit quantum features in collective macroscopic variables. Quantum optomechanics and superconducting quantum circuits (also discussed in this book) are examples.

A key feature of the theory of engineered quantum systems is how the quantum description is given. One does not solve the Schrödinger equation for every atomic or molecular constituent of the macroscopic system. On the contrary, one begins with a classical description of the relevant macroscopic degrees of freedom — elastic deformations in the case of mechanics and classical current and flux in the case of quantum circuits — and quantises these collective degrees of freedom directly. This works if the macroscopic system can be so designed that the relevant collective degrees of freedom largely decouple from the microscopic degrees of freedom, which remain only as a source of noise and dissipation.

This approach was first pioneered in the field of superconductor circuits and championed by Tony Leggett in the 1980s. He pointed out that the quantum theory of such circuits could begin with a direct quantisation of the classical circuit equations. This approach was vindicated by the landmark experiment of Martinis, Devoret and Clarke in 1987 [194]. In quantum optomechanics one often begins with the classical continuum mechanics of an elastic solid, for example, a toroidal shaped piece of silica or the collective excitations of a fluid, as in the case of superfluid helium. This continuum field is then quantised directly. Essentially this is an "effective quantum field theory."

Engineered quantum systems hold the promise of new technologies based on designing complex systems to exhibit functional quantum behaviour. In the case of optomechanics, this might include a gain in sensitivity to external forces and fields that arises when these influence quantum interference phenomena. An intriguing possibility arises when the mechanical system becomes sufficiently large that gravitational interactions need to be included. This holds the promise of better determinations of the Newtonian gravitational constant and even the possibility of experimental evidence for how quantum theory and gravity might be reconciled (see Chapter 10).

Structure of the book

The book begins in Chapter 1 with the basic physics of quantum harmonic oscillators and how they interact with their environment. This chapter is targeted towards readers who are unfamiliar with the physics of open quantum systems, and aims to provide a basic understanding, useful tools that are applicable to quantum optomechanics, and an appreciation of how quantum forcing from the environment dictates both the dissipation and fluctuations experienced by quantum harmonic oscillators – an important concept for processes such as optomechanical cooling. In Chapter 2 we introduce the radiation pressure interaction between light and matter, along with the basic Hamiltonians used in quantum optomechanics. The semi-classical dynamics of a quantum optomechanical system are solved in the steady state leading to the concept of optomechanical bistability.

Chapters 3 to 5 focus on the linearised regime of quantum optomechanics, where the effect of a single photon on the optomechanical system is negligible and instead a bright coherent optical driving field is used. We begin in Chapter 3 with the radiation pressure noise that is imparted on a mechanical oscillator due to momentum kicks from light and the precision limits this introduces on measurements of mechanical motion. In Chapter 4 radiation pressure–mediated coherent interactions between light within an optical cavity and a mechanical oscillator are introduced, including strong coupling where energy is coherently exchanged between the light and the mechanical oscillator, optomechanical cooling and entanglement, optomechanically induced transparency, where a coherent optical driving introduces a sharp transparency window for the light, and ponderomotive squeezing, where the optomechanical system acts like an effective optical Kerr nonlinearity for the light and thereby squeezes it. Chapter 5 deals with linear measurement-based control of the mechanical system, introducing feedback cooling, as well as quantum measurement techniques such as back-action evading measurements, which allow the radiation pressure noise introduced in Chapter 3 to be avoided.

In Chapters 6 and 7 we consider scenarios where the simple linearised picture of quantum optomechanics no longer holds. Chapter 6 introduces the physics of quantum optomechanics in the single photon strong coupling regime, where a single photon is sufficient to substantially affect the system dynamics, for instance, blocking the absorption of subsequent photons by the cavity or allowing the generation of macroscopic quantum superposition states of the mechanical oscillator. Chapter 7 deals with situations where an additional nonlinearity is introduced to the optomechanical system over and above the usual nonlinearity of the radiation pressure interaction. This includes intrinsic mechanical nonlinearities such as the Duffing nonlinearity and nonlinear damping, and nonlinearities introduced by engineering the optomechanical interaction.

Chapter 8 introduces hybrid optomechanical systems, where the canonical quantum optomechanical system is coupled to another quantum object such as a superconducting circuit, microwave field, or two-level system. Chapter 9 considers an alternative form of hybrid optomechanical system, where multiple optomechanical systems interact between themselves. This leads to the phenomenon of synchronisation where, through their interaction with light, multiple mechanical oscillators lock into phase with each other. Finally, in Chapter 10 we consider the possible ramification of quantum optomechanics for tests of gravitational physics.

Author biographies

Warwick Bowen earned a PhD in experimental physics from the Australian National University in 2004 for work on continuous variable quantum information systems. He is currently an associate professor in physics and Australian Research Council Future Fellow at the University of Queensland. He manages the Quantum Opto- and Nano-Mechanics Program of the Australian Research Council Centre of Excellence for Engineered Quantum Systems. His current research interests include quantum optomechanics, precision metrology and sensing, and biological applications of quantum measurement.

Gerard Milburn earned a PhD in theoretical physics from the University of Waikato in 1982 for work on squeezed states of light and quantum nondemolition measurements. He is currently professor of physics at the University of Queensland and director of the Australian Research Council Centre of Excellence for Engineered Quantum Systems. His current research interests include quantum optomechanics, superconducting quantum circuits, and quantum control.

Acknowledgements

The authors have benefited from a great many interactions with colleagues in the fields of quantum optics and control, quantum optomechanics, and precision measurement over the years. While we have the deepest gratitude, we cannot — and therefore do not attempt to — acknowledge all of you. You know who you are, and should know that this book would not be what it is without you. We take this opportunity to acknowledge the people who specifically helped in the process of writing — from wrestling with the communication of difficult concepts, to many hours of proofreading and feedback. We are deeply indebted to Markus Aspelmeyer, Sahar Basiri-Esfahani, James Bennett, Andrew Doherty, Adil Gangat, Kaerin Gardner, Simon Haine, Glen Harris, Catherine Holmes, Chitanya Joshi, Dvir Kafr, Nir Kampel, Florian Marquart, David McAuslan, Casey Myers, Cindy Regal, Stewart Szigeti, Jake Taylor, and John Teufel.

CHAPTER 1

Quantum harmonic oscillators

CONTENTS

1.1   Quantising the harmonic oscillator   2
     1.1.1   Ladder operators   3
     1.1.2   Thermal equilibrium statistics   4
     1.1.3   Zero-point motion and the uncertainty principle ...   5
     1.1.4   Dimensionless position and momentum operators ..   6
     1.1.5   Phasor diagrams   6
     1.1.6   Unitary dynamics in the Heisenberg picture   7
1.2   Fluctuations and dissipation in a quantum harmonic
     oscillator   8
     1.2.1   Classical fluctuation–dissipation theorem   9
     1.2.2   Quantum forcing of a harmonic oscillator   10
         1.2.2.1   Upwards going transition probability ..   10
         1.2.2.2   Quantum power spectral density   12
         1.2.2.3   Transition rates between energy levels .   13
     1.2.3   Relationship between $S_{FF}(\Omega)$ and $S_{FF}(-\Omega)$ in
         thermal equilibrium   14
     1.2.4   Quantum fluctuation–dissipation theorem   15
1.3   Modelling open system dynamics via quantum Langevin
     equations   18
     1.3.1   Quantum Langevin equation for a particle in a
         potential   19
         1.3.1.1   Quantum Langevin equation for a
             general system operator   22
         1.3.1.2   Markovian limit   22
     1.3.2   Harmonic oscillator dynamics   23
         1.3.2.1   Susceptibility of an oscillator   25
     1.3.3   Determining the temperature of an oscillator via
         sideband asymmetry   26

    1.3.4   Determining the effective oscillator temperature
           from the integral of $\bar{S}_{QQ}(\omega)$ 26

1.4   Quantum Langevin equation within the rotating wave
    approximation 28
    1.4.1   Commutation and correlation relations 29
    1.4.2   Power spectral densities and uncertainty relations
           of the bath within the rotating wave approximation 30
    1.4.3   Input-output relations 31
    1.4.4   Harmonic oscillator dynamics within the rotating
           wave approximation 32
           1.4.4.1   Mechanical susceptibility within the
                      rotating wave approximation 32
    1.4.5   Transformation into a rotating frame 34
    1.4.6   Quadrature operators 34
    1.4.7   Energy decay with and without the rotating wave
           approximation 35

In this chapter we introduce the basic concept of a quantum harmonic oscillator, and how such oscillators couple to their environment. Our purpose is not to exhaustively cover all aspects of such systems, but rather to provide the foundation material required to treat the quantum mechanics of mechanical resonators and optical cavities – and their interactions – in later chapters. The quantum fluctuation–dissipation theorem is derived relating the power spectral density of environmental forcing of an oscillator to the rate of energy decay from the oscillator into the environment. This is in direct analogy to the familiar fluctuation–dissipation theorem in classical physics. However, in contrast to the classical case, the noncommutation of position and momentum introduces an asymmetry between the positive and negative frequency components of the power spectrum. This provides a natural definition for the effective temperature of the oscillator. The quantum Langevin approach is introduced as a method to model the open dynamics of quantum systems, both with and without recourse to the rotating wave approximation.

1.1 QUANTISING THE HARMONIC OSCILLATOR

Here we provide a brief overview of quantum harmonic oscillators, before discussing the effects of environmental coupling on their dynamics. A more comprehensive discussion of quantum harmonic oscillators can be found in most introductory quantum mechanics textbooks (e.g., [129]).

As illustrated in Fig. 1.1, the essential quantum features of a harmonic oscillator are quantised energy levels spaced evenly by $\hbar\Omega$, and finite *zero-point energy* of $\hbar\Omega/2$ when cooled to their ground state at zero kelvin. The

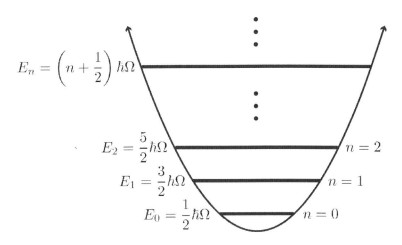

$$E_n = \left(n + \frac{1}{2}\right)\hbar\Omega$$

$$E_2 = \frac{5}{2}\hbar\Omega \qquad n = 2$$

$$E_1 = \frac{3}{2}\hbar\Omega \qquad n = 1$$

$$E_0 = \frac{1}{2}\hbar\Omega \qquad n = 0$$

FIGURE 1.1 Quantised energy levels E_n of an harmonic oscillator.

quantised excitations of a mechanical oscillator are termed *phonons*, while those of the optical field are, of course, termed *photons*.

The Hamiltonian describing an isolated quantum harmonic oscillator with mass m and frequency Ω is

$$\hat{H} = \frac{k\hat{q}^2}{2} + \frac{\hat{p}^2}{2m}, \tag{1.1}$$

where $\hat{q}$ and $\hat{p}$ are the position and momentum operators, respectively, and $k \equiv m\Omega^2$ is the *spring constant*.

1.1.1 Ladder operators

Quantum harmonic oscillators are naturally described in terms of ladder operators that act to add or subtract phonons from the oscillator. The creation $a^\dagger$ and annihilation a operators obey the *Boson commutation relation*

$$[a, a^\dagger] = 1. \tag{1.2}$$

If a harmonic oscillator is initially prepared in the n phonon state $|n\rangle$, the action of these operators is

$$
\begin{aligned}
a|n\rangle &= \sqrt{n}|n-1\rangle & \text{(1.3a)}\\
a^\dagger|n\rangle &= \sqrt{n+1}|n+1\rangle, & \text{(1.3b)}
\end{aligned}
$$

with the exception that $a|0\rangle \equiv 0$, since it is not possible to subtract a phonon from an oscillator that is already in its ground state. It is then immediately obvious why $a^\dagger$ and a are termed ladder operators, since they cause transitions up and down on level on the harmonic oscillator ladder, respectively.

Furthermore, since

$$a^\dagger a |n\rangle = n|n\rangle, \qquad (1.4)$$

we see that

$$\hat{n} \equiv a^\dagger a \qquad (1.5)$$

is the phonon number operator.

1.1.2 Thermal equilibrium statistics

Since phonons (and photons) are bosons, when a harmonic oscillator is in equilibrium with its environment at temperature T, Bose–Einstein statistics determine the occupancy probability $p(n)$ of each energy level. Specifically,

$$p(n) = \exp\left(-\frac{\hbar\Omega n}{k_B T}\right)\left[1 - \exp\left(-\frac{\hbar\Omega}{k_B T}\right)\right], \qquad (1.6)$$

where $k_B = 1.381 \times 10^{-23}$ m^2 kg s^{-1} K^{-1} is Boltzmann's constant.

The mean occupancy of the oscillator is

$$\bar{n} = \langle\hat{n}\rangle = \sum_{n=0}^{\infty} n p(n) = \left[\exp\left(\frac{\hbar\Omega}{k_B T}\right) - 1\right]^{-1}, \qquad (1.7)$$

where, here and often we use the accent ¯ for the expectation value of an observable. In the classical limit where $k_B T \gg \hbar\Omega$, this simplifies to the familiar expression

$$\bar{n}_{k_B T \gg \hbar\Omega} = \frac{k_B T}{\hbar\Omega}. \qquad (1.8)$$

That is, classically, the mean occupancy of a harmonic oscillator is equal to the thermal energy $k_B T$ divided by the energy of a phonon. For a typical micro- or nano-mechanical oscillator with resonance frequency in the range $\Omega/2\pi = 1 - 1000$ MHz at 300 K, $k_B T/\hbar\Omega = 10^4 - 10^7$. We see, therefore, that, without cooling of some kind, in most cases such oscillators can be viewed as purely classical. As oscillators become smaller, their resonance frequency typically increases. Consequently, the mechanical vibrations of atoms and molecules must generally be viewed quantum mechanically. Since optical fields are generated by transitions between atomic energy levels, it is unsurprising that also must be treated quantum mechanically. Visible light has a frequency of $\Omega/2\pi \sim 5 \times 10^{14}$. Equation (1.7) then gives a mean thermal occupancy at 300 K of $\bar{n} \sim 10^{-35}$. Therefore, optical fields in thermal equilibrium at room temperature can be safely considered to be in their ground state. This has important consequences in quantum optomechanics. Most particularly, as discussed in Chapter 4, when coupled to a mechanical oscillator appropriately, optical fields act as a cold bath that can cool the oscillator into its ground state.

1.1.3 Zero-point motion and the uncertainty principle

The position and momentum operators are expressed in terms of the ladder operators as

$$\hat{q} = x_{zp} \left(a^\dagger + a\right) \tag{1.9a}$$

$$\hat{p} = i p_{zp} \left(a^\dagger - a\right), \tag{1.9b}$$

where x_{zp} and p_{zp} are, respectively, the standard deviation of the zero-point motion and zero-point momentum of the oscillator, given by

$$x_{zp} = \sqrt{\frac{\hbar}{2m\Omega}} \tag{1.10a}$$

$$p_{zp} = \sqrt{\frac{\hbar m\Omega}{2}}. \tag{1.10b}$$

The existence of zero-point motion and momentum is a direct consequence of the zero-point energy of a harmonic oscillator. As we will see in later chapters, many of the quantum features of harmonic oscillators only becomes evident if it is possible to measure the motion of the oscillator with a precision surpassing the level of the zero-point motion. While mechanical oscillators used in quantum optomechanics experiments vary in size and frequency through many orders of magnitude, a typical micromechanical oscillator might have mass and resonance frequency on the order of $m \sim 1$ μg and $\Omega/2\pi \sim 1$ MHz, respectively, while the reduced $m \sim 1$ pg mass of nanomechanical oscillators results in generally higher resonance frequencies on the order of 1 GHz. Equation (1.10a) then results in typical order-of-magnitude estimates for the zero-point motion of $x_{zp}^{\text{micro}} \sim 10^{-17}$ m and $x_{zp}^{\text{nano}} \sim 10^{-14}$ m.

It can be straightforwardly shown using Eqs. (1.9), Eqs. (1.10), and the Boson commutation relation of Eq. (1.2) that the position and momentum operators obey the commutation relation

$$[\hat{q}, \hat{p}] = i\hbar. \tag{1.11}$$

This nonzero commutation relation between position and momentum leads directly to the famous Heisenberg uncertainty relation

$$\sigma(\hat{q})\sigma(\hat{p}) \geq \frac{1}{2} |\langle [\hat{q}, \hat{p}] \rangle| = \frac{\hbar}{2}, \tag{1.12}$$

where $\sigma(\hat{O}) = [\langle \hat{O}^2 \rangle - \langle \hat{O} \rangle^2]^{1/2}$ is the standard deviation of operator $\hat{O}$.

1.1.4 Dimensionless position and momentum operators

Throughout this text, we will generally work with dimensionless position $\hat{Q}$ and momentum $\hat{P}$ operators

$$\hat{Q} = \frac{1}{\sqrt{2}}\frac{\hat{q}}{x_{zp}} = \frac{1}{\sqrt{2}}\left(a^\dagger + a\right) \tag{1.13a}$$

$$\hat{P} = \frac{1}{\sqrt{2}}\frac{\hat{p}}{p_{zp}} = \frac{i}{\sqrt{2}}\left(a^\dagger - a\right). \tag{1.13b}$$

We will often refer to these operators as simply the position and the momentum, or position and momentum operators. Using Eq. (1.11), it is immediately apparent that

$$[\hat{Q}, \hat{P}] = i, \tag{1.14}$$

and that

$$\sigma(\hat{Q})\sigma(\hat{P}) \geq \frac{1}{2}\left|\left\langle [\hat{Q}, \hat{P}] \right\rangle\right| = \frac{1}{2}. \tag{1.15}$$

The variances of the dimensionless position and momentum operators of an oscillator in thermal equilibrium with its environment are linearly dependent on the mean phonon number[1]

$$\sigma^2(\hat{Q}) = \sigma^2(\hat{P}) = \bar{n} + \frac{1}{2}. \tag{1.16}$$

As expected, the variances do not approach zero, at zero kelvin, and instead saturate the uncertainty principle (see Eq. (1.15)).

We may also define dimensionless operators $\hat{Q}^\theta$ and P^θ rotated by a phase angle θ from the position and momentum as

$$\hat{Q}^\theta = \frac{1}{\sqrt{2}}\left(a^\dagger e^{i\theta} + ae^{-i\theta}\right) = \hat{Q}\cos\theta + \hat{P}\sin\theta \tag{1.17a}$$

$$\hat{P}^\theta = \frac{i}{\sqrt{2}}\left(a^\dagger e^{i\theta} - ae^{-i\theta}\right) = \hat{P}\cos\theta - \hat{Q}\sin\theta, \tag{1.17b}$$

which also satisfy the commutation relation $[\hat{Q}^\theta, \hat{P}^\theta] = i$.

1.1.5 Phasor diagrams

Similarly to a classical harmonic oscillator, it is natural to represent the state of a quantum harmonic oscillator as a vector on a phasor diagram of position versus momentum. However, unlike a classical oscillator, the uncertainty principle dictates that the direction and length of the vector cannot be wholly

[1] These results can be shown – for example, for the position operator – from the definition of the variance $\sigma^2(\hat{Q}) = \langle \hat{Q}^2 \rangle - \langle \hat{Q} \rangle^2$. One expands $\langle \hat{Q} \rangle = \langle a + a^\dagger \rangle/\sqrt{2}$, and $\langle \hat{Q}^2 \rangle = \langle a^\dagger a + aa^\dagger + a^2 + a^{\dagger 2} \rangle/2$. Recognising that the density matrix of a thermal state has no off-diagonal terms in the phonon-number basis, we have $\langle a \rangle = \langle a^\dagger \rangle = \langle a^2 \rangle = \langle a^{\dagger 2} \rangle = 0$, so that, using the Boson commutation relations in Eq. (1.2), $\sigma^2(\hat{Q}) = \langle a^\dagger a \rangle + 1/2$.

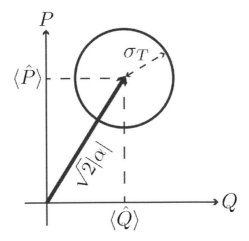

FIGURE 1.2 Phasor diagram of a displaced thermal state. $\alpha = \langle a \rangle$. $\sigma_T = \sigma(\hat{Q}) = \sigma(\hat{P}) = [\bar{n} + 1/2]^{1/2}$.

deterministic. There are several commonly used quantum phase space distributions, including the P, Q, and Wigner functions [305]. Of these, the *Wigner function* or *Wigner distribution* provides the closest quantum analogy to a classical phase space probability distribution of position and momentum. It should be thought of as a quasi-probability distribution; quasi, because the uncertainty principle allows localised regions of negative quasi-probabilities. While this may appear unphysical, one of the key properties of the Wigner function is that, when projected onto any marginal of the distribution, that marginal reproduces the correct positive-definite quantum distributions given by the Born rule. Furthermore, quantum states whose Wigner functions exhibit no negativity are indistinguishable from a classical state with classical probability distribution matching the shape of the Wigner function. It is often useful to visualise the state of an oscillator on a "ball and stick" diagram, as shown, for example, for a displaced thermal state in Fig. 1.2. This is a rough pictorial representation of a Wigner function, generally only including a vector (or stick) from the origin to the point $\{\langle \hat{Q} \rangle, \langle \hat{P} \rangle\}$ in the Q-P plane, and an uncertainty contour (or ball) illustrating the extent of the state.

1.1.6 Unitary dynamics in the Heisenberg picture

In terms of the dimensionless position and momentum operators, the Hamiltonian Eq. (1.1) becomes

$$\hat{H} = \frac{\hbar\Omega}{2}\left(\hat{Q}^2 + \hat{P}^2\right);\tag{1.18}$$

while substituting Eqs. (1.9a) and Eqs. (1.9b) into Eq. (1.1), we retrieve the canonical quantum harmonic oscillator Hamiltonian

$$\hat{H} = \hbar\Omega\, a^\dagger a = \hbar\Omega\, \hat{n}, \tag{1.19}$$

where we have neglected the half-quanta of zero-point energy.

In the absence of coupling to the environment, the dynamics of any operator $\hat{O}$ of a quantum system can be determined in the Heisenberg picture via the Heisenberg equation of motion

$$\dot{\hat{O}}(t) = \frac{1}{i\hbar}\left[\hat{O}(t), \hat{H}(t)\right]. \tag{1.20}$$

Using the canonical Hamiltonian of Eq. (1.19), the dynamics of the annihilation operator $a(t)$ are then

$$\begin{aligned}
\dot{a}(t) &= -i\Omega\left[\hat{a}(t), a^\dagger(t)a(t)\right] \tag{1.21}\\
&= -i\Omega a(t) \tag{1.22}
\end{aligned}$$

where we include the time dependence explicitly to remind the reader that, in the Heisenberg picture, the operators evolve in time, and we have made use of the Boson commutation relation (Eq. (1.2)).

We see that the annihilation operator obeys the usual equation for the dynamics of a simple harmonic oscillator, with the solution

$$\hat{a}(t) = \hat{a}(0)\exp(-i\Omega t). \tag{1.23}$$

Equations (1.13) then give the equations of motion for the position and momentum:

$$\begin{aligned}
\hat{Q}(t) &= \frac{1}{\sqrt{2}}\left(a^\dagger(t) + a(t)\right) = \cos(\Omega t)\hat{Q}(0) + \sin(\Omega t)\hat{P}(0) \tag{1.24a}\\
\hat{P}(t) &= \frac{i}{\sqrt{2}}\left(a^\dagger(t) - a(t)\right) = \cos(\Omega t)\hat{P}(0) - \sin(\Omega t)\hat{Q}(0). \tag{1.24b}
\end{aligned}$$

As might be expected, the position and momentum oscillate with frequency Ω.

1.2 FLUCTUATIONS AND DISSIPATION IN A QUANTUM HARMONIC OSCILLATOR

It is important to include coupling to the environment in any realistic model of an open quantum system. This coupling both introduces a damping channel for the system and allows environmental noise to perturb the oscillator. Furthermore, it provides a channel through which measurements of the state of the system may be performed.

1.2.1 Classical fluctuation–dissipation theorem

Before treating the connection between damping and perturbation in a quantum context, it is worthwhile to recollect the classical relationship. For a classical oscillator in thermal equilibrium with a bath at temperature T, the *fluctuation–dissipation theorem* establishes a formal link between the energy decay rate of the oscillator γ and the power spectral density $S_{FF}(\omega)$ of the thermal force $F(t)$ exerted on it by the bath via [166]

$$S_{FF}(\omega) = 2\gamma m k_B T, \tag{1.25}$$

where, in general, the power spectral density $S_{hh}(\omega)$ of a complex classical variable $h(t)$ is

$$S_{hh}(\omega) = \lim_{\tau \to \infty} \frac{1}{\tau} \langle h_\tau^*(\omega) h_\tau(\omega) \rangle . \tag{1.26}$$

Here, $h_\tau(\omega)$ is the Fourier transform of $h(t)$ sampled over the time period $-\tau/2 < t < \tau/2$. Throughout this text the Fourier transform and its inverse are defined as

$$h(\omega) = \mathcal{F}(h(t)) = \int_{-\infty}^{\infty} h(t) e^{i\omega t} dt \tag{1.27a}$$

$$h(t) = \mathcal{F}^{-1}(h(\omega)) = \frac{1}{2\pi} \int_{-\infty}^{\infty} h(\omega) e^{-i\omega t} d\omega, \tag{1.27b}$$

so that $h_\tau(\omega) = \int_{-\tau/2}^{\tau/2} h(t) e^{i\omega t} dt$.

The Wiener–Khinchin theorem provides a direct relationship between the power spectral density of a variable and its autocorrelation function in the time domain. For a stationary time–independent variable[2]

$$S_{hh}(\omega) = \int_{-\infty}^{\infty} d\tau \, e^{i\omega\tau} \langle h^*(t+\tau) h(t) \rangle_{t=0} = \int_{-\infty}^{\infty} d\omega' \, \langle h^*(-\omega) h(\omega') \rangle , \tag{1.28}$$

where the first relation is the Wiener–Khinchin theorem. The reader is referred, for example, to [68] for its derivation. The second relation can be derived straightforwardly from the first using the properties of the Fourier transform. While this second relation bears some resemblance to the standard definition of the power spectral density in Eq. (1.26), it is generally more convenient to use since it is expressed in terms of the full Fourier transform $h(\omega)$ rather than the windowed Fourier transform $h_\tau(\omega)$. Note that throughout this textbook we use the standard convention that $h^*(\omega) = \mathcal{F}\{h(t)\}^*$, and the same convention for operators. The alternative convention $h^*(\omega) = \mathcal{F}\{h^*(t)\}$ is frequently found in the quantum optomechanics literature.

Exercise 1.1 *Derive the second relation in Eq. (1.28).*

---

[2]Note, by stationary here, we do not mean that the variable itself is a constant, but rather that its statistical properties are unchanging with time.

1.2.2   Quantum forcing of a harmonic oscillator

To introduce damping and dissipation to a quantum harmonic oscillator, we begin by including a forcing term

$$\hat{V} = \hat{q}\hat{F} \tag{1.29}$$

in the Hamiltonian of Eq. (1.19), where $\hat{F}$ is a force exerted on the oscillator due to coupling to a bath consisting of another quantum system – or an ensemble of such systems – which commutes with the position $\hat{q}$. The resulting Hamiltonian is

$$\hat{H} = \hat{H}_0 + \hat{V}, \tag{1.30}$$

where the bare Hamiltonian $\hat{H}_0 = \hbar\Omega a^\dagger a + \hat{H}_{\text{bath}}$ and $\hat{H}_{\text{bath}}$ is the Hamiltonian of the bath. Using this Hamiltonian, a quantum version of the fluctuation–dissipation theorem for a quantum harmonic oscillator can be readily derived. Here we follow the approach outlined in [73]. The derivation turns out to be highly illuminating for the understanding of optical cooling and heating of mechanical systems.

1.2.2.1   Upwards going transition probability

The combined oscillator-bath system evolves as dictated by the Hamiltonian of Eq. (1.30) and the Schrödinger equation. During the evolution, the force $\hat{F}(t)$ drives transitions between the energy levels of the oscillator. We seek to determine the probability of a transition from some initial energy eigenstate $|\psi(0)\rangle$ to an orthogonal final energy eigenstate $|\psi_f\rangle$ after some time t, and follow a Fermi-golden rule-style derivation.

It is convenient to remove the bare dynamics of both bath and oscillator by moving into an interaction picture. The state of the combined system then evolves as $|\psi_I(t)\rangle \equiv \hat{U}_0^\dagger(t)|\psi(t)\rangle = \hat{U}_I(t)|\psi(0)\rangle$, where the unitary time evolution operators are $\hat{U}_0(t) = e^{-i\hat{H}_0 t/\hbar}$ and $\hat{U}_I(t) = e^{-i\hat{V}t/\hbar}$ with the interaction picture labelled here, and often throughout the textbook, with the subscript I. The probability amplitude that the system is found in state $|\psi_f\rangle$ at time t is given in terms of $|\psi_I(t)\rangle$ by

$$\begin{aligned}
A_{i\to f}(t) &= \langle\psi_f|\psi(t)\rangle \tag{1.31} \\
&= \langle\psi_f|\hat{U}_0(t)\hat{U}_I(t)|\psi(0)\rangle \tag{1.32} \\
&= e^{-iE_f t/\hbar}\langle\psi_f|\psi_I(t)\rangle, \tag{1.33}
\end{aligned}$$

where we have used the property that since $|\psi_f\rangle$ is an energy eigenstate $\hat{U}_0(t)|\psi_f\rangle = e^{-iE_f t/\hbar}|\psi_f\rangle$ with E_f being the energy of the final state.

In the interaction picture the Hamiltonian of Eq. (1.30) becomes $\hat{H}_I(t) = \hat{U}_0^\dagger(t)\hat{H}\hat{U}_0(t) = \hat{U}_0^\dagger(t)\hat{V}\hat{U}_0(t) = \hat{V}_I(t)$. Formally integrating the Schrödinger equation, the evolution of the full system can then be expressed exactly as the

Dyson series

$$
\begin{aligned}
|\psi_I(t)\rangle &= |\psi_I(0)\rangle + \frac{1}{i\hbar} \int_0^t d\tau_1 \hat{V}_I(\tau_1) |\psi_I(\tau_1)\rangle \qquad (1.34) \\
&= |\psi(0)\rangle + \frac{1}{i\hbar} \int_0^t d\tau_1 \hat{V}_I(\tau_1) |\psi(0)\rangle \\
&\quad + \frac{1}{(i\hbar)^2} \int_0^t \int_0^{\tau_1} d\tau_1 d\tau_2 \hat{V}_I(\tau_1) \hat{V}_I(\tau_2) |\psi_I(\tau_2)\rangle, \qquad (1.35)
\end{aligned}
$$

where in the second line we have substituted $|\psi_I(0)\rangle$ for the Schrödinger picture ket $|\psi(0)\rangle$ since $\hat{U}_0(0) = \mathbb{1}$, with $\mathbb{1}$ being the identity operator.

We wish to determine the transition rate between oscillator energy levels in the short-time limit. We therefore perform standard time-dependent first-order perturbation theory, neglecting the last term in Eq. (1.35). We take the initial states of the bath and oscillator to be separable, with the oscillator in an energy eigenstate $|n\rangle$ and the bath in some eigenstate $|j\rangle$ so that $|\psi(0)\rangle = |\psi_{\text{sys}}(0)\rangle \otimes |\psi_{\text{bath}}(0)\rangle \equiv |n, j\rangle$, and assume that the interaction is sufficiently weak and the bath sufficiently large that they remain separable throughout the evolution.[3] Recognising that the initial and final states are orthogonal so that $\langle \psi_f | \psi(0) \rangle = 0$ and utilising Eq. (1.33), the probability amplitude that the interaction causes an upwards going transition in the oscillator to $|n + 1\rangle$ and leaves the bath in some eigenstate $|k\rangle$ is then

$$
\begin{aligned}
A_{i \to f}(t) &= \frac{1}{i\hbar} \int_0^t d\tau_1 \langle n+1, k | \hat{V}_I(\tau_1) | n, j \rangle \qquad (1.36) \\
&= \frac{1}{i\hbar} \int_0^t d\tau_1 \langle n+1 | \hat{q}_I | n \rangle \langle k | \hat{F}_I(\tau_1) | j \rangle \qquad (1.37) \\
&= \frac{x_{zp}}{i\hbar} \int_0^t d\tau_1 e^{i\Omega\tau_1} \langle n+1 | a^\dagger + a | n \rangle \langle k | \hat{F}_I(\tau_1) | j \rangle \qquad (1.38) \\
&= \frac{x_{zp}\sqrt{n+1}}{i\hbar} \int_0^t d\tau_1 e^{i\Omega\tau_1} \langle k | \hat{F}_I(\tau_1) | j \rangle, \qquad (1.39)
\end{aligned}
$$

where $\hat{q}_I \equiv \hat{U}_0^\dagger(t) \hat{q} \hat{U}_0(t)$ and $\hat{F}_I \equiv \hat{U}_0^\dagger(t) \hat{F} \hat{U}_0(t)$, we have neglected the overall phase factor $e^{-iE_f t/\hbar}$ which cancels in the final transition probability, and we have used the properties of the annihilation and creation operators in Eqs. (1.3).

Ultimately, we are not interested in, or indeed able to determine, the final state of the bath. Summing over all possible bath final states, the probability

---

[3] This is a form of *Born approximation*.

of the oscillator transitioning from state $|n\rangle$ to state $|n + 1\rangle$ is then

$$
\begin{aligned}
P_{n\to n+1} &= \sum_k |A_{i\to f}(t)|^2 \tag{1.40} \\
&= \frac{x_{zp}^2(n+1)}{\hbar^2} \int\!\!\int_0^t d\tau_1 d\tau_2 e^{i\Omega(\tau_2 - \tau_1)} \sum_k \langle j|\hat{F}_I(\tau_1)|k\rangle \langle k|\hat{F}_I(\tau_2)|j\rangle \\
&= \frac{x_{zp}^2(n+1)}{\hbar^2} \int\!\!\int_0^t d\tau_1 d\tau_2 e^{i\Omega(\tau_2 - \tau_1)} \left\langle \hat{F}_I(\tau_1)\hat{F}_I(\tau_2) \right\rangle, \tag{1.41}
\end{aligned}
$$

where we have used the fact that $\hat{F}$ is Hermitian ($\hat{F}^\dagger = \hat{F}$), as well as the completeness and orthonormality of the bath modes so that $\sum_k |k\rangle\langle k| = \mathbb{1}$.

1.2.2.2 Quantum power spectral density

The integral in Eq. (1.41) is a form of autocorrelation function and can therefore be related to the force power spectral density via the Wiener–Khinchin theorem. The power spectral density of a general operator $\hat{O}$ is defined as

$$
S_{OO}(\omega) \equiv \lim_{\tau\to\infty} \frac{1}{\tau} \left\langle \hat{O}_\tau^\dagger(\omega)\hat{O}_\tau(\omega) \right\rangle, \tag{1.42}
$$

in direct analogy to the classical equivalent (Eq. (1.26)), where, as in the classical case, $\hat{O}_\tau(\omega)$ is the Fourier transform of $\hat{O}(t)$ windowed over the time period $-\tau/2 < t < \tau/2$. The Wiener–Khinchin theorem then states that, for an operator with stationary statistics,

$$
S_{OO}(\omega) = \int_{-\infty}^{\infty} d\tau\, e^{i\omega\tau} \left\langle \hat{O}^\dagger(t+\tau)\hat{O}(t) \right\rangle_{t=0} = \int_{-\infty}^{\infty} d\omega' \left\langle \hat{O}^\dagger(-\omega)\hat{O}(\omega') \right\rangle, \tag{1.43}
$$

while, for the conjugate operator $\hat{O}^\dagger$,

$$
S_{O^\dagger O^\dagger}(\omega) = \int_{-\infty}^{\infty} d\tau\, e^{i\omega\tau} \left\langle \hat{O}(t+\tau)\hat{O}^\dagger(t) \right\rangle_{t=0} = \int_{-\infty}^{\infty} d\omega' \left\langle \hat{O}(\omega)\hat{O}^\dagger(\omega') \right\rangle. \tag{1.44}
$$

While, like a classical power spectral density, $S_{OO}(\omega)$ and $S_{O^\dagger O^\dagger}(\omega)$ are always real, one striking difference between the quantum and classical power spectral densities is immediately apparent. For a classical variable $h(t)$, the product $h^*(t+\tau)h(t) = h(t)h^*(t+\tau)$. However, for a quantum operator $[\hat{O}^\dagger(t+\tau), \hat{O}(t)] \neq 0$ in general. Consider, for example, the case where $\hat{O}$ is the position $\hat{q}$ of an isolated quantum harmonic oscillator. Then, after a time delay $\tau = \pi/2\Omega$ of a quarter oscillator period $\hat{q}(t+\tau) = \hat{p}(t)$ and therefore clearly does not commute with $\hat{q}(t)$.

One consequence of this difference between the power spectral densities of classical and quantum variables becomes particularly evident when $\hat{O}(t)$ (and therefore $\hat{O}_\tau$) is a Hermitian observable such as force, position, or momentum.[4] For any Hermitian operator, it is straightforward to show from the definition of the Fourier transform that, similar to a real classical variable, $\hat{O}^\dagger(\omega) = \hat{O}(-\omega)$. From Eq. (1.42) we then immediately find that $S_{\hat{O}\hat{O}}(\omega) = \lim_{\tau\to\infty}\langle\hat{O}_\tau(-\omega)\hat{O}_\tau(\omega)\rangle/\tau$, while $S_{\hat{O}\hat{O}}(-\omega)$ is identical but with swapped operator ordering. Since $\hat{O}_\tau(\omega)$ and $\hat{O}_\tau(\omega')$ do not necessarily commute, in general, for a quantum observable, $S_{\hat{O}\hat{O}}(\omega) \neq S_{\hat{O}\hat{O}}(-\omega)$. That is, the power spectral density of an operator is generally not symmetric in frequency. The power spectral density of a classical variable, on the other hand, is always frequency symmetric ($S_{hh}(\omega) = S_{hh}(-\omega)$). As we will see in this chapter, this difference between the positive and negative parts of the power spectral density of a quantum observable has important consequences for our understanding of fluctuation and dissipation in quantum systems and is particularly useful in defining their temperature.

1.2.2.3 Transition rates between energy levels

With the quantum power spectral density defined, we now return to the problem of understanding the transition rates between levels in a quantum harmonic oscillator driven by an external force. To transform Eq. (1.41) into a form more closely resembling a power spectral density, we make the substitutions $\tau_1 = t' + \tau$ and $\tau_2 = t'$ so that

$$P_{n\to n+1} = \frac{x_{zp}^2(n+1)}{\hbar^2} \int_0^t dt' \int_{-t'}^{t-t'} d\tau e^{-i\Omega\tau} \left\langle \hat{F}_I(t'+\tau)\hat{F}_I(t') \right\rangle. \quad (1.45)$$

If the integration is taken over times long compared to the autocorrelation time of the bath,[5] the limits on the second integral may be approximated to $\pm\infty$. The average probability of an upwards going transition can then be expressed in terms of the power spectral density as

$$P_{n\to n+1} = \frac{x_{zp}^2(n+1)}{\hbar^2} \int_0^t dt' S_{FF}(-\Omega) \quad (1.46)$$

$$= \frac{x_{zp}^2(n+1)}{\hbar^2} t\, S_{FF}(-\Omega), \quad (1.47)$$

where, while we have dropped the subscript I's, it should be remembered that the force $\hat{F}_I(t)$ is within an interaction picture. We see that the transition probability increases linearly with time and is proportional to the force power spectral density at frequency $-\Omega$. Note that, since we have taken a first-order perturbative approach to derive this expression, it is only valid at times

---

[4]In general, an operator is Hermitian if the property $\hat{O}^\dagger(t) = \hat{O}(t)$ holds.
[5]Typically, the bath is taken to be Markovian and therefore delta correlated such that this approximation is eminently reasonably (see Section 1.3.1.2). In the case of radiation pressure forcing within an optical cavity treated later in this textbook, the autocorrelation time is determined by the optical cavity decay time, which is typically much shorter than the mechanical decay time, in which case the approximation is also reasonable.

sufficiently short that $P_{n \to n+1} \ll 1$. Taking the time derivative gives the upwards going transition rate from $|n\rangle$ to $|n+1\rangle$,

$$\gamma_{n \to n+1} = \frac{x_{zp}^2}{\hbar^2}(n+1)\, S_{FF}(-\Omega). \tag{1.48}$$

A similar calculation for downwards going transitions from $|n\rangle$ to $|n-1\rangle$ yields

$$\gamma_{n \to n-1} = \frac{x_{zp}^2}{\hbar^2}n\, S_{FF}(\Omega). \tag{1.49}$$

Exercise 1.2 *Derive Eq. (1.49) for yourself. Start by calculating the probability amplitude $A_{i \to f}(t)$ of a transition $|n, j\rangle \to |n - 1, k\rangle$ where, as before, the first and second arguments in the kets respectively label energy eigenstates of the harmonic oscillator and the bath (see Section 1.2.2.1).*

As one might expect, both upwards and downwards going transitions are Bose enhanced by the occupation number n, with the downwards going transition rate equal to zero when $n = 0$. Notice also that upwards going transitions are driven by the negative frequency part of the force power spectral density, while downwards going transitions are driven by the positive frequency part of the spectrum. This gives a strong indication that heating or cooling of a mechanical oscillator could be achieved by controlling the spectrum of the force noise impinging upon it and turns out to elegantly explain how the effect of an optical cavity on the optical power spectral density can result in resolved sideband cooling. We investigate resolved sideband cooling in detail in Chapter 4.

1.2.3 Relationship between $S_{FF}(\Omega)$ and $S_{FF}(-\Omega)$ in thermal equilibrium

Combined with the concept of detailed balance and the Bose–Einstein distribution, the upwards and downwards going transition rates in Eqs. (1.48) and (1.49) allow a strict relationship to be established between the positive and negative frequency parts of the force power spectrum in thermal equilibrium. Detailed balance is a concept in thermodynamics which states that in equilibrium the total rate of transitions from one state to another must exactly balance the reverse rate [290]. That is, in thermal equilibrium

$$p(1)\gamma_{1 \to 2} = p(2)\gamma_{2 \to 1}, \tag{1.50}$$

where $p(1)$ and $p(2)$ are the occupation probabilities for states 1 and 2. Applying detailed balance to the harmonic oscillator transitions $n \to n + 1$ and $n + 1 \to n$ using Eqs. (1.48) and (1.49), we find

$$\frac{p(n+1)}{p(n)} = \frac{\gamma_{n \to n+1}}{\gamma_{n+1 \to n}} = \frac{S_{FF}(-\Omega)}{S_{FF}(\Omega)}. \tag{1.51}$$

The occupation probabilities for a mechanical oscillator in thermal equilibrium are given by the Bose–Einstein distribution in Eq. (1.6). Substituting for $p(n)$ and $p(n + 1)$, we immediately find

$$\frac{S_{FF}(\Omega)}{S_{FF}(-\Omega)} = \exp\left(\frac{\hbar\Omega}{k_B T}\right) \tag{1.52}$$

$$= 1 + \frac{1}{\bar{n}}. \tag{1.53}$$

We see that, in thermal equilibrium, the positive and negative frequency parts of the quantum force power spectral density are related by the Boltzmann factor [254]. At sufficiently high temperatures, such that $\bar{n} \gg 1$, the classical result that the force power spectral density is independent of frequency Ω is retrieved (see Eq. (1.25)). However, at low temperatures this is no longer true, with the ratio diverging as $\bar{n} \to 0$.

Inverting Eqs. (1.52) and (1.53) provides general definitions of the temperature, and mean phonon number, of an oscillator in thermal equilibrium.

$$T = \frac{\hbar\Omega}{k_B}\left[\ln\left(\frac{S_{FF}(\Omega)}{S_{FF}(-\Omega)}\right)\right]^{-1} \tag{1.54a}$$

$$\bar{n} = \frac{S_{FF}(-\Omega)}{S_{FF}(\Omega) - S_{FF}(-\Omega)} \tag{1.54b}$$

Commonly, we will deal with circumstances where the oscillator has reached steady state but is not in thermal equilibrium due, for example, to optical cooling or heating. In such circumstances, it is natural to define the frequency–dependent *effective temperature*,

$$T_{\text{eff}}(\omega) = \frac{\hbar\omega}{k_B}\left[\ln\left(\frac{S_{FF}(\omega)}{S_{FF}(-\omega)}\right)\right]^{-1}, \tag{1.55}$$

in analogy to Eq. (1.52).

1.2.4 Quantum fluctuation–dissipation theorem

Now that we have linked the transition rates between oscillator energy levels to the power spectral density of the driving force from a bath, and linked this power spectral density to the effective temperature of the bath, it is possible to derive a quantum version of the fluctuation–dissipation theorem we introduced in Eq. (1.25).

Again taking the approach in [73], let us consider an oscillator which, at some initial time, is out of equilibrium, with mean phonon occupancy

$$\bar{n}_b = \sum_{n=0}^{\infty} np_n, \tag{1.56}$$

where p_n is the probability of occupancy of state $|n\rangle$. The rate of change of the mean occupancy is

$$\dot{\bar{n}}_b = \sum_{n=0}^{\infty} n \dot{p}_n. \tag{1.57}$$

Following [73], we may derive rate equations for the occupancy of each level of the harmonic oscillator using the transition rates in Eqs. (1.48) and (1.49),

$$\dot{p}_0 = \gamma_\downarrow p_1 - \gamma_\uparrow p_0 \tag{1.58}$$

$$\dot{p}_1 = \gamma_\uparrow p_1 + 2\gamma_\downarrow p_2 - [\gamma_\downarrow + 2\gamma_\uparrow] p_1 \tag{1.59}$$

$$\vdots$$

$$\dot{p}_n = n\gamma_\uparrow p_{n-1} + (n+1)\gamma_\downarrow p_{n+1} - [n\gamma_\downarrow + (n+1)\gamma_\uparrow] p_n \tag{1.60}$$

$$\vdots \quad ,$$

where we have defined $\gamma_{n \to n+1} \equiv (n+1)\gamma_\uparrow$ and $\gamma_{n \to n-1} \equiv n\gamma_\downarrow$.

Exercise 1.3 *By substituting Eq. (1.60) into Eq. (1.57) show that*

$$\dot{\bar{n}}_b = \gamma_\uparrow - \gamma \bar{n}_b, \tag{1.61}$$

where the dissipation rate $\gamma \equiv \gamma_\downarrow - \gamma_\uparrow$. Then use the definitions of $\gamma_\downarrow$ and $\gamma_\uparrow$ above along with Eqs. (1.48) and (1.49) to show that, for linear forcing $\hat{F}$ as described by the Hamiltonian in Eq. (1.30),

$$\gamma = \frac{x_{zp}^2}{\hbar^2} \left(S_{FF}(\Omega) - S_{FF}(-\Omega) \right). \tag{1.62}$$

The first term in Eq. (1.61) is due to fluctuations that heat the oscillator, while the second term describes an exponential decay (or dissipation) of energy out of the oscillator, the rate of which is determined by the difference between the positive and negative frequency force power spectral densities (Eq. (1.62)).

Reexpressing the equation of motion in Eq. (1.61) in terms of energy by substituting $\bar{n}_b = \bar{E}/\hbar\Omega - 1/2$, we have

$$\dot{\bar{E}} = \frac{\hbar\Omega}{2} (\gamma_\uparrow + \gamma_\downarrow) - \gamma \bar{E} \tag{1.63}$$

$$= \frac{1}{2m} \bar{S}_{FF}(\Omega) - \gamma \bar{E}, \tag{1.64}$$

where $\bar{S}_{FF}(\Omega)$ is the symmetrised force power spectral density, with the *symmetrised power spectral density* of an operator $\hat{O}$ defined in general as

$$\bar{S}_{OO}(\Omega) \equiv \frac{S_{OO}(\Omega) + S_{OO}(-\Omega)}{2}, \tag{1.65}$$

and to arrive at the final expression we have used the definitions of $\gamma_\downarrow$ and

$\gamma_\uparrow$ along with Eqs. (1.48) and (1.49). We see that, while the dissipation γ introduced by the force $\hat{F}$ arises from the difference of the positive and negative frequency force power spectral densities, the fluctuations arise from the sum. Symmetrised power spectral densities have broad physical significance. For instance, they not only quantify the fluctuations introduced to a harmonic oscillator from the environment but also, as we will see in Section 3.3.4, the spectrum obtained from an ideal homodyne measurement of an optical field.

Since, in thermal equilibrium, $S_{FF}(\Omega)$ and $S_{FF}(-\Omega)$ are related by detailed balance (Eq. (1.52)), it is possible to link the magnitude of the fluctuations to the dissipation rate. Using Eq. (1.52), the symmetrised force spectral density can be reexpressed as

$$\bar{S}_{FF}(\Omega) = m\gamma\,\hbar\Omega\,(2\bar{n} + 1). \tag{1.66}$$

Exercise 1.4 *Show this result.*

This is the *quantum fluctuation–dissipation theorem*, analogous to the classical equivalent given in Eq. (1.25) [166]. The equivalence can be seen by taking the high temperature classical limit of Eq. (1.66). If $k_B T \gg \hbar\Omega$, $\coth(\hbar\Omega/2k_B T) \approx 2k_B T/\hbar\Omega$ so that

$$\bar{S}_{FF}(\Omega \ll k_B T/\hbar) = 2\gamma m\, k_B T. \tag{1.67}$$

Using Eq. (1.66), we can rewrite the equation of motion for the energy of an oscillator that is coupled to a thermal environment as

$$\dot{\bar{E}} = \gamma\hbar\Omega\left(\bar{n} + \frac{1}{2}\right) - \gamma\bar{E}, \tag{1.68}$$

which shows that if an oscillator damps through coupling to a thermal environment at a rate γ, it will also heat at a rate of $\gamma\hbar\Omega(\bar{n} + 1/2)$ J s^{-1}.

One conundrum that the perceptive reader may have observed relating to the equations of motion for energy and phonon decay given in Eqs. (1.68) and (1.61) is that the decay is independent of oscillation phase. On the other hand, since a real mechanical oscillator experiences damping proportional to its velocity, one would expect the energy decay to oscillate between a maximum value when the oscillator has maximum velocity, and zero when it is stationary. The cause of this discrepancy is that, strictly, the upwards and downwards going transition rates derived in Section 1.2.2.1 only apply to a mechanical oscillator initially prepared in a mechanical Fock state ($|n\rangle$), and only apply at short times where transitions back into state $|n\rangle$ – and therefore interference phenomena – may be neglected. If the mechanical oscillator is prepared in a different state, such as a coherent or thermal state, interference will occur between the different possible paths into each energy level. This interference results in the observed oscillation in energy decay. We will show in Section 1.4.7 that Eqs. (1.68) and (1.61) are valid generally, within the regime of validity of the rotating wave approximation. The essential physics behind this is that to perform the rotating wave approximation requires both

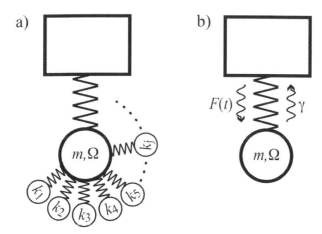

FIGURE 1.3 Illustration of a quantum system (in this case a mechanical oscillator with mass m and resonance frequency Ω) coupled to its environment. (a) Independent-oscillator model, where the quantum system is coupled via springs with spring constant k_j to a large ensemble of environmental oscillators. (b) Reduced model where the net result of the environmental coupling is to introduce forcing $\hat{F}(t)$ and damping γ.

that the oscillator is high quality and that the relevant time scales are long compared to the mechanical period. In this case, the oscillations in energy decay average out.

1.3 MODELLING OPEN SYSTEM DYNAMICS VIA QUANTUM LANGEVIN EQUATIONS

In Section 1.2 we studied the effect of an external force on the dynamics of a mechanical oscillator, arriving at a quantum version of the fluctuation–dissipation theorem. Here, we consider the open dynamics of a quantum system more generally, introducing the quantum Langevin equation as a tool to model such systems. The essential idea is to couple the system to a bath consisting of a large number (eventually infinite) of oscillators, as shown schematically in Fig. 1.3a. In this way, the system dynamics may be substantially affected by the presence of the bath, while the effect of the system on each individual bath oscillator remains negligible. It is then possible to derive equations describing the system dynamics, without the requirement of also solving for the dynamics of the bath. This approach was introduced by Langevin in 1908 to study the classical statistical mechanics of Brownian motion [170]. Several methods were developed in the 1980s and 1990s to extend the approach to open quantum processes and dynamics (see, for example, [54, 55, 105, 219, 206, 113]) motivated by experimental progress in the field of quantum optics. Due to their classical statistical mechanics origin, quantum Langevin methods are typically more physically intuitive than other methods that have been developed to model open quantum systems such as master equations (see Section 5.1).

Quantum Langevin equations have been thoroughly treated in many textbooks, for example, [111, 310]. Often, these treatments make use of the rotating wave approximation, which is generally valid for quantum optics problems where the optical frequency is typically much faster than any other frequency of relevance. The rotating wave approximation is often valid for mechanical oscillators. However, since their frequencies are typically some seven orders of magnitude lower than optical frequencies, this is is not always the case. Here, we briefly introduce a nonrotating wave quantum Langevin equation suitable for modelling quantum optomechanical systems, following [105] and [118].

1.3.1 Quantum Langevin equation for a particle in a potential

We consider a quantum system in a potential $\hat{V}(\hat{q})$ that is spring-coupled to each of an ensemble of independent bath oscillators, as shown in Fig. 1.3a, in what is termed the *independent-oscillator model* [105]. This model is closely related to the perhaps more familiar, Caldeira–Leggett model [54, 55], where a linear position-position coupling interaction is introduced to an ensemble of bath oscillators. While both models yield similar results, here we use the independent-oscillator model since the Caldeira–Leggett model requires the introduction of a non-unique correction term to correct for frequency renormalisation effects that can lead to erroneous predictions [55, 105]. In the independent-oscillator model the total Hamiltonian is

$$\hat{H} = \hat{H}_{\text{sys}} + \hat{H}_{\text{sys}-\text{bath}}, \tag{1.69}$$

where

$$\hat{H}_{\text{sys}} = \frac{\hat{p}^2}{2m} + \hat{V}(\hat{q}) \tag{1.70}$$

and

$$\hat{H}_{\text{sys}-\text{bath}} = \sum_j \left[\frac{\hat{p}_j^2}{2m_j} + \frac{k_j}{2}(\hat{q}_j - \hat{q})^2 \right], \tag{1.71}$$

with $\hat{q}_j$, $\hat{p}_j$, Ω_j, m_j, and $k_j = m_j \Omega_j^2$ being the position, momentum, resonance frequency, mass, and spring constant of bath oscillator j, respectively.

Exercise 1.5 *Derive Heisenberg equations of motion for the the operators $\hat{q}$, $\hat{p}$, $\hat{q}_j$, and $\hat{p}_j$ using Eqs. (1.20) and (1.69). Then eliminate the momentum operators to obtain the second-order differential equations*

$$m\ddot{\hat{q}} = -\frac{\partial \hat{V}(\hat{q})}{\partial \hat{q}} + \sum_j k_j(\hat{q}_j - \hat{q}) \tag{1.72a}$$

$$m_j \ddot{\hat{q}}_j = -k_j(\hat{q}_j - \hat{q}). \tag{1.72b}$$

The general solution to Eq. (1.72b) is

$$\hat{q}_j = \hat{q}_j^h(t) + \hat{q}(t) - \int_{-\infty}^{t} dt' \cos[\Omega_j(t-t')]\dot{\hat{q}}(t'), \tag{1.73}$$

where

$$\hat{q}_j^h(t) = \hat{q}_j(0)\cos(\Omega_j t) + \frac{\hat{p}_j(0)}{\Omega_j m_j}\sin(\Omega_j t) \tag{1.74}$$

is the solution to the homogeneous differential equations obtained by omitting the $\hat{q}$ terms in Eq. (1.72b). As detailed in [105], substitution of this general solution into Eq. (1.72a) yields the quantum Langevin equation

$$m\ddot{\hat{q}} + \int_{-\infty}^{t} dt' \mu(t-t')\dot{\hat{q}}(t') + \frac{\partial \hat{V}(\hat{q})}{\partial \hat{q}} = \hat{F}(t), \tag{1.75}$$

where $\mu(t)$ is a memory kernel, and

$$\hat{F}(t) = \sum_j k_j \hat{q}_j^h(t) \tag{1.76}$$

is an operator-valued stochastic force with zero expectation value.

The memory kernel $\mu(t)$ is given explicitly by

$$\mu(t) = \sum_j k_j \cos(\Omega_j t) \tag{1.77}$$

$$= \int_0^\infty d\Omega\, \rho(\Omega) k(\Omega) \cos(\Omega t), \tag{1.78}$$

and characterises the time scale over which the bath force decorrelates, where we have taken the continuum limit in the second equation expressing the sum as an integral over an arbitrary density of oscillators $\rho(\Omega)$. Its effect is to both damp the oscillator and shift its resonance frequency. This can be seen by taking the Fourier transform of the relevant term in Eq. (1.75).

Exercise 1.6 *Neglecting transients that arise from initial conditions, show that*

$$\mathcal{F}\left\{\int_{-\infty}^{t} dt' \mu(t-t')\dot{\hat{q}}(t')\right\} = \mathcal{F}\left\{\int_{-\infty}^{\infty} dt'\, \Theta(t-t')\mu(t-t')\dot{\hat{q}}(t')\right\}$$

$$= (k_{\text{bath}} - i\omega m\gamma(\omega))\,\hat{q}(\omega) \tag{1.79}$$

where the coefficients

$$k_{\text{bath}} = \int_0^\infty d\Omega\, \rho(\Omega) k(\Omega) \tag{1.80a}$$

$$\gamma(\omega) = \frac{k(\omega)\rho(\omega)}{4m}, \tag{1.80b}$$

and $\Theta(t)$ *is the Heaviside step function with Fourier transform* $\Theta(\omega) = \delta(\omega)/2 + i/\omega$. *The Fourier transform properties* $\mathcal{F}\{\dot{f}(t)\} = -i\omega f(\omega)$ *and* $\mathcal{F}\{\cos\Omega t\} = [\delta(\omega - \Omega) + \delta(\omega + \Omega)]/2$ *may be useful. Then take the inverse Fourier transform of Eq. (1.79) to show that*

$$\int_{-\infty}^{t} dt' \, \mu(t - t')\dot{\hat{q}}(t') = k_{\text{bath}}\hat{q}(t) + m \int_{-\infty}^{\infty} dt' \, \gamma(t - t')\dot{\hat{q}}(t'). \qquad (1.81)$$

From this result it is clear that the system-bath interaction introduces both the expected frequency-dependent dissipation $\gamma(\omega)$ and a harmonic potential with spring constant k_{bath} in addition to the bare system potential $\hat{V}(\hat{q})$. This additional potential is somewhat analogous to the Lamb shift exhibited by atomic energy levels due to interaction with the vacuum.

Setting the upper limit of the integral in Eq. (1.81) to t as required to maintain causality, the quantum Langevin equation for the system (Eq. (1.75)) becomes

$$m\ddot{\hat{q}} + m \int_{-\infty}^{t} dt' \, \gamma(t - t')\dot{\hat{q}}(t') + \frac{\partial \hat{V}(\hat{q})}{\partial \hat{q}} = \hat{F}(t), \qquad (1.82)$$

where we have incorporated the bath-induced spring constant into the potential as

$$\hat{V}(\hat{q}) \rightarrow \hat{V}(\hat{q}) + k_{\text{bath}}\hat{q}^2. \qquad (1.83)$$

Thus we see that the complex problem initially introduced of solving the dynamics of a compound system consisting of the quantum system of interest coupled to a large ensemble of bath oscillators is reduced to the problem of solving the dynamics of the system of interest in the presence of damping γ and environmental forcing $\hat{F}(t)$. This is shown schematically in Fig. 1.3b.

Exercise 1.7 *Assuming the bath oscillators are in thermal equilibrium, use the expression in Eq. (1.76) for the bath forcing of the quantum system to show that the force power spectral densities* $S_{FF}(\omega)$ *and* $S_{FF}(-\omega)$ *are*

$$S_{FF}(\omega) = 2m\gamma(\omega)\hbar\omega\,(\bar{n}(\omega) + 1) \qquad (1.84a)$$
$$S_{FF}(-\omega) = 2m\gamma(\omega)\hbar\omega\,\bar{n}(\omega), \qquad (1.84b)$$

where

$$\bar{n}(\omega) = \frac{1}{e^{\hbar\omega/k_B T} - 1}. \qquad (1.85)$$

Use these expressions to verify that the independent-oscillator model satisfies the quantum fluctuation–dissipation theorem of Eq. (1.66), and that the $\gamma(\omega)$ *you obtain is consistent with Eq. (1.80b).*

1.3.1.1 Quantum Langevin equation for a general system operator

In the previous section we derived a quantum Langevin equation for the position of a quantum particle in a potential $\hat{V}(\hat{q})$ coupled to a bath. It is straightforward to extend this result to an arbitrary system operator $\hat{O}$. From the Hamiltonian, Eq. (1.69), the Heisenberg equation of motion is

$$\dot{\hat{O}} = \frac{1}{i\hbar}\left[\hat{O}, \hat{H}_{\text{sys}}\right] + \frac{1}{i\hbar}\left[\hat{O}, \hat{H}_{\text{sys-bath}}\right] \tag{1.86}$$

$$= \frac{1}{i\hbar}\left[\hat{O}, \hat{H}_{\text{sys}}\right] - \frac{1}{2i\hbar}\sum_j k_j \left\{\left[\hat{O}, \hat{q}\right], \hat{q}_j - \hat{q}\right\}_+ \tag{1.87}$$

where the anticommutator is defined as $\{\hat{A}, \hat{B}\}_+ = \hat{A}\hat{B} + \hat{B}\hat{A}$, and we have used the commutation relation property $[\hat{A}, \hat{B}^2] = \{[\hat{A}, \hat{B}], \hat{B}\}_+$ and made the explicit assumption that all system operators commute with the bath ($[\hat{O}, \hat{q}_j] = 0$).

Exercise 1.8 *By substituting in for $\hat{q}_j - \hat{q}$ using Eq. (1.73), show that*

$$\dot{\hat{O}} = \frac{1}{i\hbar}\left[\hat{O}, \hat{H}_{\text{sys}}\right] - \frac{1}{i\hbar}\left[\hat{O}, \hat{q}\right]\hat{F}(t) + \frac{m}{2i\hbar}\left\{\left[\hat{O}, \hat{q}\right], \int_{-\infty}^{t} dt'\, \gamma(t - t')\dot{\hat{q}}(t')\right\}_+ . \tag{1.88}$$

You should use the definitions of $\hat{F}(t)$ and $\mu(t)$ in Eqs. (1.76) and (1.77) as well as Eq. (1.81), and again incorporate the bath-induced spring constant k_{bath} into the system Hamiltonian.

1.3.1.2 Markovian limit

The limit of linear damping due to constant friction is particularly relevant for the dynamics of quantum oscillators and many other quantum systems. Substituting $\gamma(t) = \gamma\delta(t)$ in Eq. (1.82), in what is known as the *first Markov approximation* [113], we obtain the familiar quantum Markovian Langevin equation

$$m\ddot{\hat{q}} + m\gamma\dot{\hat{q}} + \frac{\partial \hat{V}(\hat{q})}{\partial \hat{q}} = \hat{F}(t). \tag{1.89}$$

The same substitution into Eq. (1.88) yields the quantum Markovian Langevin equation for the general observable $\hat{O}$:

$$\dot{\hat{O}} = \frac{1}{i\hbar}\left[\hat{O}, \hat{H}_{\text{sys}}\right] - \frac{1}{i\hbar}\left[\hat{O}, \hat{q}\right]\hat{F}(t) + \frac{m}{2i\hbar}\left\{\left[\hat{O}, \hat{q}\right], \gamma\dot{\hat{q}}(t)\right\}_+ . \tag{1.90}$$

For classical processes, the Markov approximation results in memoryless (or *Markovian*) bath coupling, identified by delta-correlated bath forcing. However, this is not generally the case for quantum processes. In our particular case, within the first Markov approximation the symmetrised force autocorrelation function is [105]

$$\frac{1}{2}\left\langle \hat{F}(t)\hat{F}(t') + \hat{F}(t')\hat{F}(t) \right\rangle = m\gamma kT \frac{d}{dt}\left\{ \coth\left[\frac{\pi k_B T}{\hbar}(t - t') \right] \right\}. \qquad (1.91)$$

This is clearly not a delta function. By contrast, taking the classical limit where $\hbar \to 0$, the autocorrelation function becomes

$$\frac{1}{2}\left\langle \hat{F}(t)\hat{F}(t') + \hat{F}(t')\hat{F}(t) \right\rangle_{\hbar \to 0} = 2m\gamma k_B T \delta(t - t'), \qquad (1.92)$$

as expected for a memoryless process. We see, therefore, that linear coupling between a bath and a quantum oscillator is not, strictly speaking, Markovian. However, the function in Eq. (1.91) features a sharp peak at $t - t'$ with characteristic time scale $\hbar/k_B T$. When the uncertainty $\hbar\gamma$ in the energy of quanta decaying from the oscillator is much less than the thermal energy $k_B T$, the oscillator decay time $2\pi/\gamma$ is much less than this characteristic time scale and the force $\hat{F}(t)$ can, for all intents and purposes, be treated as delta correlated and the bath coupling as Markovian. For an oscillator at room temperature $\hbar/k_B T \sim 10$ fs, while at 10 mK $\hbar/k_B T \sim 1$ ns, in both cases significantly shorter than the decay time of oscillators used in quantum optomechanics experiments. We therefore refer to Eq. (1.90) as a *Markov quantum Langevin equation*, though it should be appreciated that this is only approximately valid.

It is generally convenient to reexpress the Markov quantum Langevin equation of Eq. (1.90) in terms of the dimensionless position operator $\hat{Q}$ (see Eqs. (1.13)) and the dimensionless input momentum fluctuations

$$\hat{P}_{\text{in}}(t) = \frac{i}{\sqrt{2}}\left(a^{\dagger}_{\text{in}}(t) - a_{\text{in}}(t) \right) \qquad (1.93)$$

$$\equiv \frac{x_{zp}\hat{F}(t)}{\hbar\sqrt{\gamma}}. \qquad (1.94)$$

The Markovian quantum Langevin equation is then

$$\dot{\hat{O}} = \frac{1}{i\hbar}\left[\hat{O}, \hat{H}_{\text{sys}} \right] + i\sqrt{2\gamma}\left[\hat{O}, \hat{Q} \right]\hat{P}_{\text{in}}(t) + \frac{1}{2iQ}\left\{ \left[\hat{O}, \hat{Q} \right], \dot{\hat{Q}}(t) \right\}_{+}, \qquad (1.95)$$

where $Q = \Omega/\gamma$ is the oscillator's quality factor, not to be confused with $\hat{Q}$, its dimensionless position operator.

1.3.2   Harmonic oscillator dynamics

Langevin equations provide a method to study the dynamics of general quantum systems interacting linearly with their environment. Here, as a particularly relevant example, we use Eq. (1.90) to study the simple case of a simple harmonic oscillator interacting with a Markovian bath. The dynamics of the

Heisenberg picture annihilation operator $a(t)$ can be determined by substituting the Hamiltonian of a simple harmonic oscillator (Eq (1.19)) into the quantum Langevin equation of Eq. (1.90) and setting $\hat{\mathcal{O}} = a$. This results in the dynamical stochastic equation of motion

$$\dot{a} = -i\Omega a + \frac{\gamma}{2}\left(a^\dagger - a\right) + \frac{ix_{zp}\hat{F}(t)}{\hbar}, \tag{1.96}$$

Exercise 1.9 *Show the above result.*

The dynamical equations governing the position and momentum operators are then straightforwardly found to be

$$\dot{\hat{Q}} = \Omega\hat{P} \tag{1.97a}$$
$$\dot{\hat{P}} = -\Omega\hat{Q} - \gamma\hat{P} + \sqrt{2\gamma}\hat{P}_{\text{in}} \tag{1.97b}$$

using the relations in Eqs. (1.13). From these equations of motion we immediately observe, as expected for viscous memoryless forcing, the damping and fluctuations only appear on the momentum of the oscillator.

The positive and negative frequency power spectral densities of $\hat{P}_{\text{in}}(t)$ can be found by taking the Fourier transform of Eq. (1.94) and using Eqs. (1.43) and (1.84), with the result

$$S_{P_{\text{in}}P_{\text{in}}}(\omega) = \frac{\omega}{\Omega}\left(\bar{n}(\omega) + 1\right) \tag{1.98a}$$
$$S_{P_{\text{in}}P_{\text{in}}}(-\omega) = \frac{\omega}{\Omega}\bar{n}(\omega). \tag{1.98b}$$

The symmetrised bath power spectral density $\bar{S}_{P_{\text{in}}P_{\text{in}}}(\omega) = (S_{P_{\text{in}}P_{\text{in}}}(-\omega) + S_{P_{\text{in}}P_{\text{in}}}(\omega))/2$ is plotted as a function of frequency ω in Fig. 1.4. As can be seen, at frequencies for which $\hbar\omega/k_BT \ll 1$ the bath power spectral density is flat and roughly equal to $\hbar\Omega/k_BT$, while, in contrast to classical Markovian processes that exhibit a spectrally flat power spectrum, at higher frequencies the bath power spectral density grows linearly with frequency.

It is often convenient to make the so-called *quantum optics approximation*, treating the bath power spectral densities in Eqs. (1.98) to be spectrally flat and equal to

$$S_{P_{\text{in}}P_{\text{in}}}(\omega) = \bar{n} + 1 \tag{1.99a}$$
$$S_{P_{\text{in}}P_{\text{in}}}(-\omega) = \bar{n}, \tag{1.99b}$$

where henceforth in the textbook we define $\bar{n} \equiv \bar{n}(\Omega)$, i.e., whenever the ω argument is missing, the mean photon number should be taken as the value at the oscillator resonance frequency Ω. This approximation is accurate in two regimes. The first regime is a semiclassical regime where $\hbar\omega/k_BT \ll 1$ and the bath fluctuations are dominated by thermal noise. As discussed in the previous paragraph, the power spectral density is flat in this regime. The second regime

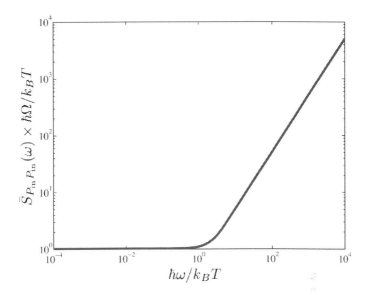

FIGURE 1.4 Symmetrised bath power spectral density, $\bar{S}_{P_{\text{in}} P_{\text{in}}}(\omega) = (S_{P_{\text{in}} P_{\text{in}}}(-\omega) + S_{P_{\text{in}} P_{\text{in}}}(\omega))/2$.

is one where the oscillator is of sufficiently high quality that all frequencies of interest lie within a narrow band around Ω over which, even in the limit $\hbar\omega/k_B T \gtrsim 1$, the power spectral density is essentially flat. Generally, micro- and nano-mechanical oscillators lie within the first regime, while optical cavity fields lie within the second.

1.3.2.1 Susceptibility of an oscillator

The mechanical dynamics of the oscillator can be written fully in terms of position by substituting into Eq. (1.97b) for $\hat{P}$ from Eq. (1.97a). This yields

$$\ddot{\hat{Q}} + \gamma\dot{\hat{Q}} + \Omega^2\hat{Q} = \sqrt{2\gamma}\Omega\hat{P}_{\text{in}}. \tag{1.100}$$

This equation can be conveniently solved in the frequency domain. Taking the Fourier transform, we find

$$\hat{Q}(\omega) = \sqrt{2\gamma}\chi(\omega)\hat{P}_{\text{in}}(\omega), \tag{1.101}$$

where $\chi(\omega)$ is the *susceptibility* which quantifies the response of the oscillator to external forces as a function of frequency and is defined here as

$$\chi(\omega) \equiv \frac{\Omega}{\Omega^2 - \omega^2 - i\omega\gamma}. \tag{1.102}$$

This susceptibility is a generalised Lorentzian function that for a high quality oscillator ($Q \equiv \Omega/\gamma \gg 1$) exhibits a sharp peak at the resonance frequency of the oscillator, decays to a non-zero plateau near $\omega = 0$ with $|\chi(0)/\chi(\Omega)| = Q^{-1}$, and decays towards zero at sufficiently large frequencies (see Fig 1.5).

1.3.3 Determining the temperature of an oscillator via sideband asymmetry

In Section 1.2.3 we introduced a definition of the temperature of an oscillator based on the ratio of the positive and negative force power spectral densities at the oscillator frequency. This definition comes from the fact that, in thermal equilibrium, the force power spectral density at Ω is greater than at $-\Omega$ (see Eqs. (1.99)). This *sideband asymmetry* in forcing results in an equivalent sideband asymmetry in the directly measurable position power spectral density of the oscillator. Equation (1.101) connects the mechanical position to the force noise in the frequency domain. The position power spectral density at $\pm\Omega$ can then be determined using Eq. (1.43) to be

$$S_{QQ}(\Omega) = 2\gamma |\chi(\Omega)|^2 S_{P_{\text{in}}P_{\text{in}}}(\Omega) \tag{1.103a}$$

$$S_{QQ}(-\Omega) = 2\gamma |\chi(-\Omega)|^2 S_{P_{\text{in}}P_{\text{in}}}(-\Omega). \tag{1.103b}$$

From Eq. (1.102) we observe that $|\chi(-\Omega)|^2 = |\chi(\Omega)|^2$, and therefore that the temperature and phonon number of an oscillator in thermal equilibrium given in Eq. (1.54) can be reexpressed directly in terms of the measurable mechanical position power spectral density as

$$T = \frac{\hbar\Omega}{k_B} \left[\ln \left(\frac{S_{QQ}(\Omega)}{S_{QQ}(-\Omega)} \right) \right]^{-1} \tag{1.104a}$$

$$\bar{n} = \left(\frac{S_{QQ}(\Omega)}{S_{QQ}(-\Omega)} - 1 \right)^{-1}. \tag{1.104b}$$

A particularly attractive feature of using the ratio of positive and negative frequency power spectral densities to determine the temperature of an oscillator is that in principle it is an absolute, calibration-free, measurement. As such, Eqs. (1.104) are routinely used for this purpose in ion trapping experiments [145, 86, 202] and more recently have been applied in several quantum optomechanics experiments (see, for example, [244, 48, 308]).

1.3.4 Determining the effective oscillator temperature from the integral of $\bar{S}_{QQ}(\omega)$

There are some circumstances where the positive and negative frequency components of the position power spectral density are not independently directly measurable – most notably in a homodyne measurement of the phase of an optical or microwave field after interaction with the oscillator (see Section 3.3.3).

Such measurements, instead, determine the symmetrised power spectral density $\bar{S}_{QQ}(\omega) = [S_{QQ}(\omega) + S_{QQ}(-\omega)]/2$. The symmetrised power spectral density also allows the temperature of the oscillator to be determined, although an accurate calibration is required.

An oscillator with spring constant k has potential energy $\hat{U} = k\hat{q}^2/2$. Its mean potential energy averaged over a long time is therefore

$$\bar{U} = \frac{1}{2}k \lim_{\tau \to \infty} \frac{1}{\tau} \int_{-\tau/2}^{\tau/2} \langle \hat{q}^2(t) \rangle dt. \qquad (1.105)$$

Using Parseval's theorem,[6] this can be expressed in terms of the position power spectral density of the oscillator as

$$\bar{U} = \frac{k}{4\pi} \int_{-\infty}^{\infty} d\omega \, S_{qq}(\omega) \qquad (1.106)$$

$$= \frac{k}{4\pi} \int_{0}^{\infty} d\omega \, [S_{qq}(\omega) + S_{qq}(-\omega)] \qquad (1.107)$$

$$= \frac{k}{2\pi} \int_{0}^{\infty} d\omega \, \bar{S}_{qq}(\omega). \qquad (1.108)$$

We see from this that the symmetrised power spectral density $\bar{S}_{qq}(\omega)$ defined in Section 1.2.4 is intimately connected to the energy of the oscillator. In thermal equilibrium, the mean kinetic and potential energies of the oscillator are equal so that the time–averaged total energy is

$$\bar{E} = \hbar\Omega\left(\bar{n} + \frac{1}{2}\right) = \frac{k}{\pi} \int_{0}^{\infty} d\omega \, \bar{S}_{qq}(\omega). \qquad (1.109)$$

The allows the mean number of quanta in the oscillator to be experimentally determined from the position power spectral density. The temperature of the oscillator can then be easily found via Eqs. (1.54). Written in terms of the dimensionless position operator, Eq. (1.109) simplifies to

$$\frac{1}{\pi} \int_{0}^{\infty} d\omega \, \bar{S}_{QQ}(\omega) = \bar{n} + \frac{1}{2}. \qquad (1.110)$$

Exercise 1.10 *Analytically calculate the integral in Eq. (1.110) in the limit $\Omega \gg \gamma$ to confirm that the independent-oscillator model returns the correct prediction for $\bar{n}$ in that limit. You should use the mechanical bath power spectral densities in Eqs. (1.99) and the frequency domain expression for $\hat{Q}(\omega)$ in Eq. (1.101).*

For an oscillator that is out of thermal equilibrium, it is natural to define an effective temperature in the same spirit as Eq. (1.55), with Eq. (1.110) used to determine the effective mean number of quanta in the oscillator.

---

[6]For an operator $\hat{O}$, Parseval's theorem states that $\lim_{\tau \to \infty} \frac{1}{\tau} \int_{-\tau/2}^{\tau/2} dt \, \langle \hat{O}^\dagger(t)\hat{O}(t) \rangle = \frac{1}{2\pi} \int_{-\infty}^{\infty} d\omega \, S_{OO}(\omega) = \frac{1}{\pi} \int_{0}^{\infty} d\omega \, \bar{S}_{OO}(\omega)$.

1.4 QUANTUM LANGEVIN EQUATION WITHIN THE ROTATING WAVE APPROXIMATION

Generally in quantum optics, and often in optomechanics, the bath coupling rate is much smaller than other relevant rates in the system. In this case, it is convenient to perform a rotating wave approximation on the system-bath Hamiltonian of Eq. (1.69), neglecting non-energy-conserving terms, and derive a new rotating wave quantum Langevin equation. This proceeds by substituting $\hat{q} = x_{zp}(a^\dagger + a)$, $\hat{p} = ip_{zp}(a^\dagger - a)$, $\hat{q}_j = x_{zp,j}(a_j^\dagger + a_j)$, $\hat{p}_j = ip_{zp,j}(a_j^\dagger - a_j)$ into Eq. (1.69) and neglecting the counterrotating terms aa, $a^\dagger a^\dagger$, $a_j^\dagger a_j^\dagger$, and $a_j a_j$. We further, make the canonical transformation $a_j \to ia_j$ to maintain consistency with the literature and for notational convenience later on. This results in a new Hamiltonian of the form

$$\hat{H}_{\text{RWA}} = \hat{H}_{\text{sys}} + \sum_j \left[\hbar\Omega_j a_j^\dagger a_j + i\hbar\gamma_j \left(a_j a^\dagger - a_j^\dagger a \right) \right], \qquad (1.111)$$

where γ_j defines the strength of coupling to each bath oscillator j, and we have neglected the zero-point energy of the bath, which has no effect on the system dynamics. A rotating wave quantum Langevin equation can then be derived in a similar approach to that followed above. We refer the reader to [113] for details of the procedure. In the Markovian limit, the resulting equation is

$$\dot{\hat{O}} = \frac{1}{i\hbar}[\hat{O}, \hat{H}_{\text{sys}}] - \left[\hat{O}, a^\dagger\right] \left(\frac{\gamma}{2}a - \sqrt{\gamma}a_{\text{in}}(t)\right) + \left(\frac{\gamma}{2}a^\dagger - \sqrt{\gamma}a_{\text{in}}^\dagger(t)\right) \left[\hat{O}, a\right]. \qquad (1.112)$$

Here the forcing term is expressed as an input noise operator

$$a_{\text{in}}(t) \equiv \frac{1}{2\pi} \int_{-\infty}^{\infty} d\omega \, e^{-i\omega(t-t_-)} a_-(\omega), \qquad (1.113)$$

where $a_-(\omega)$ is the annihilation operator of the bath oscillator with resonance frequency ω at some initial time $t_- < t$ and arises from taking a continuum limit of the sum of bath operators a_j. Note that the taking to negative infinity of the integral in Eq. (1.113) is clearly unphysical since it then includes bath oscillators with negative frequencies. However, as long as the system dynamics are constrained to a narrow band of frequencies that are well separated from zero, as is the case for a high-quality harmonic oscillator, the approximation is reasonable. Within this approximation, Eq. (1.113) is simply an inverse Fourier transform (c.f. Eq. (1.27b)), such that, in the frequency domain, the input noise operator is $a_{\text{in}}(\omega) = a_-(\omega)e^{i\omega t_-}$.

By inspection of Eq. (1.113) we see immediately that the units of $a_{\text{in}}(t)$ and $a_-(\omega)$ are not the same. Indeed, $\hat{n}_{\text{in}}(t) = a_{\text{in}}^\dagger(t)a_{\text{in}}(t)$ is the flux of quanta (e.g., phonons or photons) incident on the system from the environment at time t, with units of s^{-1}. In this picture, the system can be thought of as interacting at each time t with a delta-like bath mode defined by Eq. (1.113) that is independent of other bath modes at earlier or later times.

1.4.1 Commutation and correlation relations

The input noise operator $a_{\text{in}}(t)$ and its adjoint $a_{\text{in}}^{\dagger}(t)$ obey the commutation and correlation relations

$$[a_{\text{in}}(t), a_{\text{in}}^{\dagger}(t')] = \delta(t - t') \tag{1.114a}$$

$$[a_{\text{in}}(t), a_{\text{in}}(t')] = [a_{\text{in}}^{\dagger}(t), a_{\text{in}}^{\dagger}(t')] = 0, \tag{1.114b}$$

with $\delta(t)$ being the Dirac delta function, and, assuming that the input is in a thermal state, the correlation relations

$$\left\langle a_{\text{in}}^{\dagger}(t) a_{\text{in}}(t') \right\rangle = \bar{n}\delta(t - t') \tag{1.115a}$$

$$\left\langle a_{\text{in}}(t) a_{\text{in}}^{\dagger}(t') \right\rangle = (\bar{n} + 1)\,\delta(t - t') \tag{1.115b}$$

$$\langle a_{\text{in}}(t) a_{\text{in}}(t') \rangle = \left\langle a_{\text{in}}^{\dagger}(t) a_{\text{in}}^{\dagger}(t') \right\rangle = 0. \tag{1.115c}$$

The corresponding frequency domain commutation and correlation relations may be straightforwardly derived from the respective time domain expressions, and are identical, except with the substitution $t \to \omega$ throughout.

Exercise 1.11 *Given that $[a_{-}(\omega), a_{-}^{\dagger}(\omega')] = \delta(\omega - \omega')$, derive the commutation relations in Eq. (1.114).*

From Eqs. (1.114) and (1.115) it can be directly shown that the dimensionless bath position and momentum operators $\hat{Q}_{\text{in}} = (a_{\text{in}}^{\dagger} + a_{\text{in}})/\sqrt{2}$ and $\hat{P}_{\text{in}} = i(a_{\text{in}}^{\dagger} - a_{\text{in}})/\sqrt{2}$ satisfy the commutation and correlation relations

$$[\hat{Q}_{\text{in}}(t), \hat{P}_{\text{in}}(t')] = i\delta(t - t') \tag{1.116a}$$

$$[\hat{Q}_{\text{in}}(t), \hat{Q}_{\text{in}}(t')] = [\hat{P}_{\text{in}}(t), \hat{P}_{\text{in}}(t')] = 0 \tag{1.116b}$$

$$\langle \hat{Q}_{\text{in}}(t)\hat{Q}_{\text{in}}(t') \rangle = \langle \hat{P}_{\text{in}}(t)\hat{P}_{\text{in}}(t') \rangle = \left(\bar{n} + \frac{1}{2}\right)\delta(t - t') \tag{1.116c}$$

$$\langle \hat{Q}_{\text{in}}(t)\hat{P}_{\text{in}}(t') \rangle = -\langle \hat{P}_{\text{in}}(t)\hat{Q}_{\text{in}}(t') \rangle = \frac{i}{2}\delta(t - t'), \tag{1.116d}$$

while the frequency domain position and momentum operators $\hat{Q}_{\text{in}}(\omega) = (a_{\text{in}}^{\dagger}(-\omega) + a_{\text{in}}(\omega))/\sqrt{2}$ and $\hat{P}_{\text{in}}(\omega) = i(a_{\text{in}}^{\dagger}(-\omega) - a_{\text{in}}(\omega))/\sqrt{2}$ satisfy

$$[\hat{Q}_{\text{in}}^{\dagger}(\omega), \hat{P}_{\text{in}}(\omega')] = [\hat{Q}_{\text{in}}(\omega), \hat{P}_{\text{in}}^{\dagger}(\omega')] = i\delta(\omega - \omega') \tag{1.117a}$$

$$[\hat{Q}_{\text{in}}(\omega), \hat{Q}_{\text{in}}(\omega')] = [\hat{P}_{\text{in}}(\omega), \hat{P}_{\text{in}}(\omega')] = 0 \tag{1.117b}$$

$$\langle \hat{Q}_{\text{in}}^{\dagger}(\omega)\hat{Q}_{\text{in}}(\omega') \rangle = \langle \hat{P}_{\text{in}}^{\dagger}(\omega)\hat{P}_{\text{in}}(\omega') \rangle = \left(\bar{n} + \frac{1}{2}\right)\delta(\omega - \omega') \tag{1.117c}$$

$$\langle \hat{Q}_{\text{in}}^{\dagger}(\omega)\hat{P}_{\text{in}}(\omega') \rangle = -\langle \hat{P}_{\text{in}}^{\dagger}(\omega)\hat{Q}_{\text{in}}(\omega') \rangle = \frac{i}{2}\delta(\omega - \omega'). \tag{1.117d}$$

1.4.2 Power spectral densities and uncertainty relations of the bath within the rotating wave approximation

From the correlation relations in Eqs. (1.117c) and (1.117d) the power spectral densities of the dimensionless bath position and momentum operators can be shown within the rotating wave approximation to be

$$S_{Q_{in}Q_{in}}(\pm\omega) = S_{P_{in}P_{in}}(\pm\omega) = \bar{n} + \frac{1}{2} \qquad (1.118a)$$

$$S_{Q_{in}P_{in}}(\pm\omega) = -S_{P_{in}Q_{in}}(\pm\omega) = \frac{i}{2}, \qquad (1.118b)$$

where $S_{Q_{in}P_{in}}(\pm\omega)$ and $S_{P_{in}Q_{in}}(\pm\omega)$ are cross-spectral densities between the bath position and momentum operators. To arrive at these expressions, we have used Eq. (1.43) and the cross-spectral density which is defined in direct extension of Eq. (1.43) as

$$S_{AB}(\omega) = \int_{-\infty}^{\infty} d\tau\, e^{i\omega\tau} \left\langle \hat{A}^\dagger(t+\tau)\hat{B}(t) \right\rangle_{t=0} = \int_{-\infty}^{\infty} d\omega' \left\langle \hat{A}^\dagger(-\omega)\hat{B}(\omega') \right\rangle \tag{1.119}$$

for two arbitrary operators $\hat{A}$ and $\hat{B}$. Unlike the self-power spectral density $S_{AA}(\omega)$, the cross-power spectral density can be complex.

Exercise 1.12 *Using Eq. (1.119), show that*

$$S_{AB}(\omega) = S_{BA}^*(\omega). \tag{1.120}$$

We see from Eqs. (1.118) that, in contrast to the nonrotating frame bath power spectral densities (see Eqs. (1.99)), within the rotating frame the bath power spectral densities are symmetric in frequency.

The commutation relations in Eqs. (1.117a) result in an uncertainty principle between the position and momentum power spectra of the bath [258],

$$S_{Q_{in}Q_{in}}(\omega)S_{P_{in}P_{in}}(\omega) \geq \frac{1}{4}, \tag{1.121}$$

which can be seen from Eqs. (1.118a) to be saturated for a vacuum state $(\bar{n} = 0)$.[7]

---

[7]This can be derived using a similar approach to that which we adopt in Section 4.4.8 to treat entanglement between a mechanical oscillator and an external optical field. Briefly, by defining an appropriately normalised bath spectral mode $u(\omega)$ with dimensions position and momentum operators $\hat{Q}_{u,in}(\omega) = u(\omega)\hat{Q}_{in}(\omega)$ and $\hat{P}_{u,in}(\omega) = u(\omega)\hat{P}_{in}(\omega)$; calculating the power spectral densities of these quadratures in the case where $u(\omega) = \delta(\omega')$, i.e. for a mode that corresponds to the bath fluctuations at some frequency ω'; using the power spectral densities to determine the quadrature variances of the mode u; and thereby showing that $\sigma^2(\hat{Q}_{u,in}) = S_{Q_{in}Q_{in}}(\omega')$ and $\sigma^2(\hat{P}_{u,in}) = S_{P_{in}P_{in}}(\omega')$. The result then follows from Eq. (1.117a).

Exercise 1.13 *Using Eqs. (1.118) show that the symmetrised power spectral density of a Hermitian operator of the form $\hat{A}(t) = f(t) * \hat{Q}_{\text{in}}(t) + g(t) * \hat{P}_{\text{in}}(t)$ is*

$$\bar{S}_{\mathcal{A}\mathcal{A}}(\omega) = |f(\omega)|^2 S_{Q_{\text{in}}Q_{\text{in}}}(\omega) + |g(\omega)|^2 S_{P_{\text{in}}P_{\text{in}}}(\omega) \quad (1.122)$$

$$= \left(|f(\omega)|^2 + |g(\omega)|^2\right)\left(\bar{n} + \frac{1}{2}\right), \quad (1.123)$$

where $f(t)$ and $g(t)$ are arbitrary functions which, if $\hat{A}(t)$ is Hermitian, must be real.

The above exercise provides a useful result, that in the measured symmetrised power spectral density of any observable, the bath cross-correlation terms $S_{Q_{\text{in}}P_{\text{in}}}(\pm\omega)$ and $S_{P_{\text{in}}Q_{\text{in}}}(\pm\omega)$ cancel out.

1.4.3   Input-output relations

A similar procedure to that described above but defining the initial condition of the bath at a time $t_+ > t$ allows output modes to be defined analogously to the input modes of Eq. (1.113) [113],

$$a_{\text{out}}(t) \equiv \frac{1}{2\pi} \int_{-\infty}^{\infty} d\omega\, e^{-i\omega(t-t_+)} a_+(\omega), \quad (1.124)$$

where $a_+(\omega)$ is the annihilation operator of the bath oscillator with resonance frequency ω at time t_+. It is straightforward to show that a_{out} satisfies the same commutation relations as a_{in}. One can then derive a time-reversed quantum Langevin equation that, with the requirement of consistency with Eq. (1.112), enforces the input-output relation [113]

$$a_{\text{out}}(t) = a_{\text{in}}(t) - \sqrt{\gamma}a(t). \quad (1.125)$$

Since this is a linear relation, it translates directly into the position and momentum operators

$$\hat{Q}_{\text{out}}(t) = \hat{Q}_{\text{in}}(t) - \sqrt{\gamma}\hat{Q}(t) \quad (1.126\text{a})$$
$$\hat{P}_{\text{out}}(t) = \hat{P}_{\text{in}}(t) - \sqrt{\gamma}\hat{P}(t). \quad (1.126\text{b})$$

In many circumstances, it is difficult to experimentally access or control the input and output from quantum systems. For instance, the motion of a mechanical oscillator drives energy into its environment, introducing phonon losses. Typically, these phonon losses go into many vibrational modes of the environment and cannot be isolated to enable them to be measured or injected into another quantum system. However, in some circumstances it is possible to engineer the coupling rate to an isolated channel so that it is comparable to, or exceeds, the sum of all uncontrolled coupling rates. A good example of this is an optical cavity, where the transmission of the input/output coupler can generally be designed to achieve a coupling rate greater than uncontrolled coupling rates to the environment due to photon scattering and absorption. A similar approach is possible for a mechanical oscillator by engineering its coupling to a phonon waveguide. In situations where an appreciable coupling rate can be achieved to a known isolated channel, Eq. (1.125) is exceptionally useful to determine the properties of the field within the channel after interacting with the system.

1.4.4 Harmonic oscillator dynamics within the rotating wave approximation

In Section 1.3.2 we examined the dynamics of a harmonic oscillator without taking the rotating wave approximation. While the rotating wave approximation is a powerful tool that allows many problems to be greatly simplified, it is important to understand the limits of its applicability. To this end, we now examine the dynamics that are predicted when the rotating wave approximation is used.

The dynamics of the oscillator within the rotating wave approximation can be determined analogously to our previous results for the general dynamics. Using the rotating wave quantum Langevin equation of Eq. (1.112) we find the equations of motion

$$\dot{a} = -i\Omega a - \frac{\gamma}{2}a + \sqrt{\gamma}a_{\text{in}} \tag{1.127a}$$

$$\dot{\hat{Q}} = \Omega\hat{P} - \frac{\gamma}{2}\hat{Q} + \sqrt{\gamma}\hat{Q}_{\text{in}} \tag{1.127b}$$

$$\dot{\hat{P}} = -\Omega\hat{Q} - \frac{\gamma}{2}\hat{P} + \sqrt{\gamma}\hat{P}_{\text{in}}. \tag{1.127c}$$

We see that the counterrotating term from Eq. (1.96) is missing in Eq. (1.127a), as one might expect from the rotating wave approximation, and the dissipation and fluctuations are now equally spread between the position and momentum. This makes some sense, since the rotating wave approximation can be thought of as averaging out the bath interactions over a full period of oscillation.

1.4.4.1 Mechanical susceptibility within the rotating wave approximation

Taking the Fourier transform of Eqs. (1.127b) and (1.127c) and neglecting transients due to the initial conditions yields the simultaneous equations

$$\hat{Q}(\omega) = \left(\frac{1}{\gamma/2 - i\omega}\right)\left(\Omega\hat{P}(\omega) + \sqrt{\gamma}\hat{Q}_{\text{in}}(\omega)\right) \tag{1.128a}$$

$$\hat{P}(\omega) = \left(\frac{1}{\gamma/2 - i\omega}\right)\left(-\Omega\hat{Q}(\omega) + \sqrt{\gamma}\hat{P}_{\text{in}}(\omega)\right). \tag{1.128b}$$

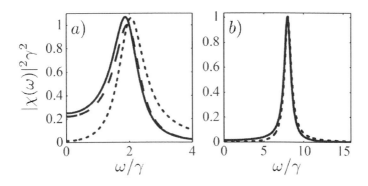

FIGURE 1.5 Susceptibility of an oscillator with and without the rotating wave approximation as a function of frequency. Solid lines: $\chi(\omega)$, dotted lines: $\chi_Q^{\mathrm{RWA}}(\omega)$, dashed lines: $\chi_P^{\mathrm{RWA}}(\omega)$. (a) $Q = \Omega/\gamma = 2$. (b) $Q = \Omega/\gamma = 8$.

These can be solved simultaneously for $\hat{Q}(\omega)$ with the result

$$\hat{Q}(\omega) = \sqrt{\gamma}\chi_P^{\mathrm{RWA}}(\omega)\hat{P}_{\mathrm{in}}(\omega) + \sqrt{\gamma}\chi_Q^{\mathrm{RWA}}(\omega)\hat{Q}_{\mathrm{in}}(\omega), \qquad (1.129)$$

where we see that, within the rotating wave approximation, the mechanical position experiences forcing from both the position and momentum of the bath, with susceptibilities

$$\chi_Q^{\mathrm{RWA}}(\omega) = \frac{\gamma/2 - i\omega}{\Omega^2 - \omega^2 - i\gamma\omega + (\gamma/2)^2} \qquad (1.130a)$$

$$\chi_P^{\mathrm{RWA}}(\omega) = \frac{\Omega}{\Omega^2 - \omega^2 - i\gamma\omega + (\gamma/2)^2}. \qquad (1.130b)$$

This is clearly quite different from the non-approximated solution (Eq. (1.101)), both in the functional form of the susceptibility and in the introduction of position driving. To illustrate the differences, the susceptibilities with and without the rotating wave approximation are plotted for two different oscillator quality factors in Fig. 1.5. It is evident that, for low frequency oscillators (Fig. 1.5a), the momentum forcing is underestimated at low frequencies,[8] while the additional position forcing is negligible at frequencies $\omega \ll \Omega$ but is comparable to the momentum forcing at $\omega = \Omega$, and dominates at $\omega \gg \Omega$. However, even for moderate quality factors of $Q = \Omega/\gamma = 8$ (Fig. 1.5b), the rotating wave approximation accurately reproduces the correct susceptibility. Indeed, in the limit of a high-quality oscillator, where both

---

[8]This is underrepresented in the figure since Eq. (1.129) is also missing a factor of $\sqrt{2}$ present in Eq. (1.101).

the detuning $\Delta = \Omega - \omega$ and decay rate γ are much smaller than the resonance frequency Ω, it can be seen that both terms in brackets in Eq. (1.130) approach $\chi(\omega)$ (c.f. Eq. (1.102)).

Generally, as long as $\{\Delta, \gamma\} \ll \Omega$, the rotating wave approximation is a good approximation for the dynamics of a harmonic oscillator, although it may even break down in this regime depending on the specific application – for instance, if the dynamics of interest occur at time scales fast compared with Ω^{-1}.

1.4.5 Transformation into a rotating frame

When using the rotating wave approximation, it is natural to also move into a frame (or interaction picture) rotating at a frequency close to the resonance frequency of the harmonic oscillator. To do this, we perform the Hamiltonian transformation

$$\tilde{H} = \hat{U}^\dagger \hat{H} \hat{U} - \hat{A} \tag{1.131}$$

where the $\tilde{}$ accent is used to identify that the new Hamiltonian is in the rotating frame,

$$\hat{A} = \hbar \Omega_r a^\dagger a, \tag{1.132}$$

and

$$\hat{U} = e^{-i\hat{A}t/\hbar}, \tag{1.133}$$

with Ω_r being the frequency of rotation. This transformation can be derived straightforwardly by applying the unitary $\hat{U}^\dagger$ to the Schrödinger equation and is valid as long as the operator a has no explicit time dependence.

Exercise 1.14 *Convince yourself of this.*

Within the rotating frame, the annihilation operator $\tilde{a}$ has the explicit time dependence

$$\tilde{a}(t) = \hat{U}^\dagger a \hat{U} = ae^{-i\Omega_r t}, \tag{1.134}$$

where here, for clarity, we have identified annihilation operators in the rotating frame with $\tilde{}$ accents. Clearly, within the rotating frame the annihilation and creation operators retain their usual Boson commutation relation (Eq. (1.2)). Generally, within this text the context within which the operators are used is sufficient to identify whether they are rotating or stationary, and unless the situation is particularly unclear, we will typically drop the accent and use a in both cases.

1.4.6 Quadrature operators

From the rotating-frame annihilation operator, we may define rotating-frame operators $\hat{X}$ and $\hat{Y}$ analogous to the dimensionless position $\hat{Q}$ and momentum

$\hat{P}$ operators

$$\hat{X}(t) = \frac{1}{\sqrt{2}} \left(\tilde{a}^\dagger + \tilde{a} \right) = \frac{1}{\sqrt{2}} \left(a^\dagger e^{i\Omega_r t} + a e^{-i\Omega_r t} \right) = \hat{Q}^{\Omega_r t} \quad (1.135a)$$

$$\hat{Y}(t) = \frac{i}{\sqrt{2}} \left(\tilde{a}^\dagger - \tilde{a} \right) = \frac{i}{\sqrt{2}} \left(a^\dagger e^{i\Omega_r t} - a e^{-i\Omega_r t} \right) = \hat{P}^{\Omega_r t}, \quad (1.135b)$$

where $\hat{Q}^{\Omega_r t}$ and $\hat{P}^{\Omega_r t}$ are position and momentum operators rotated by an angle $\theta = \Omega_r t$ (see Eqs. (1.17)), and $[\hat{X}, \hat{Y}] = i$ due to the Boson commutation relation between $\tilde{a}$ and $\tilde{a}^\dagger$. We see that the operators $\hat{X}(t)$ and $\hat{Y}(t)$ alternate in time between the nonrotating position and momentum at frequency Ω_r. These rotating-frame dimensionless position and momentum operators are often referred to as *quadrature operators*. In the specific case of a mechanical oscillator they are termed the *position and momentum quadratures*, while, for an optical field, they are termed the *amplitude and phase quadratures*.

It is straightforward to show that within the rotating wave approximation the commutation relations, correlation relations, and power spectral densities of the rotating frame bath operators are identical to those given in Eqs. (1.114) to (1.121), with the substitutions $\hat{Q} \to \hat{X}$ and $\hat{P} \to \hat{Y}$. The input-output relations of Eqs. (1.125) and (1.126) also directly apply with the same substitution.

1.4.7 Energy decay with and without the rotating wave approximation

In Section 1.2 we derived an equation for the damping rate of the mean number of quanta in a harmonic oscillator by considering the upwards and downwards going transition rates driven by an external force. However, this derivation was performed for the special case of an initial Fock state $|n\rangle$. Here we briefly revisit the calculation using quantum Langevin equations. To proceed, we substitute $\hat{O} = \hat{n} = \hat{b}^\dagger \hat{b}$ into both Eq. (1.90) for the full Markov quantum Langevin equation and Eq. (1.112) for the rotating wave approximation quantum Langevin equations in the Markov regime. This results in the equations of motion

$$\dot{\hat{n}} = \sqrt{\gamma} \hat{P}_{\text{in}} \hat{P} - \gamma \hat{P}^2 \quad (1.136a)$$

$$\dot{\hat{n}}_{\text{RWA}} = \sqrt{\gamma} \left(a_{\text{in}} a^\dagger + a_{\text{in}}^\dagger a \right) - \gamma \hat{n}_{\text{RWA}}, \quad (1.136b)$$

where $\hat{n}_{\text{RWA}}$ denotes the phonon number operator in the rotating wave approximation.

Exercise 1.15 *Derive these expressions.*

We observe that, while the rotating wave approximation predicts a damping rate proportional to the number of quanta in the oscillator, consistent with Eq. (1.61) earlier, the full quantum Langevin equation predicts the correct momentum-squared damping. Figure 1.6 shows the mean energy decay of a mechanical oscillator initially excited with a large coherent displacement

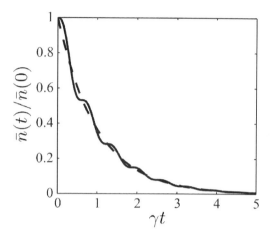

FIGURE 1.6 Decay of the mean energy in a damped harmonic oscillator with quality factor $Q = \Omega/\gamma = 5$ using the complete model (solid line) and the rotating wave approximation (dashed line).

calculated from both Eqs. (1.136). As can be seen, the overall exponential decay is in agreement. However, the full dynamics exhibit oscillations between zero damping when $\langle \hat{P} \rangle = 0$ to maximum damping when $\langle \hat{P} \rangle$ is maximised that are not present when using the rotating wave approximation. This discrepancy becomes less severe as the quality factor of the oscillator increases, and in most cases has a negligible effect on calculations for $Q = \Omega/\gamma \gtrsim 10$.

CHAPTER 2

Radiation pressure interaction

CONTENTS

2.1   Basic radiation pressure interaction 37
2.2   Effective quantisation 39
2.3   Dispersive optomechanics 40
    2.3.1   Hamiltonian in the rotating frame for the light 43
2.4   Electro- and opto-mechanical systems 43
2.5   Mechanical and optical decoherence rates 45
2.6   Dynamics of dispersive optomechanical systems 47
    2.6.1   Semiclassical dynamics 48
        2.6.1.1   Steady-state displacements of
                  mechanical oscillator and field 49
    2.6.2   Optomechanical bistability 50
2.7   Linearisation of the optomechanical Hamiltonian 51
2.8   Dissipative optomechanics 55

In this chapter we introduce the fundamental radiation pressure interaction between a mechanical element and a cavity field that lies at the core of cavity optomechanics. We begin by deriving the full interaction Hamiltonian for dispersive optomechanics within an optical cavity, and briefly introducing the range of opto- and electomechanical implementation of cavity optomechanics and their relevant decoherence rates. We then examine the semiclassical dynamics of cavity optomechanical systems touching on optomechanical bistability that is exhibited by them. We derive the linearised optomechanical Hamiltonian and finally introduce the concept of dissipative optomechanics.

2.1   BASIC RADIATION PRESSURE INTERACTION

At its most basic level, the radiation pressure interaction between light and matter involves an exchange of momentum between light and a mechanical degree of freedom. Before examining this interaction in detail, it is worthwhile to consider the basic example of a single photon reflecting from the mirrored surface of a mechanical harmonic oscillator. The impact of the photon on the oscillator will cause a momentum kick and start it oscillating. From a quantum perspective, the regime in which the maximum displacement of the oscillator is larger than its zero-point motion $x_{zp} = (\hbar/2m\Omega)^{1/2}$ is clearly interesting. In this regime, measurements on a mechanical oscillator that is initially in its ground state can reveal that the photon impact occurred, acting as a quantum nondemolition measurement of the number of photons in the field. Since the momentum of a photon with wavelength λ is h/λ, in the simple case where the photon reflects from the mechanical oscillator at normal incidence and with unit efficiency, it imparts a $2h/\lambda$ momentum kick to the oscillator. Assuming that the oscillator is well underdamped, after a quarter period this momentum kick results in a displacement of $\Delta q = 2h/\lambda m\Omega = 8\pi x_{zp}^2/\lambda$, as can be shown straightforwardly, for example, using Eqs. (1.13) and (1.24). Rearranging, we arrive immediately at the condition

$$\frac{x_{zp}}{\lambda} > \frac{1}{8\pi} \tag{2.1}$$

for the recoil to exceed the oscillator zero-point motion. Since, for typical micro-mechanical oscillators, x_{zp} is beneath a femtometer, while optical wavelengths are in the range of a micrometer and microwave wavelengths even larger, this condition poses a significant challenge. As a result, to reach the quantum regime mechanisms are required to enhance the interaction between the field and the oscillator, with the standard approaches being to confine the field in an optical or microwave cavity to boost the number of times the photon interacts with the oscillator, increase the intensity of the optical field, though, as we will see in Section 2.7, this also alters the form of the interaction, washing out some of the most interesting physics, and reduce the mass of the oscillator to increase its zero-point motion. This motivates the treatment given in this chapter of the cavity-enhanced optomechanical interaction between a mechanical oscillator and a bright optical field.

Note that Eq. (2.1) is only intended to give a rough indication of a regime where quantum mechanics plays a role in the interaction between a field and a mechanical oscillator. A more rigorous treatment would include the additional constraint that, even when placed in the best refrigerators, mechanical oscillators with typical resonance frequencies in the megahertz to gigahertz range remain far from their ground state in thermal equilibrium. This motivates the use of cryogenic techniques in quantum optomechanics experiments. Furthermore, it should be recognised that, for a high-quality mechanical oscillator, measurements can be made over many cycles of the oscillator, rather than just one, as considered above, which has the effect of relaxing the criterion in Eq. (2.1). A more rigorous treatment of this problem is given in Chapter 3.

2.2 EFFECTIVE QUANTISATION

In the previous section we introduced the idea that quantum mechanics can play a significant role in the interaction between light and a macroscopic material object, without concerning ourselves with the microscopic degrees of freedom of the material object. This object might be a bulk elastic material, a harmonically bound mirror, an ensemble of atoms, or even a single atom. We will be exclusively concerned with the first two of these. Of course a bulk elastic material is made up of atoms, but we do not describe the optomechanical interaction in terms of the atomic constituents of bulk matter and their corresponding phonon modes. Instead the theory is given in terms of an effective quantum field theory.

The response of an elastic material to the applied optical stress is modelled initially in terms of a strain field in the bulk described classically by continuum mechanics. Usually we treat the strain field in terms of a simple scalar displacement field $u(\vec{x}, t)$, where $\vec{x}$ labels a point in the bulk. Typically the response is elastic and the displacement field obeys a linear wave equation. In that case we expand the scalar field in terms of harmonic modes

$$u(\vec{x}, t) = \sum_k \alpha_k(t)\phi_k^*(\vec{x}) + \alpha_k^*(t)\phi_k^*(\vec{x}) \tag{2.2}$$

where $\phi_k^*(\vec{x})$ are a suitable set of spatial mode functions. The amplitudes $\alpha_k(t)$ obey equations of motion equivalent to those of independent harmonic oscillators

$$\dot{\alpha}_k = -i\omega_k \alpha_k. \tag{2.3}$$

In passing to the quantum description, we simply replace each of these harmonic oscillators by a quantum harmonic oscillator using

$$\alpha_k(t) \quad \rightarrow \quad a_k(t) \tag{2.4}$$

$$\alpha_k^*(t) \quad \rightarrow \quad a_k^\dagger(t) \tag{2.5}$$

with canonical commutation relations $[a_k(t), a_k'^\dagger(t)] = \delta_{k,k'}$. In most cases we can restrict the discussion to the response of a single mode a_0 with frequency Ω_0 to the applied optical force. Equivalently we may introduce canonical position and momentum operators for this mode by

$$\hat{q} = \frac{\hbar}{2m_0\Omega_0}(a_0 + a_0^\dagger) \tag{2.6}$$

$$\hat{p} = -i\frac{\hbar m_0\Omega_0}{2}(a_0 - a_0^\dagger) \tag{2.7}$$

where m_0 is the effective mass of this particular harmonic mode. In this book we almost always use this single-mode treatment and so we drop the subscript on the mass m and frequency Ω. For more details on this approach, see [201].

That this effective quantisation procedure works is abundantly clear from

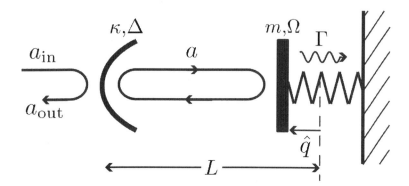

FIGURE 2.1 Schematic of a canonical cavity optomechanical system. κ, Δ, and L, are the cavity decay rate, detuning, and bare length, respectively; while m, Ω, and Γ are respectively the effective mass, resonance frequency, and damping rate of the mechanical oscillator. a and $\hat{q}$ are the annihilation operator for the intracavity field and the position operator for the mechanical oscillator, respectively; and a_{in} and a_{out} are the annihilation operators for the incident and output fields.

the many experiments we will discuss in this book. The reason it works is that, under the right conditions (often very low temperature), these collective degrees of freedom interact only very weakly with the microscopic atomic degrees of freedom (in this case short wavelength phononic modes). The relevant macroscopic displacement coordinate then largely factors out of the many microscopic degrees of freedom, which remain as a source of dissipation and noise [313]. This kind of effective quantisation of collective macroscopic degrees of freedom is typical of how we describe a wide class of engineered quantum systems. We will see another example in Chapter 8 when we discuss superconducting quantum circuits.

2.3 DISPERSIVE OPTOMECHANICS

In dispersive optomechanics, the position of the mechanical oscillator is parametrically coupled to the resonance frequency of an optical cavity or microwave resonator. Radiation pressure imparted by the optical or microwave field allows the motion of the oscillator to be controlled, while the field leaving the cavity or resonator carries information about the resonance frequency and therefore the mechanical position.

The interaction Hamiltonian describing the parametric interaction between the mechanical oscillator and the field can be easily established by considering the change in optical cavity[1] resonance frequency of a Fabry–Pérot cavity induced by a change in mechanical position.

Let us consider a Fabry–Pérot cavity of bare length L with one end mirror forming part of a mechanical oscillator, as shown in Fig. 2.1. Motion of the mechanical oscillator shifts the length of the cavity with $L(q) = L - q$, where q is the displacement of the mechanical oscillator away from its equilibrium position and the minus sign is arbitrary, defining only the positive direction of mechanical motion. A Fabry–Pérot cavity of length $L(q)$ sustains a comb of longitudinal optical modes with wavelengths

$$\lambda_j = \frac{2(L-q)}{j}, \tag{2.8}$$

where j is the mode number. The mode frequencies are therefore

$$\Omega_{c,j}(q) = \frac{2\pi c}{\lambda_j} = \frac{\pi c j}{L - q} \approx \Omega_{c,j}\left(1 + \frac{q}{L}\right), \tag{2.9}$$

where the approximation is valid as long as $q \ll L$, which is appropriate for the vast majority of cavity optomechanical systems, c is the speed of light, and $\Omega_{c,j} \equiv \Omega_{c,j}(0)$ is the bare optical resonance frequency of the cavity, with the subscript c used to label the resonance frequency of the cavity/microwave resonator throughout this textbook. We see, unsurprisingly, that to first order the motion of the mechanical oscillator acts to linearly shift the optical resonance frequency. The frequency shift per meter is quantified by the *optomechanical coupling strength G*

$$G \equiv \frac{\delta\Omega_c(q)}{\delta q} \tag{2.10}$$

$$= \frac{\Omega_c}{L} \quad \text{(Fabry–Pérot cavity)} \tag{2.11}$$

where we have omitted the mode number subscript j. While we arrived at Eq. (2.10) by considering the specific case of a Fabry–Pérot cavity, and the relation $G = \Omega_c/L$ is only relevant to that geometry, the optomechanical coupling strength is an important and widely applicable parameter.

We can now proceed to the description of the interaction between the cavity field and the mechanical element. The energy of the light in a particular cavity mode is given by the energy per quantum, $\hbar\Omega_c$, times the number, $n(t)$, of photons in the cavity

$$H_L = \hbar\Omega_c(q)\, n, \tag{2.12}$$

where we explicitly include the dependence of the cavity frequency on the

---

[1]Note that in this text we will usually discuss the physics of cavity optomechanical systems with reference to optical cavities. However, it should be kept in mind that the physics is – at least at the fundamental level – identical and equally applicable for microwave resonators (see Section 8.1).

classical displacement of the mechanical oscillator described above. In the
absence of the radiation pressure interaction, the energy of the mechanical
oscillator is

$$H_M = \frac{p^2}{2m} + \frac{m\Omega^2}{2}q^2, \tag{2.13}$$

where, here and henceforth, Ω denotes the resonance frequency of the mechanical oscillator, and as usual, p and m are its momentum and mass, respectively.
The total classical Hamiltonian is $H = H_L + H_M$. From the perspective of
the mechanical element, the interaction with the light appears as a potential
energy term and a corresponding force per photon given by $F_1 = -\hbar\frac{d\Omega_c(q)}{dq}$. If
we expand $\Omega_c(q)$ to linear order in q as in Eq. (2.9), we see that this force is
simply $F_1 \approx \hbar G$.

As outlined for the mechanical oscillator in Section 2.2, the quantum description of the system is then given by replacing the classical variables with
the corresponding operators $q \to \hat{q}$, $p \to \hat{p}$, $n \to a^\dagger a$ which satisfy the canonical computation relations given in Eqs. (1.2) and (1.11). The quantum Hamil-
tonian is then

$$\hat{H} = \frac{\hat{p}^2}{2m} + \frac{m\Omega^2}{2}\hat{q}^2 + \hbar\Omega_c(\hat{q})a^\dagger a. \tag{2.14}$$

Introducing the raising and lowering operators for the mechanical excitations
as $b^\dagger$ and b, which also satisfy the Boson commutation relation in Eq. (1.2),
and expanding the frequency to linear order in $\hat{q}$ enables us to write

$$\hat{H} = \hbar(\Omega_c + G\hat{q})a^\dagger a + \hbar\Omega b^\dagger b \tag{2.15}$$
$$= \hbar\Omega_c a^\dagger a + \hbar\Omega b^\dagger b + \hbar g_0 a^\dagger a \left(b^\dagger + b\right), \tag{2.16}$$

where to arrive at Eq. (2.16) we have used the definition of $\hat{q}$ in Eq. (1.9a),
and we have defined the *vacuum optomechanical coupling rate*

$$g_0 \equiv Gx_{zp}, \tag{2.17}$$

which has units of [rad s^{-1}]. From the point of view of the cavity field, for
any cavity optomechanical geometry, the optomechanical coupling strength
quantifies the linear dispersive shift in the optical resonance frequency induced
by the mechanical motion.

The vacuum optomechanical coupling rate, g_0, is one of the central parameters in the field of quantum optomechanics. It can be interpreted as the
optical frequency shift induced by a mechanical displacement equal to the
mechanical zero-point motion, as can be seen from inspection of Eq. (2.16)
(remember that $b^\dagger + b = \hat{q}/x_{zp}$). Conversely, the radiation pressure from a
single photon within the cavity acts to displace the mechanical oscillator. The
ratio g_0/Ω quantifies this displacement in units of the mechanical zero-point
motion. Throughout the remainder of this textbook we reserve the annihilation
operators a and b, respectively, for the cavity field and mechanical oscillator,
and position and momentum operators $\hat{p}$ and $\hat{q}$ for the mechanical oscillator.

2.3.1   Hamiltonian in the rotating frame for the light

In the majority of cavity optomechanics experiments, the optical cavity (or microwave resonator) resonance frequency Ω_c is much larger than all other system rates. For instance, compare a typical visible laser frequency of $\sim 5 \times 10^{14}$ Hz to typical mechanical oscillator frequencies and optical decay rates in the range of $10^3 - 10^9$ Hz. In this regime, it is convenient to move into a rotating frame at the incident laser frequency Ω_L and thereby remove the fast oscillations of the optical field. Since the unitary used to transform into the rotating frame (Eq. (1.133)) commutes with all terms in Eq. (2.16), the transformation is particularly simple, with the result

$$\hat{H} = \hbar\Delta a^\dagger a + \hbar\Omega b^\dagger b + \hbar g_0 a^\dagger a \left(b^\dagger + b \right), \qquad (2.18)$$

where here a is now in the rotating frame, and we have defined the detuning Δ between the optical cavity and the incident laser frequency as

$$\Delta \equiv \Omega_c - \Omega_L. \qquad (2.19)$$

2.4   ELECTRO- AND OPTO-MECHANICAL SYSTEMS

While for the most part in this textbook we use the term *quantum optomechanics*, it should not be misconstrued that the physics described is limited to the optical domain. Indeed, quantum optomechanics experiments are performed not only at optical frequencies, but also in the microwave domain (see Fig. 2.2).

The essential interaction in the optical domain is as described in the previous section – motion of the mechanical oscillator induces a dispersive shift in an optical resonance frequency. As discussed below, this domain has the primary advantage that the thermal equilibrium bath for the field is essentially in a vacuum state even at room temperature due to the high frequency of optical fields. A second crucial advantage is that, since the optical wavelength is in the vicinity of 1 μm, it is possible to fabricate small optical cavities on micrometer and nanometer scales, and thereby enhance the optomechanical coupling strengths (see Eq. (2.11)), and at the same time utilise small low mass mechanical with large zero-point motion. The optical domain, further, offers well-developed high-efficiency detectors functioning at all intensities from watts to single photon levels, and natural integration into fibre optic telecommunication systems.

Electromechanical systems typically operate using microwave fields with frequencies around 10 GHz. At first sight, such systems appear to have several significant disadvantages compared with optomechanical systems. Apart from obvious challenges with freezing out the thermal noise in the environment (see the next section), since the photon momentum scales as λ^{-1}, the momentum imparted by a single microwave photon upon reflection is some four orders of magnitude smaller than that of an optical photon. Furthermore, microwave

Electromechanics

Optomechanics

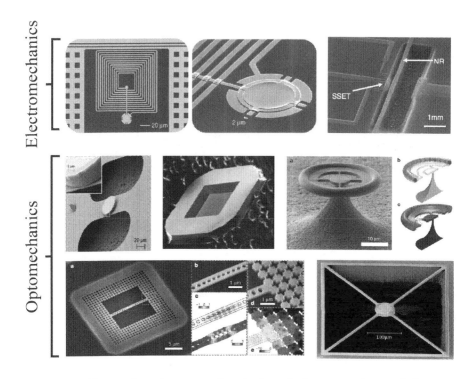

FIGURE 2.2 Examples of optomechanical and electromechanical devices. **Electromechanical devices.** *Left:* Aluminium drum mechanical resonator coupled to a superconducting lumped element microwave resonator [adapted by permission from Macmillan Publishers Ltd: *Nature* [282], copyright 2011]. *Right:* Nanomechanical beam coupled to a superconducting single-electron transistor [adapted by permission from Macmillan Publishers Ltd: *Nature* [208], copyright 2006]. **Optomechanical devices.** *Clockwise from top left:* Microscale end-mirror from a Fabry–Pérot cavity optomechanical system [adapted by permission from Macmillan Publishers Ltd: *Nature Physics* [131], copyright 2009]; silicon nitride membrane used in membrane-in-the-middle-type Fabry–Pérot cavity optomechanical system [adapted by permission from Macmillan Publishers Ltd: *Nature* [283], copyright 2008]; microtoroidal integrated cavity optomechanical system [adapted by permission from Macmillan Publishers Ltd: *Nature* [295], copyright 2012]; phononic-photonic crystal integrated optomechanical system [adapted by permission from Macmillan Publishers Ltd: *Nature* [66], copyright 2011]; trampoline microscale end-mirror from a Fabry–Pérot cavity optomechanical system [162].

cavities are limited to sizes in the centimetre range and larger due to the centimetre wavelength of microwave fields. Combined, these two effects would appear to severely constrain the optomechanical coupling strengths achievable in the microwave domain (this is immediately evident from Eq. (2.11)). To overcome these constants, electromechanical systems confine the microwave field using transmission line resonators or lumped element circuits rather than microwave cavities. Transmission line resonators can have one-dimensional confinement at a level beneath 10 μm [32]. Lumped element circuits consist of separate inductive and capacitive elements that respectively contain the magnetic and electric fields, and can have volumes many orders of magnitude smaller than a cubic wavelength. The capacitor in the lumped element circuit shown at the top left of Fig. 2.2, for instance, confines the circuit's electric field to around 15 μm in two dimensions and 50 nm in the third dimension [282]. A cavity optomechanical system is formed by making one plate of the capacitor mechanically compliant. Mechanical drum modes alter the plate separation and through this the capacitance and LC resonance frequency of the circuit. A substantial advantage of electromechanical systems is that they can be directly coupled to superconducting qubits which can be used to control and read out the state of the mechanical oscillator, as will be discussed further in Section 8.2.

2.5 MECHANICAL AND OPTICAL DECOHERENCE RATES

The rates at which the optical and mechanical degrees of freedom decohere are critical parameters for optomechanical systems that, along with the op-tomechanical coupling rate, play a central role in determining the types of physics they display. In Section 1.2.4 we showed that, within the rotating wave approximation, the mean heating rate of a quantum oscillator coupled to a thermal environment is $\gamma \hbar \Omega (\bar{n} + 1/2)$ J s^{-1} (see Eq. (1.68)). This heating introduces quantum decoherence to the state of the oscillator. The natural time scale for this decoherence is the time over which half a quanta of energy enters the oscillator from the environment – i.e., the time scale over which the environment introduces an amount of energy equal to the oscillators ground state (or zero-point) energy. A characteristic thermal decoherence rate can therefore be defined for the oscillator

$$\gamma_{\text{decoh}} = \gamma \left(2\bar{n} + 1 \right). \tag{2.20}$$

While this rate has wide significance in quantum optomechanics, it should of course be noted that, strictly speaking, quantum decoherence is a state-dependent phenomenon. For instance, it is well known that the decoherence rate of Schrödinger's cat states increases as the separation of the superposition becomes larger, while for a number state $|n\rangle$ the decoherence rate is $\gamma_{\text{decoh}}^{(n)} = \gamma[(n+1)\bar{n} + n(\bar{n}+1)]$ and therefore increases with increasing n.

Exercise 2.1 *Derive the above expression for the decoherence rate of a number state $\gamma_{\mathrm{decoh}}^{(n)}$ using Eqs. (1.48), (1.49), (1.53), and (1.62).*

As you would expect, the thermal decoherence rate of Eq. (2.20) contains a vacuum term and a term that scales linearly with the temperature of the bath. This introduces two natural regimes for decoherence, a thermal noise dominated regime and a vacuum noise dominated regime. Even at room temperature, the high resonance frequencies of optical cavities put them very much in the vacuum dominated regime, with $\bar{n} \sim 10^{-36}$ (see Eq. (1.7)).[2] On the other hand, a microwave resonator with $\Omega = 10$ GHz has $\bar{n} \approx 600$ at room temperature and is therefore in the thermal noise dominated regime. Consequently, all microwave quantum optomechanics experiments are operated in cryogenic conditions.[3] For a 10 GHz resonance frequency the cross-over between thermal and vacuum noise dominated regimes occurs at $T \approx \hbar\Omega/k_B = 0.5$ K, which introduces the requirement to work in ultracryogenic conditions provided by dilution refrigerators and adiabatic demagnetisation refrigerators. Henceforth in this textbook, unless explicitly stated otherwise, we will treat the optical cavity/microwave resonator as being within the vacuum noise dominated regime, with decoherence rate

$$\kappa_{\mathrm{decoh}} = \kappa, \tag{2.21}$$

where we have made the substitution $\gamma \to \kappa$, and throughout the textbook reserve the notation κ for the decay rate of the cavity.

Mechanical oscillators typically have resonance frequencies well below the microwave resonance frequencies considered in the previous paragraph. As a result, their decoherence rate is generally thermal noise limited, such that

$$\Gamma_{\mathrm{decoh}} = \Gamma \left(2\bar{n} + 1 \right) \approx 2\Gamma\bar{n}, \tag{2.22}$$

where we have made the substitution $\gamma \to \Gamma$, henceforth reserving the notation Γ for the mechanical decay rate and $\bar{n}$ for the mean phonon number of the mechanical bath.

An important minimum requirement for most quantum optomechanics experiments is that the individual oscillations of both the optical cavity field and the mechanical resonator are quantum coherent. By this we mean that the fluctuations introduced within one period of the oscillator are smaller than the zero-point fluctuations of the oscillator, or equivalently that less than half a quanta of heating is introduced over one oscillation period. In terms of the resonant frequency of the oscillator, this criterion can be expressed simply as

$$\Omega > \gamma_{\mathrm{decoh}}. \tag{2.23}$$

---

[2]Note that it is not unusual for technical laser noise to raise the effective noise floor above the vacuum noise level.

[3]This has the added benefit of allowing the use of superconducting materials that result in greatly reduced resonator decay rates.

For the optical cavity field with resonance frequency Ω_c this amounts to the condition that $\Omega_c > \kappa$, or in other words that the cavity is not overdamped ($Q_c = \Omega_c/\kappa > 1$), which is essentially always the case. For the mechanical oscillator the nonzero bath phonon number leads to the more stringent condition

$$\Omega > \Gamma\,(2\bar{n}+1).\qquad(2.24)$$

Under the assumption that $\bar{n} \gg 1$, we can approximate $2\bar{n}+1 \approx k_B T/\hbar\Omega$. Rearranging Eq. (2.24), we then arrive at the condition for quantum coherent oscillation of the mechanical oscillator

$$Q\Omega > \frac{2k_B T}{\hbar},\qquad(2.25)$$

where $Q = \Omega/\Gamma$ is the quality factor of the mechanical oscillator. Since the right-hand side of this expression is a function of only temperature and fundamental constants, we see that the so-called Q–f product is an important metric to quantify whether a particular mechanical oscillator is suitable for quantum mechanical experiments or applications. Consequently, much effort has been made to increase the Q–f product by increasing the frequency of the mechanical oscillator while maintaining or reducing the mechanical dissipation rate [125]. It is worth noting that, in the nano- and micro-electromechanics community, it has been found phenomenologically that in situations where the mechanical dissipation rate is dominated by internal material losses, the Q–f product is roughly invariant for a given material and temperature over a wide range of mechanical oscillator geometries [117]. Thus material choice is crucial for quantum optomechanics experiments.

2.6 DYNAMICS OF DISPERSIVE OPTOMECHANICAL SYSTEMS

Using the Hamiltonian of Eq. (2.18), it is possible to use the tools developed in Chapter 1 to determine the Langevin equations of motion that describe the open system dynamics of a cavity optomechanical system. As discussed in Sections 1.3.1.2 and 1.3.2, it is generally appropriate to treat both optical and mechanical baths as Markovian, and we do so here. We, further, make a rotating wave approximation for the optical field which is reasonable in almost all circumstances due to its high frequency. The rotating wave approximation is often also appropriate for high-quality mechanical oscillators; however, we choose not to make that approximation to keep the generality of the derivation.

Exercise 2.2 *Using the general Markov Langevin equation of Eq. (1.95) for the mechanical oscillator variables and the rotating wave Langevin equation of Eq. (1.112) for the optical field, show that the Hamiltonian Eq. (2.18) yields* *the cavity optomechanical open-system dynamics*

$$\dot{a} = -\left[\frac{\kappa}{2} + i\left(\Delta + \sqrt{2}g_0\hat{Q}\right)\right]a + \sqrt{\kappa}\,a_{\text{in}} \tag{2.26a}$$

$$\dot{\hat{Q}} = \Omega\hat{P} \tag{2.26b}$$

$$\dot{\hat{P}} = -\Omega\hat{Q} + \sqrt{2\Gamma}\,\hat{P}_{\text{in}} - \Gamma\hat{P} - \sqrt{2}g_0 a^\dagger a, \tag{2.26c}$$

where we have identified $\hat{Q}$ and $\hat{P}$ with the dimensionless position and momentum quadratures of the mechanical oscillator, and as usual κ and Γ are the decay rates of the optical cavity and mechanical oscillator.

As expected, we can observe from these expressions that the optomechanical interaction applies a radiation pressure force to the mechanical oscillator, driving its momentum, and modifies the effective detuning of the optical cavity.

The field a_{in} represents an input field to the optical cavity which drives the optical cavity and is often coherently populated so that $\alpha_{\text{in}} \equiv \langle a_{\text{in}} \rangle \neq 0$. However, generally, the dissipation from the cavity (and fluctuations driving it) occurs through several channels, including an experimentally accessible coherent driving channel which we term the *input port*, and other loss channels due, for instance, to absorption, scattering, or imperfect mirror coatings, which we treat as a combined *loss port*. In practise it is important to account separately for these two different types of channel. This can be done via the substitution

$$\sqrt{\kappa}\,a_{\text{in}} \rightarrow \sqrt{\kappa_{\text{in}}}\,a_{\text{in}} + \sqrt{\kappa_{\text{loss}}}\,a_{\text{loss}}, \tag{2.27}$$

where κ_{in} and κ_{loss} are the dissipation rates through the input and loss ports, respectively, with the total dissipation rate $\kappa = \kappa_{\text{in}} + \kappa_{\text{loss}}$, and in the substitution a_{in} and a_{loss} are the fields entering through each port. For an optical cavity at room temperature, the field entering through the loss port can be well treated as a vacuum state so that $\langle a_{\text{loss}} \rangle = 0$.

Equations (2.26) are nonlinear coupled quantum differential equations, whose moments – in general – cannot be found analytically. We refer the reader to Chapters 6, 7, and 9 for treatments applicable in some specific cases. For the remainder of this Chapter we consider their semiclassical solutions and the linearised equations that arise from considering small fluctuations around the semiclassical steady state.

2.6.1 Semiclassical dynamics

The semiclassical dynamics of a cavity optomechanical system can be obtained by taking the expectation values of Eqs. (2.26) and approximating operator product terms of the form $\langle \hat{A}\hat{B} \rangle$ as $\langle \hat{A} \rangle \langle \hat{B} \rangle$. This approximation can be justified as long as for each operator product the fluctuation term $\langle (\hat{A} - \langle \hat{A} \rangle)(\hat{B} - \langle \hat{B} \rangle) \rangle$ is small compared to $\langle \hat{A} \rangle \langle \hat{B} \rangle$. This is generally, though not always, the case in cavity optomechanical systems that have significantly more than one intracavity photon on average – particular care should be taken when the system is driven close to an instability where the fluctuation terms can be greatly amplified (see, for example, [324]). The result of this process is the following set of coupled classical differential equations of motion:

$$\dot{\alpha} = -\left[\frac{\kappa}{2} + i\left(\Delta + \sqrt{2}g_0\langle\hat{Q}\rangle\right)\right]\alpha + \sqrt{\kappa_{\text{in}}}\,\alpha_{\text{in}} \quad (2.28a)$$

$$\langle\dot{\hat{Q}}\rangle = \Omega\langle\hat{P}\rangle \quad (2.28b)$$

$$\langle\dot{\hat{P}}\rangle = -\Omega\langle\hat{Q}\rangle - \Gamma\langle\hat{P}\rangle - \sqrt{2}g_0|\alpha|^2, \quad (2.28c)$$

where we use the substitution in Eq. (2.27), assume the mechanical oscillator is driven incoherently by its bath ($\langle\hat{P}_{\text{in}}\rangle = 0$), and define the coherent amplitude in the cavity $\alpha \equiv \langle a\rangle$ which is related to the mean occupancy N of the optical cavity introduced by coherent driving via $N = |\alpha|^2$. These equations exhibit a nonlinear coupling between the mechanical oscillator and the optical field that results in a rich range of classical behaviour, including regimes of parametric instability characterised by exponential growth in the amplitude of mechanical oscillation (see, for example, [61, 159]) as well as chaotic regimes [60]. They are therefore generally difficult to solve analytically.

2.6.1.1 Steady-state displacements of mechanical oscillator and field

Here, since we are primarily interested in small quantum fluctuations around some classical steady state we consider only the steady-state solutions in stable regimes where the oscillation of the mechanical position is small enough to only weakly modulate the optical field. We envision that the system has reached a steady-state at some time t_0 and average Eqs. (2.28b) and (2.28c) over a subsequent time period of length τ much longer than the mechanical oscillation period and time scale of any oscillations of the optical intensity. In this case each of the derivatives on the left-hand side of Eqs. (2.28b) and (2.28c) average to zero

$$\frac{1}{\tau}\int_{t_0}^{t_0+\tau}\langle\dot{\hat{Q}}\rangle\,dt = \frac{1}{\tau}\int_{t_0}^{t_0+\tau}\langle\dot{\hat{P}}\rangle\,dt = 0 \quad (2.29)$$

so that

$$\bar{Q} = -\frac{\sqrt{2}g_0}{\Omega\tau}\int_{t_0}^{t_0+\tau}|\alpha|^2\,d\tau, \quad \bar{P} = 0, \quad (2.30)$$

where

$$\bar{Q} \equiv \frac{1}{\tau}\int_{t_0}^{t_0+\tau}\langle\hat{Q}\rangle\,dt \quad (2.31)$$

is the time average of $\langle\hat{Q}\rangle$, with a similar definition for $\bar{P}$.

We wish to make the approximation $\langle\hat{Q}\rangle \to \bar{Q}$ in Eq. (2.28a) so that only the time-averaged mechanical position contributes to the dynamics of α. This approximation is reasonable as long as the fluctuating part of $\langle\hat{Q}\rangle$ is sufficiently small, specifically $|\langle\hat{Q}\rangle - \bar{Q}| \ll \kappa/[2\sqrt{2}g_0]$, which is the case for typical cavity optomechanical systems when not operated close to an instability. Equation (2.28a) then becomes

$$\dot{\alpha} = -\alpha\left[\frac{\kappa}{2} + i\left(\Delta + \sqrt{2}g_0\bar{Q}_{ss}\right)\right] + \sqrt{\kappa_{\mathrm{in}}}\alpha_{\mathrm{in}}. \tag{2.32}$$

Let us now take the specific case where the coherent optical driving α_{in} is a monochromatic tone which oscillates at the laser frequency Ω_L in the laboratory frame (i.e., the nonrotating frame). This is the typical scenario relevant to most quantum optomechanics experiments.[4] In the rotating frame α_{in} is then a constant. Since the term in square brackets is also a constant, in the long time limit the coherent amplitude in the cavity is also constant, and the derivative $\dot{\alpha} = 0$. We therefore find from Eqs. (2.28a) and (2.30) that, in the steady-state, the intracavity coherent amplitude α_{ss} and mechanical displacement $\bar{Q}_{ss}$ are

$$\alpha_{ss} = \sqrt{\kappa_{\mathrm{in}}}\alpha_{\mathrm{in}}\left[\frac{\kappa}{2} + i\left(\Delta + \sqrt{2}g_0\bar{Q}\right)\right]^{-1} \tag{2.33a}$$

$$\bar{Q}_{ss} = -\frac{\sqrt{2}g_0|\alpha_{ss}|^2}{\Omega}. \tag{2.33b}$$

2.6.2   Optomechanical bistability

From Eqs. (2.33) we see that the mean effect of the optical field is to displace the position of the mechanical oscillator by an amount dependent on the optomechanical coupling strength, the mechanical resonance frequency, and the optical intensity. As can be seen in Eq. (2.33a), a displacement of the mechanical oscillator itself displaces the resonance frequency of the optical cavity and through this the intracavity intensity. This gives rise to the prospect of multistability, where a given intensity of incident light can produce multiple different steady states in both intracavity photon number and mechanical position. These ideas were first explored experimentally by the laboratory of Herbert Walther in 1983 using a Fabry–Pérot-type optomechanical system in the optical domain [93].

Substituting Eq. (2.33b) into Eq. (2.33a) results in a cubic equation for $N = |\alpha_{ss}|^2$. Cubic equations exhibit regimes with one real root, and separate regimes with three real roots. As such, the classical dynamics of a cavity optomechanical system will exhibit regimes where only one real solution exists

---

[4]Note, there is a class of cavity optomechanics protocols termed *pulsed optomechanics* [292], where the input field is pulsed, typically with pulse length short compared to the mechanical frequency such that the optomechanical interaction occurs over a time frame in which the mechanical oscillator can essentially be treated as stationary. While in these protocols α_{in} is, by definition, changing in time, often the time scale is chosen to be slower than the decay time of the optical cavity, in which case the steady-state solution to Eq. (2.28a), taking the mechanical oscillator as stationary, provides the appropriate α at each time.

for N in the steady state and other regimes where three possible solutions exist.

Exercise 2.3 *Show that*

$$N\left[\frac{\kappa^2}{4} + \left(\Delta - \frac{2g_0^2}{\Omega}N\right)^2\right] = \kappa_{\mathrm{in}}N_{\mathrm{in}}, \qquad (2.34)$$

where $N_{\mathrm{in}} = |\alpha_{\mathrm{in}}|^2$ is the input photon flux in units of photons per second. Find analytical expressions for the roots of this expression and determine the regimes of mono- and multi-stability.

The real solutions for the steady-state intracavity photon number N are illustrated as a function of optical cavity detuning Δ in Fig. 2.3. Fig. 2.3a shows the case where there is no optomechanical interaction ($g_0 = 0$). As can be seen – and as should be expected – the intracavity photon number is a symmetric Lorenzian centred at zero detuning. As the optomechanical interaction increases (Fig. 2.3b) the Lorenzian skews towards higher detunings, with the radiation pressure force on the mechanical oscillator shifting the resonance frequency of the cavity. At sufficiently high optomechanical interaction strengths (Fig. 2.3c and d) a tristability appears, with three possible intracavity photon numbers in the steady state. While three solutions exist, it is experimentally difficult to access the central solution. If the cavity detuning is adiabatically swept up in frequency, the intracavity photon number follows the upper solution until it no longer exists, and then discontinuously jumps to the lowest solution (see the down arrows on Fig. 2.3c and d). On the other hand, in a hysteretic type of behaviour, if the cavity is adiabatically swept down in frequency, the intracavity photon number follows the lower branch before eventually jumping to the upper branch (see the up arrows on Fig. 2.3c and d). Consequently, while strictly speaking optomechanical systems are tristable, for most practical intents and purposes their behaviour is bistable.

2.7 LINEARISATION OF THE OPTOMECHANICAL HAMILTONIAN

The vacuum optomechanical coupling rate g_0 is usually much smaller than the optical and mechanical decoherence rates of Eqs. (2.21) and (2.22).[5] A common approach to address this issue is to – rather than rely on the coupling between a single intracavity photon and the mechanical oscillator – coherently drive the optical cavity by injecting a bright coherent field, as we discussed in the semiclassical limit earlier in this chapter. This has the effect of greatly increasing the radiation pressure force, and therefore the optomechanical coupling rate, and also changes the essential character of the interaction.

---

[5]This is not always the case, as is discussed in more detail in Chapter 6.

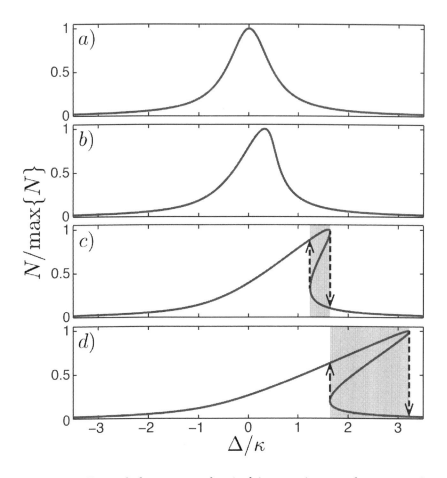

FIGURE 2.3 Effect of the optomechanical interaction on the mean photon number $N = |\alpha|^2$ within an optical resonator, displaying optomechanical bistability. Horizontal axis: optical detuning Δ normalised by the optical decay rate κ. Vertical axis: mean intracavity photon number N normalised to the peak N achieved when the cavity is on resonance and there is no optomechanical interaction ($\Delta = g_0 = 0$). Shaded regions: regions of multistability. Arrows: discontinuous jumps in the mean photon number that occur as the detuning is increased and decreased, showing hysteresis. For these models the incident photon flux $N_{\text{in}} = |\alpha_{\text{in}}|^2$ is $N_{\text{in}} = 40\kappa/\eta$ where $\eta = \kappa_{\text{in}}/\kappa$ is the cavity escape efficiency, while for a–d the optomechanical interaction strength is characterised by $g_0^2/\kappa\Omega = \{0, 1, 5, 10\} \times 10^{-3}$, respectively.

The tri- and bi-stability that arise from the optomechanical interaction which we discussed in the previous section – while interesting in their own right – are quite often not important for quantum optomechanics experiments (see Chapters 7 and 9 for some exceptions). When the coherent driving is suitably strong, the system dynamics can generally be well approximated by a linearised description considering only small fluctuations around the semiclassical steady state.

Coherent optical driving can be introduced at the level of the Hamiltonian by including a driving term in Eq. (2.18)

$$\hat{H} = \hbar\Delta a^\dagger a + \hbar\Omega b^\dagger b + \hbar g_0 a^\dagger a \left(b^\dagger + b \right) + \hbar\epsilon \left(a^\dagger + a \right), \qquad (2.35)$$

where ϵ is the drive strength which we have defined to be real without loss of generality and is linearly proportional to the input optical field amplitude $\alpha_{\rm in}$ used in previous sections. We saw in Section 2.6.1.1 that the effect of this drive is to displace the steady states of both the intracavity field and – through the optomechanical interaction – the mechanical position. To linearise the Hamiltonian, we wish to remove these displacements, transforming into a frame in which $\alpha_{ss} = \bar{Q}_{ss} = 0$; i.e., we wish to perform the displacements

$$a \rightarrow \alpha + a \qquad (2.36a)$$

$$b \rightarrow \beta + b, \qquad (2.36b)$$

where $\beta \equiv \langle b \rangle = \bar{Q}_{ss}/\sqrt{2}$ is the optically induced steady-state displacement of the mechanical state and is real, and we omit the subscript on α_{ss} for succinctness.[6] Henceforth we choose the intracavity optical coherent amplitude α to also be real. This can be done without loss of generality, with the intracavity field then providing a phase reference for the input and output fields, as well as any other fields introduced to the system.

Exercise 2.4 *By substituting Eqs. (2.36) into the Hamiltonian, show that the choice of optical displacement*

$$\alpha = -\frac{\epsilon}{\Delta + 2\beta g_0} \qquad (2.37)$$

cancels the coherent optical driving (terms proportional to $a^\dagger + a$), while with $\bar{Q}$ equal to the steady-state mechanical displacement which we found earlier (Eq. (2.33b)) the coherent mechanical driving is also cancelled.

Combined, Eqs. (2.37) and (2.33) connect the drive strength ϵ introduced in this section to the coherent amplitude $\alpha_{\rm in}$ of the input optical field.

Exercise 2.5 *By neglecting the terms that do not depend on a or b – and*

---

[6] Formally, these displacements can be achieved by applying the displacement operators $\hat{D}_a(\alpha) = \exp(\alpha a^\dagger - \alpha^* a)$ and $\hat{D}_b(\beta) = \exp[\beta(b^\dagger - b)]$ to the Hamiltonian. That is, $\hat{H} \rightarrow \hat{D}_a^\dagger \hat{D}_b^\dagger \hat{H} \hat{D}_a \hat{D}_b$. This amounts to the substitution of Eqs. (2.36) into the Hamiltonian.

therefore do not contribute to the equations of motion for their dynamics – in the Hamiltonian you arrived at in the previous exercise, derive the driven-displaced optomechanical Hamiltonian

$$\hat{H} = \hbar \left(\Delta - \frac{2g_0^2 \alpha^2}{\Omega} \right) a^\dagger a + \hbar \Omega b^\dagger b + \hbar g_0 \left[\alpha \left(a^\dagger + a \right) + a^\dagger a \right] \left(b^\dagger + b \right). \quad (2.38)$$

Several observations can be made about the Hamiltonian in Eq. (2.38). The optomechanical interaction acts to change the detuning of the optical cavity consistent with our prior semiclassical analysis (see Eqs. (2.33)), with the detuning proportional to the intracavity photon number $N = \alpha^2$ and the square of the vacuum optomechanical coupling rate g_0, and inversely proportional to the mechanical frequency Ω. The first term in the square brackets is a position-position interaction term between the amplitude quadrature of the light and the dimensionless position of the mechanical oscillator. Two observations can be immediately made about this term. Firstly, it is amplified by the coherent amplitude α of the field which, for a typical cavity optomechanical system, might be of order 10^3. Secondly, while the interaction term in the original Hamiltonian involves the product of three operators, and therefore constitutes a third-order nonlinearity, this term involves the product of two operators and is therefore only a second-order nonlinear term. The second term in the square brackets retains the third-order nonlinearity but is not enhanced by the coherent amplitude of the field. Since in this displaced frame $\langle a \rangle = 0$, it is generally negligible compared to the other terms. Note that this is not always the case. Circumstances where this term cannot be neglected are examined in Chapter 6.

Motivated by the discussion in the preceding paragraph, we neglect the second term in the square brackets in Eq. (2.38). This is known as the *linearisation approximation*. Since in practise the optical detuning is generally easily controlled by either tuning the incident laser frequency or modifying the length of the optical cavity, we further subsume the optomechanical modification of the optical detuning within the overall detuning, making the substitution

$$\Delta \to \Delta + \frac{2g_0^2 \alpha^2}{\Omega}. \quad (2.39)$$

We then arrive at the linearised cavity optomechanical Hamiltonian

$$\hat{H} = \hbar \Delta a^\dagger a + \hbar \Omega b^\dagger b + \hbar g \left(a^\dagger + a \right) \left(b^\dagger + b \right) \quad (2.40)$$

$$= \frac{\hbar \Delta}{2} \left(\hat{X}^2 + \hat{Y}^2 \right) + \frac{\hbar \Omega}{2} \left(\hat{Q}^2 + \hat{P}^2 \right) + 2 \hbar g \hat{X} \hat{Q}, \quad (2.41)$$

where we have used the definitions of the dimensionless position and momentum operators from Eqs. (1.13) and quadrature operators from Eqs. (1.135) to reach the second equation, and defined the linearised *optomechanical coupling rate*

$$g \equiv \alpha g_0. \quad (2.42)$$

We have, further, identified the optical amplitude and phase quadratures with the symbols $\hat{X}$ and $\hat{Y}$, respectively, and the dimensionless mechanical position and momentum with $\hat{Q}$ and $\hat{P}$. This choice of terminology is motivated by the aim to maintain succinctness in notation, since throughout the text we will generally (though not always) deal with optical field within a rotating frame and the mechanical oscillator within a stationary frame, as is the case here. We see that the linearised optomechanical interaction is simply that of a pair of position-position coupled oscillators. This linearised interaction can be used to describe the majority of quantum optomechanics experiments to date and will form the basis of much of Chapters 3–5.

With the change in definition of the optical detuning in Eq. (2.39) above, the intracavity optical coherent amplitude from Eqs. (2.33) becomes

$$\alpha = \left(\frac{\sqrt{\kappa_{\text{in}}}}{\kappa/2 + i\Delta} \right) \alpha_{\text{in}}. \qquad (2.43)$$

Using the input-output relations (Eq. (1.125)), the coherent amplitude of the output-coupled field can then be shown to be

$$\alpha_{\text{out}} = \left(1 - \frac{\kappa_{\text{in}}}{\kappa/2 + i\Delta} \right) \alpha_{\text{in}}. \qquad (2.44)$$

Differences in action of the full and linearised optomechanical Hamiltonians are illustrated via the phasor diagrams in Fig. 2.4. To first order when the optical cavity detuning is zero, the full Hamiltonian shifts the phase of the optical field in proportion to $\hat{Q}$, as shown in Fig. 2.4 (*top left*), and displaces the momentum of the mechanical oscillator in proportion to the photon number operator $a^{\dagger}a$. By comparison, the linearised Hamiltonian displaces the phase *quadrature* $\hat{Y}$ of the optical field in proportion to $\alpha\hat{Q}$ and displaces the momentum of the mechanical oscillator in proportion to $\alpha\hat{X}$.

Exercise 2.6 *Convince yourself of this.*

2.8 DISSIPATIVE OPTOMECHANICS

An alternative approach to cavity optomechanics is to engineer a mechanical position dependent coupling rate between the intracavity optical field and the optical bath. Since this optomechanical interaction modifies the optical dissipation rather than the cavity resonance frequency, it is termed *dissipative optomechanics* [102]. In the usual linear dissipative optomechanics, the optical decay rate κ is replaced with

$$\kappa \to \kappa + H\hat{q} = \kappa + h_0\hat{Q}, \qquad (2.45)$$

where H is the dissipative optomechanical coupling strength with units of Hz per metre, and $h_0 \equiv x_{zp}H$ is the dissipative vacuum optomechanical coupling rate. Dissipative optomechanical systems display many of the same behaviours as their dispersive counterparts, including bistability, instability, radiation pressure heating, etc. However, they also display significant differences.

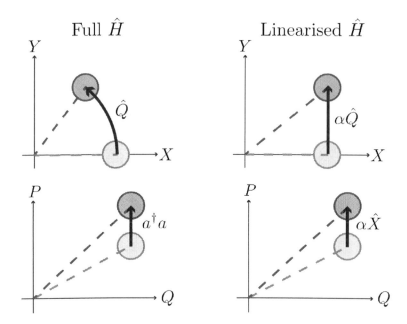

FIGURE 2.4 Phasor diagrams (or "ball and stick" diagrams; see Section 1.1.5) comparing the effects of the full (*left column*) and linearised (*right column*) optomechanical Hamiltonians on the mechanical oscillator and intracavity optical field. Top: phasor diagrams for the intracavity field. Bottom: phasor diagrams for the mechanical oscillator. The light grey phasors represent the optical and mechanical states without any optomechanical interaction. The arrows and dark grey phasors represent – illustratively – the effect of the interaction.

For instance, it has been shown theoretically that they can allow ground-state cooling of the mechanical oscillator without requiring either feedback control or that the system be in the good cavity regime ($\kappa \ll \Omega$) [102, 325]. While dispersive coupling is an interesting alternative to dispersive coupling, significantly less experimental progress has been made in this area to date. We therefore focus the remainder of this text on dispersive optomechanics.

CHAPTER 3

Linear quantum measurement of mechanical motion

CONTENTS

3.1   Free-mass standard quantum limit 58
3.2   Radiation pressure shot noise 60
      3.2.1   Optomechanical dynamics with on-resonance
              optical driving 60
      3.2.2   Effect of detuning 62
      3.2.3   Mechanical power spectral density with
              on-resonance optical driving 63
3.3   Measurement of mechanical motion 64
      3.3.1   Quantum bound on force-imprecision product 66
      3.3.2   Output field from a cavity optomechanical system . 67
      3.3.3   Linear detection of optical fields 68
      3.3.4   Photocurent power spectral density 69
      3.3.5   Phase-referenced detection 70
      3.3.6   Measured power spectral densities from a cavity
              optomechanical system 71
              3.3.6.1   Power spectral density from homodyne
                        detection 72
              3.3.6.2   Power spectral density from heterodyne
                        detection 73
              3.3.6.3   Characterisation of optomechanical
                        parameters 75
3.4   Standard quantum limit on mechanical position
      measurement 76
      3.4.1   Incident power required to reach the standard
              quantum limit 77

3.4.2 Dependence of detection noise spectrum on
incident power 78
3.4.3 Optomechanical system as a probe of the
mechanical bath 80
3.4.4 Standard quantum limit for displacement
measurement 81
3.5 Standard quantum limit for gravitational wave
interferometry 86
3.6 Standard quantum limit for force measurement 88
3.6.1 Bandwidth of optomechanical force sensing 88
3.6.2 Sensitivity of classical force sensing 91

In this chapter we use the quantum theory of continuous measurement to describe the linearised interaction between light and a mechanical oscillator in a cavity optomechanical system, and quantify how the optical output field can be used to monitor the quantum state of the mechanics. We show that continuous measurement results in unavoidable quantum back-action that heats the mechanical oscillator and introduces a standard quantum limit to force and displacement sensing, and detail how important parameters in quantum optomechanics, including optomechanical coupling rate, cooperativity, and oscillator temperature, can be characterised by optical measurement.

3.1 FREE-MASS STANDARD QUANTUM LIMIT

Before introducing the effects of quantum measurement on the dynamics and precision of cavity optomechanical systems, it is illustrative to consider the simple example of quantum measurement of the position of a free-mass. Specifically, we wish to determine the position of the mass after some time τ by performing two sequential position measurements and are interested in the effect of measurement back-action on the precision [327, 63, 215, 44]. In its continuous limit [46, 65, 64] this is a scenario that is relevant, for example, in interferometric gravitational wave observatories that seek to observe ripples in space-time via the length changes they induce in an interferometer. Current earth-based observatories target gravitational waves with frequencies in the range of a kilohertz and have kilogram-scale suspended end-mirrors with fundamental resonance frequencies in the hertz range. Consequently, on the characteristic time scale of the gravitational waves, the end-mirrors can very well be thought of as free masses.

If the initial measurement localises the position of the mass with a standard deviation $\sigma[\hat{q}(0)]$, the Heisenberg uncertainty principle (Eq. (1.12)) tells us that the measurement must also introduce quantum back-action on the momentum of the oscillator, increasing its uncertainty to at least

$$\sigma[\hat{p}(0)] = \frac{\hbar}{2\sigma[\hat{q}(0)]}. \tag{3.1}$$

The mass will then freely evolve until the second measurement is performed at time τ. This evolution is, of course, governed by the Hamiltonian

$$\hat{H} = \frac{\hat{p}^2}{2m}, \tag{3.2}$$

which, using the Heisenberg equation of motion (Eq. (1.20)), yields the evolution

$$\hat{q}(\tau) = \hat{q}(0) + \frac{\tau}{m}\hat{p}(0). \tag{3.3}$$

The second measurement can, in principle, be performed with arbitrary precision without any back-action penalty on the position estimate since any back-action effects that it introduces will only alter the dynamics at future times. The uncertainty in the measurement outcome is then entirely specified by $\sigma^2[\hat{q}(\tau)]$, which includes only the uncertainty in the initial localisation of the mass $\sigma[\hat{q}(0)]$ and the component of the mass's momentum uncertainty that has coupled into position during the evolution

$$\sigma^2[\hat{q}(\tau)] = \sigma^2\left[\hat{q}(0) + \frac{\tau}{m}\hat{p}(0)\right] \tag{3.4}$$

$$= \sigma^2[\hat{q}(0)] + \left(\frac{\tau}{m}\right)^2 \sigma^2[\hat{p}(0)] \tag{3.5}$$

$$= \sigma^2[\hat{q}(0)] + \left(\frac{\hbar\tau}{2m}\right)^2 \frac{1}{\sigma^2[\hat{q}(0)]}. \tag{3.6}$$

Here, by simply taking the sum of the uncertainty contributions from the position and momentum, we are assuming that the position and momentum of the mass are not correlated after the first measurement.[1] Indeed, no such correlations can exist if the mass is in a minimum uncertainty state, as described by Eq. (3.1). The regime in which this assumption breaks down is interesting and can allow improved measurement precision. We consider this regime in some detail in Chapter 5.

Since Eq. (3.6) contains terms that scale both as $\sigma^2[\hat{q}(0)]$ and as $\sigma^{-2}[\hat{q}(0)]$, it is clear that an optimal measurement precision exists that maximises the accuracy of the measurement. It is straightforward to show that this optimum is

$$\sigma_{sql}[\hat{q}(0)] = \sqrt{\frac{\hbar\tau}{2m}}, \tag{3.7}$$

which yields a *standard quantum limit* to position measurement of a free mass

$$\sigma_{sql}[\hat{q}(\tau)] = \sqrt{\frac{\hbar\tau}{m}}. \tag{3.8}$$

---

[1] Or, specifically, that $\langle\hat{q}(0)\hat{p}(0) + \hat{p}(0)\hat{q}(0)\rangle - 2\langle\hat{p}(0)\rangle\langle\hat{q}(0)\rangle = 0$.

We see therefore that the presence of quantum measurement back-action provides a fundamental limit to the precision of position measurements. To take an example, since gravitational waves cause an oscillation in the relative length of the two arms of an interferometer, a differential position measurement is most sensitive to them if the first measurement is made when one arm of the interferometer is fully extended and the second is made when it is fully contracted. Consequently, detection of a kilohertz gravity wave requires measurements with a delay of around a millisecond. If the end-mirrors have a kilogram mass, Eq. (3.8) gives a standard quantum limited precision of 4×10^{-19} m, within an order of magnitude of the current sensitivity of state-of-the-art ground-based gravitational wave interferometers.

3.2 RADIATION PRESSURE SHOT NOISE

For the majority of the remainder of this chapter we will examine the continuous limit of this measurement back-action and its consequences for the precision of position and force sensors. We begin in this section by considering the effect of fluctuations in radiation pressure due to optical shot noise on the dynamics of a mechanical oscillator. This radiation pressure shot noise is the necessary quantum back-action on the oscillator that compliments the information imprinted on the optical field about the position of the oscillator.

3.2.1 Optomechanical dynamics with on-resonance optical driving

Here we will consider only the linearised dynamics of the optomechanical system (as discussed in Section 2.7), taking the limit where a bright optical field is injected into the cavity and the system is far away from the single photon strong coupling regime (see Chapter 6 for a discussion of this regime). We will, further, take the Markovian limit where the bath has no memory, for both the optical field and the mechanical oscillator, and make the rotating wave approximation on the optical field, taking the cavity resonance frequency to be much higher than any other rates in the problem. In this regime, linearised equations of motion for the dimensionless position and momentum of the mechanical oscillator and optical field can be derived from the Hamiltonian of Eq. (2.41) using the quantum Langevin equations Eqs. (1.90) and (1.112), respectively, for the mechanical oscillator and optical field. The resulting equations of motion are

$$\dot{\hat{X}} = -\frac{\kappa}{2}\hat{X} + \sqrt{\kappa}\hat{X}_{\text{in}} \tag{3.9a}$$

$$\dot{\hat{Y}} = -\frac{\kappa}{2}\hat{Y} + \sqrt{\kappa}\hat{Y}_{\text{in}} - 2g\hat{Q} \tag{3.9b}$$

$$\dot{\hat{Q}} = \Omega\hat{P} \tag{3.9c}$$

$$\dot{\hat{P}} = -\Omega\hat{Q} - \Gamma\hat{P} + \sqrt{2\Gamma}\hat{P}_{\text{in}} - 2g\hat{X}, \tag{3.9d}$$

where we have taken the case of on-resonance optical driving ($\Delta = 0$) for simplicity, and we remind the reader that as usual $\hat{X}$ and $\hat{Y}$ refer to the optical amplitude and phase quadratures, $\hat{Q}$ and $\hat{P}$ refer to the dimensionless mechanical position and momentum, κ and Γ are the optical and mechanical decay rates, and g is the coherent amplitude boosted optomechanical coupling rate. The non-zero detuning case is very interesting and allows cooling, light-mechanical entanglement, and measurements beyond the standard quantum limit. We consider those scenarios in Chapters 4 and 5. Notice that, as might be expected and was discussed in the previous chapter, the position of the mechanical oscillator is imprinted on the phase (or momentum) quadrature of the intracavity field, while the amplitude (or position) quadrature of the optical field is similarly imprinted on the momentum of the mechanical oscillator.

Because we have chosen to drive the optical cavity on resonance ($\Delta = 0$), the quantum stochastic equations describing the optical amplitude and phase quadratures (Eqs. (3.9a) and (3.9b)) are independent. This is not the case for the mechanical oscillator. However, the two first-order stochastic differential equations (Eqs. (3.9c) and (3.9d)) describing the mechanical oscillator are easily combined into a single second-order differential equation

$$\ddot{\hat{Q}} + \Gamma\dot{\hat{Q}} + \Omega^2\hat{Q} = \sqrt{2\Gamma}\Omega\hat{P}_{\text{in}} - 2g\Omega\hat{X}. \tag{3.10}$$

This linear system of equations can then be solved straightforwardly in the frequency domain. Taking the Fourier transform, we obtain the steady-state solutions

$$\hat{X}(\omega) = \frac{\sqrt{\kappa}\hat{X}_{\text{in}}}{\kappa/2 - i\omega} \tag{3.11a}$$

$$\hat{Y}(\omega) = \frac{\sqrt{\kappa}\hat{Y}_{\text{in}} - 2g\hat{Q}}{\kappa/2 - i\omega} \tag{3.11b}$$

$$\hat{Q}(\omega) = \chi(\omega)\left(\sqrt{2\Gamma}\hat{P}_{\text{in}} - 2g\hat{X}\right) \tag{3.11c}$$

where the mechanical susceptibility $\chi(\omega) \equiv \Omega/(\Omega^2 - \omega^2 - i\omega\Gamma)$, and we have neglected the ω arguments on the right-hand side and in subsequent analysis. The mechanical susceptibility derived here is identical to that derived in Chapter 1 in the absence of optical driving (Eq. (1.102)), showing that for zero detuning ($\Delta = 0$) the optical field does not modify the response of the mechanical oscillator to its environment. We will see later that this is no longer true if the optical cavity is detuned.

Substituting Eq. (3.11a) for the optical amplitude quadrature into Eq. (3.11c), we arrive at the expression for the mechanical position

$$\hat{Q}(\omega) = \sqrt{2\Gamma}\chi(\omega)\left(\hat{P}_{\text{in}} - \sqrt{2C_{\text{eff}}}\hat{X}_{\text{in}}\right), \tag{3.12}$$

where we have introduced the *effective optomechanical cooperativity*

$$C_{\text{eff}}(\omega) \equiv \frac{C}{(1 - 2i\omega/\kappa)^2} \tag{3.13}$$

with C being the *optomechanical cooperativity*

$$C \equiv \frac{4g^2}{\kappa\Gamma}. \qquad (3.14)$$

We see that, through radiation pressure, the optical shot noise contributes a heating term to the mechanical oscillator dynamics, with magnitude dependent on the effective optomechanical cooperativity. As might be expected, this heating is attenuated at frequencies above the cavity bandwidth ($\omega > \kappa$) since the incident optical fluctuations at these frequencies are off-resonance and therefore partially rejected from the cavity. Particularly in the resolved sideband regime where $\Omega > \kappa$ that we consider in detail in Chapters 4 and 5, this attenuation has the effect of reducing the optomechanical interaction strength. However, this does not pose a fundamental constraint. Cavity optomechanical systems have been proposed and demonstrated that utilise multiple optical resonances to alter the optical response function and thereby ensure that the optical cavity admits optical fluctuations at both the optical carrier frequency and at sidebands spaced by the mechanical resonance frequency away from the carrier (see, for example, [89, 336, 45]).

3.2.2 Effect of detuning

We have seen above that, in the case of zero optical detuning ($\Delta = 0$), the effect of radiation pressure on the mechanical oscillator is solely to introduce additional heating. The dynamics become more complicated when a non-zero detuning is present. In this case, the optical amplitude and phase quadratures are dynamically linked. Mechanical position information encoded on the intracavity phase quadrature is then transferred to the intracavity amplitude quadrature, and can drive the momentum of the mechanical oscillator. This dynamical back-action on the mechanical oscillator acts to modify its susceptibility, and leads to the possibilities of laser cooling, parametric amplification, and other dynamical effects (see Chapter 4).

Exercise 3.1 *Derive equations of motion for the optical field and mechanical oscillator similar to Eqs. (3.9), but including an optical detuning Δ. Solving these equations in the same manner as above, show that the radiation pressure interaction modifies the mechanical susceptibility to*

$$\chi^{-1}(\omega, \Delta) = \chi^{-1}(\omega) - \frac{4g^2\Delta}{\left(\kappa/2 - i\omega\right)^2 + \Delta^2}, \qquad (3.15)$$

where $\chi(\omega)$ is the bare susceptibility in the absence of the optical field (Eq. (1.102)).

Unlike the bare mechanical susceptibility, in general, the dynamical back-action modified mechanical susceptibility of Eq. (3.15) is not a generalised Lorentzian function. However, in the low frequency limit where $\omega^2 \ll (\kappa/2)^2 +$

Δ^2 it can be shown to have approximately the same functional form as Eq. (3.15), with modified decay rate $\Gamma(\Delta)$ and frequency $\Omega(\Delta)$ given by

$$\frac{\Gamma(\Delta)}{\Gamma} = 1 + C \frac{\Delta\kappa^2\Omega}{(\kappa^2/4 + \Delta^2)^2} \tag{3.16a}$$

$$\frac{\Omega(\Delta)}{\Omega} = \left(1 - \frac{C}{Q}\frac{\kappa\Delta}{\kappa^2/4 + \Delta^2}\right)^{1/2}, \tag{3.16b}$$

where $Q \equiv \Omega/\Gamma$ is the mechanical quality factor.

Exercise 3.2 *Show these results.*

We can observe from these expressions that, with non-zero detuning, the radiation pressure interaction provides a method to both shift the mechanical resonance frequency, termed the *optical spring effect* and first demonstrated in [259], and, by changing the energy decay rate from the mechanical oscillator without changing its environmental heating rate, to parametrically heat or cool the mechanical oscillator [37, 160]. For further discussion of parametric heating and cooling, and their respective uses to reach the mechanical ground state and generate optomechanical entanglement, the reader is referred to Sections 4.2 and 4.4. From inspection of Eqs. (3.16) it is apparent that the sign of the detuning determines both which of heating or cooling occurs, and whether the optical spring effect pulls the oscillator to higher or lower frequencies. In the special case where the magnitude of detuning $|\Delta| = \kappa/2$, we find that appreciable cooling/heating occurs once the optomechanical cooperativity $C \sim \kappa/\Omega$, while the optical spring effect induces a frequency shift comparable to the mechanical resonance frequency when $C \sim Q$.

3.2.3 Mechanical power spectral density with on-resonance optical driving

Let us now return to the zero-detuning case. Calculating the position power spectral density of the mechanical oscillator using Eq. (1.43), we find

$$S_{QQ}(\omega) = 2\Gamma|\chi(\omega)|^2 \left(S_{P_{\text{in}}P_{\text{in}}}(\omega) + 2\,|C_{\text{eff}}(\omega)|\,S_{X_{\text{in}}X_{\text{in}}}(\omega)\right), \tag{3.17}$$

where we have assumed that the optical and mechanical baths are independent and uncorrelated and therefore neglected the cross terms.

Exercise 3.3 *Assuming that the optical field is of sufficiently high frequency that the optical bath is essentially in a vacuum state, show that within the quantum optics approximation (see Section 1.3.2) the power spectral densities*

$$S_{QQ}(\omega) = 2\Gamma|\chi(\omega)|^2\left(\bar{n} + |C_{\text{eff}}(\omega)| + 1\right) \tag{3.18a}$$

$$S_{QQ}(-\omega) = 2\Gamma|\chi(\omega)|^2\left(\bar{n} + |C_{\text{eff}}(\omega)|\right). \tag{3.18b}$$

Then, treating the effective optomechanical cooperativity as constant across the mechanical resonance, show that the radiation pressure shot noise raises the mean occupancy $\bar{n}_b$ of the mechanical oscillator in thermal equilibrium from $\bar{n}$ to

$$\bar{n}_b = \bar{n} + \bar{n}_{ba}, \tag{3.19}$$

where

$$\bar{n}_{ba} \equiv |C_{\text{eff}}(\Omega)| = \frac{C}{1 + 4(\Omega/\kappa)^2} \tag{3.20}$$

quantifies the increase in mechanical phonon occupancy due to radiation pressure back-action.

Hint: in the rotating frame the optical bath power spectral density $S_{X_{\text{in}}X_{\text{in}}}(\omega) = \bar{n}_L + 1/2$ (see Section (1.4.6)), with $\bar{n}_L$ being the photon occupancy of the incident optical field evaluated at the laser carrier frequency Ω_L.

Equation (3.19) makes the physical significance of the effective optomechanical cooperativity C_{eff} quite clear. It establishes the optomechanical coupling strength required for the optical back-action on the mechanical oscillator to heat the oscillator by one phonon. As we will see in later chapters, the optomechanical cooperativity has other significance as well. For instance, it is the metric that establishes the effectiveness of both resolved sideband cooling (see Section 4.2.2) and feedback cooling (see Section 5.2).

Radiation pressure effects are widely seen in experiments with cold atomic gases. For instance, the random recoil experienced by atoms upon spontaneous emission of light sets a limit on the temperature of an atomic ensemble in a magneto-optic trap [177]. However, the radiation pressure back-action heating of a single collective mode of motion that we have introduced here was first observed only in 2008 [207], in that case using a collective mechanical mode of a cloud of 9,000 ultracold atoms. Since the optomechanical cooperativity is proportional to the square of the zero-point motion of the mechanical oscillator, as the mass of the mechanical oscillator increases it becomes increasingly difficult to achieve $|C_{\text{eff}}| > \bar{n}$, as required for radiation pressure shot noise to dominate thermal heating. In 2013 radiation pressure shot noise was observed for the first time for a mechanical resonance of a macroscopic mechanical oscillator using a 7 ng silicon nitride membrane [231] (see Fig. 3.1).

3.3 MEASUREMENT OF MECHANICAL MOTION

As we have just seen, the radiation pressure interaction perturbs the motion of a mechanical oscillator, introducing noise. This noise can be understood to be a direct consequence of the Heisenberg uncertainty principle between position and momentum. The interaction encodes position information on the optical field, necessitating an increase in noise on the mechanical momentum. In this section we introduce the linear detection processes of homodyne and

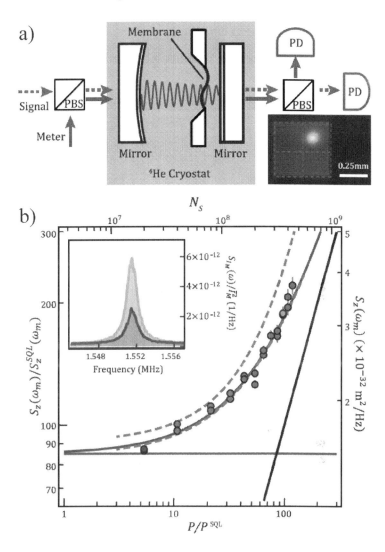

FIGURE 3.1 Observation of measurement back-action on a cryogenically cooled silicon nitride membrane with 7 ng effective mass in a Fabry–Pérot cavity. From [231]. Reprinted with permission from AAAS. (a) Experimental schematic. (b) Peak position power spectral density as a function of incident power showing that the resonant mode temperature increases with power. N_S: intracavity photon number, P/P^{SQL}: ratio of injected power to power at the standard quantum limit, $S_z(\omega_m)$: mechanical power spectral density at the mechanical resonance frequency, $S_z^{\mathrm{SQL}}(\omega_m)$: mechanical power spectral density at the standard quantum limit, $S_{I_M}(\omega)/\bar{I}_M^2$: normalised power spectral density of secondary meter beam, PBS: polarising beam splitter, PD: photodiode.

heterodyne detection commonly used to extract information about the mechanical oscillator from the field that exits the optomechanical system. This will allow us to establish, later in the chapter, standard quantum limits for both mechanical position and force measurement.

Before embarking on this discussion of detection techniques, however, we will establish a formal minimum bound on the product of the measurement imprecision and the magnitude of radiation pressure noise exerted on a mechanical oscillator valid in the case of continuous linear measurements [44].

3.3.1 Quantum bound on force-imprecision product

The force experienced by the mechanical oscillator in an optomechanical system described by the linearised Hamiltonian of Eq. (2.41) may be determined in the usual way via $\hat{F} = \partial \hat{H}/\partial \hat{q}$, where $\hat{q}$ is of course the mechanical position operator. The contribution to this force from the optical field is easily shown to be

$$\hat{F}_L = \frac{\sqrt{2}\hbar g}{x_{zp}} \hat{X}. \qquad (3.21)$$

Exercise 3.4 *Convince yourself of this.*

Here the optical amplitude quadrature $\hat{X}$ is driven by the fluctuations of the optical input field $\hat{X}_{\text{in}}$, as we already found in Eq. (3.9a). Let us consider, for now, the bad cavity limit where κ is much larger than Ω as well as all other system rates, leaving a more general treatment to later sections. In this simple case $\hat{X}$ reaches equilibrium much faster than any other system variable and Eq. (3.9a) can be adiabatically eliminated via the approximation $\dot{\hat{X}} = 0$. Equation (3.21) can then be expressed in terms of the input optical fluctuations as

$$\hat{F}_L = \sqrt{\frac{8}{\kappa}} \left(\frac{\hbar g}{x_{zp}} \right) \hat{X}_{\text{in}}, \qquad (3.22)$$

with the radiation pressure force power spectral density given by

$$S_{FF}(\omega) = \frac{8}{\kappa} \left(\frac{\hbar g}{x_{zp}} \right)^2 S_{X_{\text{in}} X_{\text{in}}}(\omega). \qquad (3.23)$$

A similar adiabatic elimination for $\hat{Y}$ in Eq. (3.9b) allows the intracavity optical phase quadrature to be related simply to the mechanical position $\hat{q}$ and input optical phase quadrature fluctuations $\hat{Y}_{\text{in}}$ by

$$\hat{Y} = \frac{2}{\kappa} \left(\sqrt{\kappa} \hat{Y}_{\text{in}} - \frac{\sqrt{2}g}{x_{zp}} \hat{q} \right). \qquad (3.24)$$

The input-output relations given in Eqs. (1.126) then allow the output optical phase quadrature fluctuations to be determined:

$$\hat{Y}_{\text{out}} = -\hat{Y}_{\text{in}} + \sqrt{\frac{8}{\kappa}} \left(\frac{g}{x_{zp}} \right) \hat{q}. \qquad (3.25)$$

The mechanical position could be determined by measuring $\hat{Y}_{\text{out}}$, with an ideal measurement only contaminated by the input phase quadrature fluctuations $\hat{Y}_{\text{in}}$. By inspection of Eq. (3.25), as long as $\hat{Y}_{\text{in}}$ is uncorrelated to the mechanical position, the minimum possible measurement (or imprecision) noise is quantified by the power spectral density

$$S_{qq}^{\text{imp}}(\omega) = \frac{\kappa}{8}\left(\frac{x_{zp}}{g}\right)^2 S_{Y_{\text{in}}Y_{\text{in}}}(\omega). \tag{3.26}$$

We therefore find that the product of imprecision and force noise power spectral densities is

$$S_{FF}(\omega)S_{qq}^{\text{imp}}(\omega) = \hbar^2 S_{X_{\text{in}}X_{\text{in}}}(\omega)S_{Y_{\text{in}}Y_{\text{in}}}(\omega). \tag{3.27}$$

A Heisenberg uncertainty principle exists between the amplitude and phase quadrature power spectral densities (see Eq. (1.121)). This results in the lower bound on the disturbance from the measurement

$$S_{FF}(\omega)S_{qq}^{\text{imp}}(\omega) \geq \left(\frac{\hbar}{2}\right)^2. \tag{3.28}$$

We therefore see that, indeed, linear measurement of mechanical motion will unavoidably result in a quantum back-action force on the mechanical oscillator. While here we arrive at Eq. (3.28) by considering the specific case of continuous measurement on a mechanical oscillator within a bad optical cavity, the expression is quite general, applying to any continuous measurement in the absence of quantum correlations [44]. This leads to standard quantum limits in both position and force measurement. Noncontinuous measurements or measurements that involve nonclassical states of the light or mechanical oscillator can exceed the limit set out in Eq. (3.28). Such schemes form much of the discussion in Chapter 5.

3.3.2 Output field from a cavity optomechanical system

We now return to the primary subject matter of this section, treating the optical field output from a cavity optomechanical system, and approaches to extract information about the mechanical oscillator from this output in a manner that is valid more generally than the bad cavity regime considered in the previous section.

The quadratures of the output optical fields from the cavity optomechanical system can be determined using Eqs. (3.11a) and (3.11b) along with the input-output relations given in Eqs. (1.126). The result is

$$\hat{X}_{\text{out}}(\omega) = -\left(\frac{\kappa/2 + i\omega}{\kappa/2 - i\omega}\right)\hat{X}_{\text{in}} \tag{3.29a}$$

$$\hat{Y}_{\text{out}}(\omega) = -\left(\frac{\kappa/2 + i\omega}{\kappa/2 - i\omega}\right)\hat{Y}_{\text{in}} + \frac{2\sqrt{\kappa}g\hat{Q}}{\kappa/2 - i\omega}$$

$$= -\left(\frac{\kappa/2 + i\omega}{\kappa/2 - i\omega}\right)\hat{Y}_{\text{in}} + 2\sqrt{\Gamma C_{\text{eff}}}\hat{Q}. \tag{3.29b}$$

As expected, we see that both quadratures experience a frequency-dependent phase rotation due to the optical resonance, and the mechanical position is imprinted on the output optical phase quadrature with a magnitude determined by the characteristic rate $\mu = \Gamma|C_{\text{eff}}|$.[2] In the nonresolved sideband limit, where $\Omega \ll \kappa$, this rate can be approximated as $\mu = 4g^2/\kappa$. On the other hand, in the resolved-sideband regime as $\Omega/\kappa \to \infty$, the measurement rate $\mu \to 0$ so that the information imprinted on the optical field becomes asymptotically small. This is important for resolved sideband cooling, as discussed in Section 4.2.2, since any information contained in the optical field necessarily introduces a complimentary back-action heating which presents a fundamental limit the cooling performance [191].

3.3.3 Linear detection of optical fields

Although nonlinear techniques such as photon counting [76] or coupling to a superconducting qubit [213] have been implemented to detect the output field of cavity optomechanical systems (see Section 8.2), by far the most common detection techniques are linear, ultimately providing signals that are proportional to the amplitude of the optical field. Such techniques can best be understood through the example of direct detection of a bright field.

When an optical field described in the Heisenberg picture by the annihilation operator $a_{\text{det}}(t)$ impinges upon a semiconductor photodiode, photons in the field are converted into photoelectrons, which generates a (nonlinearised) photocurrent that can be described by the detected field operator

$$\hat{i}(t) = \hat{n}_{\text{det}}(t) = a^\dagger_{\text{det}}(t)a_{\text{det}}(t), \tag{3.30}$$

where $\hat{n}_{\text{det}}$ is the photon number operator for the detected field and $\langle a_{\text{det}} \rangle = \alpha_{\text{det}}$ is the coherent amplitude of the field, with $\alpha^*_{\text{det}}\alpha_{\text{det}}$ being the mean photon flux per second incident on the detector. Linearising the optical field in the usual way by making the substitution $a_{\text{det}} \to \alpha_{\text{det}} + a_{\text{det}}$ and neglecting both the constant term $(\alpha^*_{\text{det}}\alpha_{\text{det}})$ and the operator product term $(a^\dagger_{\text{det}}a_{\text{det}})$

---

[2]While this rate appears to depend linearly on the mechanical decay rate Γ and therefore improve as the quality factor of the mechanical oscillator degrades, this is not the case since Γ also appears in the denominator of the expression for $|C_{\text{eff}}|$.

yields the linearised detected field operator

$$\hat{i}(t) = \alpha_{\text{det}}a_{\text{det}}^{\dagger} + \alpha_{\text{det}}^{*}a_{\text{det}} \qquad (3.31)$$

$$= |\alpha_{\text{det}}|\hat{X}_{\text{det}}^{\theta_{\text{det}}}, \qquad (3.32)$$

where here $e^{i\theta_{\text{det}}} = \alpha_{\text{det}}/|\alpha_{\text{det}}|$, and the rotated quadrature operator $\hat{X}_{\text{det}}^{\theta}$ is defined in Eq. (1.17a). We can therefore expect that direct detection of an optical field will produce a stochastic photocurrent that is directly proportional to the field quadrature which is oriented in the direction of the coherent amplitude (see Fig. 1.2). As we will see in the next section, however, one must be somewhat careful in this interpretation since $\hat{i}(t)$ is an operator, not a classical variable.

3.3.4 Photocurent power spectral density

Any real classical photocurrent must have a power spectral density that is symmetric in frequency. This is not the case, in general, for $\hat{i}(t)$. In the case of optical detection of mechanical motion considered in this chapter, for example, $\hat{i}(t)$ must necessarily contain a term proportional to $\hat{Q}(t)$, which, as we saw in Section 1.2, is asymmetric in frequency. The resolution to this apparent contradiction is that $\hat{i}(t)$ is still, of course, a quantum operator. It is transformed into a stochastic real classical variable $i(t)$ with a symmetrised power spectral density through the detection process and subsequent irreversible electronic amplification. This observed power spectral density can be calculated formally using Glauber's theory of photo-detection [122]. Using Eq. (1.43) one can immediately calculate the power spectral density $S_{\hat{i}\hat{i}}(\omega)$ of the *operator* $\hat{i}(t)$ through the substitution $\hat{\mathcal{O}}(t) = \hat{i}(t)$. The *photocurrent* $i(t)$, on the other hand, is generated by photon annihilation events at the detector, with the photocurrent power spectral density involving two-time photon coincidences. As recognised by Glauber, to accurately determine the photocurrent the annihilation and creation operators must be normally ordered,[3] so that direct detection of the field a_{det} yields the power spectral density

$$S_{ii}(\omega) = \int_{-\infty}^{\infty} d\tau\, e^{i\omega\tau} \langle a_{\text{det}}^{\dagger}(t+\tau)a_{\text{det}}^{\dagger}(t)a_{\text{det}}(t+\tau)\hat{a}_{\text{det}}(t)\rangle_{t=0}. \qquad (3.33)$$

Using the commutation relation of Eq. (1.114) this can be reexpressed in terms of the symmetrised power spectral density of the photocurrent operator $\bar{S}_{\hat{i}\hat{i}}(\omega)$ as[4]

$$S_{ii}(\omega) = \bar{S}_{\hat{i}\hat{i}}(\omega) - \langle \hat{n}_{\text{det}}\rangle \qquad (3.34)$$

$$\approx \bar{S}_{\hat{i}\hat{i}}(\omega), \qquad (3.35)$$

---

[3] That is, the creation operators should all appear to the left of the annihilation operators.
[4] Note the ^'s here to distinguish between a classical variable and the detected field operator.

where Eq. (3.35) is valid within the linearisation approximation, since linearisation of $\bar{S}_{ii}(\omega)$ results in quadrature variance terms that are amplified by the square of the coherent amplitude of the field. We see, therefore, that photodetection will always produce a symmetrised power spectral density.

Exercise 3.5 *Show this result.*

3.3.5 Phase-referenced detection

In the particular case of a cavity optomechanical system in the steady state with zero detuning $(\Delta = 0)$ considered in this chapter, we see from Eqs. (2.43) and (2.44) that the coherent amplitude of the output field α_{out} is real.[5] This means that, if the field is directly detected, $\theta_{det} = m\pi$ where m is an integer, and the measured quadrature is the optical amplitude quadrature. Inspection of Eqs. (3.29) shows that no information about the mechanical oscillator is contained on this quadrature. Hence, direct detection is ineffective for probing the mechanical oscillator when no cavity detuning is present. This necessitates a detuning to be introduced to the optical cavity, or the use of alternative forms of detection that provide a phase reference. *Homodyne* and *heterodyne* detection are two such phase-referenced detection techniques that find broad use in coherent communications, microwave electronics, and quantum optics, as well as quantum optomechanics. These techniques are contrasted to direct detection in Fig. 3.2.

The mathematics of homodyne and heterodyne detection is treated in Appendix A. For the purposes of the majority of this textbook it is only important to know that, by interfering the detected field with a bright local oscillator field of the same carrier frequency, homodyne detection provides a measurement of an arbitrary quadrature $\hat{X}_{det}^{\theta}$ of the detected optical field with normalised power spectral density

$$S_{ii}^{\text{homo}}(\omega) = \bar{S}_{X_{det}^{\theta} X_{det}^{\theta}}(\omega). \tag{3.36}$$

Heterodyne detection, on the other hand, uses a carrier frequency offset from the laser frequency Ω_L by Δ_{LO}. Making use of the rotating wave approximation (see Appendix A), this results in the normalised power spectral density

$$
\begin{aligned}
S_{ii}^{\text{het}}(\omega) = \frac{1}{4}\Big[& S_{X_{det}X_{det}}(\Delta_{LO} + \omega) + S_{Y_{det}Y_{det}}(\Delta_{LO} + \omega) \\
& + S_{X_{det}X_{det}}(\Delta_{LO} - \omega) + S_{Y_{det}Y_{det}}(\Delta_{LO} - \omega)\Big].
\end{aligned} \tag{3.37}
$$

Here, frequency components at $\Delta_{LO} \pm \omega$ have been "mixed down" to ω by the beating between the detected field and local oscillator.

---

[5] Remembering that we have defined the intracavity amplitude α to be real.

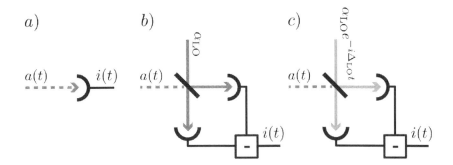

FIGURE 3.2 Schematics of direct (a), homodyne (b), and heterodyne (c) detection of an optical field. $a(t)$ is the input field to be measured. $i(t)$ is the resulting photocurrent. In homodyne detection the input field is interfered on a 50/50 beam splitter with a bright local oscillator that has the same optical carrier frequency. The phase of the local oscillator α_{LO} determines the measured optical quadrature. In heterodyne detection an offset local oscillator carrier frequency is used, resulting in a temporally oscillating phase relative to the input field. The measured optical quadrature therefore also oscillates with time.

3.3.6 Measured power spectral densities from a cavity optomechanical system

Using the result of the previous section, we can determine the power spectral densities of homodyne and heterodyne detection of the optical field output from a cavity optomechanical system. As we have observed previously, Eqs. (3.29) show that, with zero detuning, the amplitude quadrature of the optical field contains no information about the mechanical motion. Consequently, here we consider only the case where $\theta = \pi/2$ so that in the frequency domain the detected quadrature is

$$\hat{X}_{\text{det}}^{\pi/2}(\omega) = \hat{Y}_{\text{out}}(\omega) \tag{3.38}$$

$$= -\left(\frac{\kappa/2+i\omega}{\kappa/2-i\omega}\right)\hat{Y}_{\text{in}}(\omega) - 4\Gamma C_{\text{eff}}\chi(\omega)\hat{X}_{\text{in}}(\omega) + 2\sqrt{\Gamma C_{\text{eff}}}\,\hat{Q}^{(0)}(\omega) \tag{3.39}$$

$$= -\left(\frac{\kappa/2 + i\omega}{\kappa/2 - i\omega}\right)\hat{Y}_{\text{in}}(\omega)$$
$$+ 2\Gamma\sqrt{2C_{\text{eff}}}\chi(\omega)\left(\hat{P}_{\text{in}}(\omega) - \sqrt{2C_{\text{eff}}}\hat{X}_{\text{in}}(\omega)\right), \tag{3.40}$$

where

$$\hat{Q}^{(0)}(\omega) \equiv \sqrt{2\Gamma}\chi(\omega)\hat{P}_{\text{in}}(\omega) \tag{3.41}$$

is the mechanical position in the absence of measurement (i.e. with $C_{\text{eff}} = 0$). Note that, in fact, while the detected signal is maximised for $\theta_{\text{LO}} = \pi/2$, correlations between the noise on the output optical amplitude and phase quadratures due the radiation pressure interaction mean that the noise in the measurement – and therefore signal-to-noise ratio – can be reduced by shifting the local oscillator phase away from $\pi/2$. We look into these effects in Chapter 5.

From inspection of Eqs. (3.40) and (3.37) it is clear that the power spectral density of the detected phase quadrature for both homodyne and heterodyne detection contains cross-terms between the mechanical bath fluctuations and the optical input amplitude and phase quadratures. While in principle it is possible for correlations to exist between these baths – for instance, if the mechanical bath consisted of some guided phonon mode that had previously interacted strongly with the optical field – here we treat the usual case where the baths are uncorrelated, so that their cross-spectral densities are zero. This leaves only the cross-spectral densities $S_{X_{\text{in}} P_{\text{in}}}(\pm\omega)$ and $S_{P_{\text{in}} X_{\text{in}}}(\pm\omega)$ between the amplitude and phase quadratures of the input field. These are also zero for a coherent or thermal input field, as we found for a general case in Exercise 1.13.

3.3.6.1 Power spectral density from homodyne detection

With no bath correlations, the phase quadrature homodyne power spectral density is simply the sum of the symmetrised power spectral densities of each term in Eq. (3.40):

$$S_{ii}^{\text{homo}}(\omega) \;=\; \frac{1}{2} + 8\eta\Gamma^2 |C_{\text{eff}}|^2 |\chi(\omega)|^2 + 4\eta\Gamma |C_{\text{eff}}| \bar{S}_{Q^{(0)} Q^{(0)}}(\omega) \quad (3.42)$$

$$\;=\; \frac{1}{2} + 8\eta\Gamma^2 |C_{\text{eff}}| \, |\chi(\omega)|^2 \left(\bar{n}_b + \frac{1}{2} \right) \quad (3.43)$$

$$\;=\; \frac{1}{2} + 4\eta\Gamma |C_{\text{eff}}| \bar{S}_{QQ}(\omega), \quad (3.44)$$

where $S_{Q^{(0)} Q^{(0)}}(\omega)$ is the power spectral density of the mechanical position in the absence of radiation pressure. To arrive at Eq. (3.43) we have used the nonrotating and rotating frame bath power spectral densities given in Eqs. (1.98) and (1.118a), respectively, for the mechanical bath and optical fields, and have assumed that the optical field has sufficiently high frequency to be well approximated as having no thermal occupancy. To obtain the final expression we have used Eq. (3.18), recognising that the mechanical occupancy includes contributions from both the mechanical bath and the radiation pressure shot noise heating (i.e., $\bar{n}_b = \bar{n} + \bar{n}_{ba}$). The factor of $1/2$ at the front of Eqs. (3.43) and (3.44) represents a white background spectrum due to shot noise on the optical field, while the second term is contributed by the mechanical oscillator, including heating from the radiation pressure shot noise driving it.

The perceptive reader will have noticed that the symbol η has been introduced in Eqs. (3.42) to (3.44). This constitutes an overall detection efficiency that accounts for inefficiencies in the escape of the optical field from the cavity, any propagation losses due – for example – to scattering and absorption before the detectors, and inefficiencies in the detection process itself. A discussion of the appropriate approach to model detection efficiencies in the Heisenberg picture is given in Appendix A.

Exercise 3.6 *Derive Eq. (3.43) including detection efficiencies.*

The homodyne power spectral density given in Eq. (3.44) is shown as a function of frequency in Fig. 3.3 for a range of effective cooperativities C_{eff}. As can be seen, the thermal noise of the mechanical mode introduces a Lorentzian noise peak to the power spectral density, with the height of the peak increasing as the optomechanical interaction strength increases, due both to improved mechanical transduction and, for $|C_{\text{eff}}| > \bar{n}$, radiation pressure shot noise heating.

3.3.6.2 Power spectral density from heterodyne detection

The power spectral density of a heterodyne measurement can be determined in the same way as the homodyne case. Substituting the output optical amplitude and phase quadrature operators from the cavity optomechanical system (Eqs. (3.29)) for $\hat{X}_{\text{det}}$ and $\hat{Y}_{\text{det}}$ in Eq. (3.37) and using the optical bath power spectral densities defined in Eqs. (1.118a), we find

$$S_{ii}^{\text{het}}(\omega) = \frac{1}{2} + \eta\Gamma\left[\, |C_{\text{eff}}(\Delta_{\text{LO}} + \omega)|\, S_{QQ}(\Delta_{\text{LO}} + \omega)\right.$$
$$\left. + |C_{\text{eff}}(\Delta_{\text{LO}} - \omega)|\, S_{QQ}(\Delta_{\text{LO}} - \omega)\right]. \qquad (3.45)$$

Here, since the power spectrum of the mechanical oscillator $S_{QQ}(\omega)$ is sharply peaked at frequencies $\omega = \pm\Omega$, we can observe that the mechanical peak will appear at four different frequencies offset from $\pm\Delta_{\text{LO}}$ by $\pm\Omega$. As long as $\Gamma \ll \Delta_{\text{LO}}$ so that the off-mechanical-resonance term in Eq. (3.45) can be neglected, the heterodyne power spectral densities on the positive and negative frequency sides of Δ_{LO} are

$$S_{ii}^{\text{het}}(\Delta_{\text{LO}} + \omega) = \frac{1}{2} + \eta\Gamma\,|C_{\text{eff}}|\,S_{QQ}(-\omega) \qquad (3.46)$$

$$= \frac{1}{2} + 2\eta\Gamma^2\,|C_{\text{eff}}|\,|\chi(\omega)|^2\,\bar{n}_b \qquad (3.47)$$

$$S_{ii}^{\text{het}}(\Delta_{\text{LO}} - \omega) = \frac{1}{2} + \eta\Gamma\,|C_{\text{eff}}|\,S_{QQ}(\omega) \qquad (3.48)$$

$$= \frac{1}{2} + 2\eta\Gamma^2\,|C_{\text{eff}}|\,|\chi(\omega)|^2\,(\bar{n}_b + 1), \qquad (3.49)$$

where we have used Eqs. (3.18) and (3.19), and have again included the overall detection efficiency η, which appears, as in the case of homodyne detection, only as a prefactor attenuating the mechanical signal.

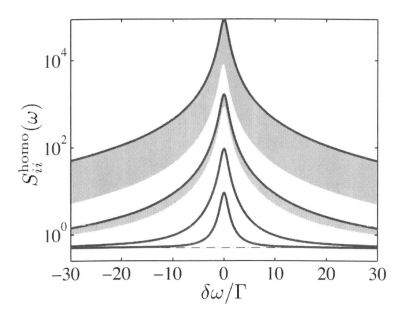

FIGURE 3.3 Photocurrent power spectral density for homodyne phase measurement of the output field from a cavity optomechanical system with cavity detuning $\Delta = 0$ as a function of frequency offset from the mechanical resonance ($\delta\omega = \omega - \Omega$). The effective cooperativities are $|C_{\text{eff}}| = \{0.1, 1, 10, 100\}$, for the bottom to top traces, respectively. The shaded regions, visible only for the top two traces, are the additional noise power introduced by radiation pressure shot noise heating. The dot-dashed line is the optical shot noise level. For each trace the thermal occupancy is $\bar{n} = 10$, the mechanical quality factor is $Q = \Omega/\Gamma = 1000$, and the measurement efficiency is taken to be unity ($\eta = 1$).

The above power spectral densities are similar to that obtained using homodyne detection in Eq. (3.43). However, there are two notable differences. Firstly, the mechanical signal here is attenuated by a factor of four; a factor of two arises from the fact that heterodyne detection is sensitive to both the amplitude and phase quadratures of the detected light field, while the mechanical signal is only encoded on the phase quadrature. The second factor of two arises because the mechanical peak appears four times in the heterodyne power spectrum, while only twice in the homodyne power spectrum. Secondly, and more importantly, where homodyne detection yields a symmetrised mechanical power spectral density, the mechanical components in the heterodyne power spectral density are asymmetric between the positive and negative sides of the local oscillator detuning Δ_{LO} and match the quantum power spectral

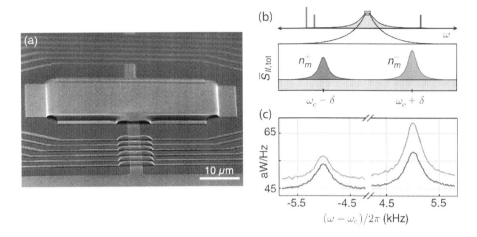

FIGURE 3.4 Sideband asymmetry experiment of [308]. (a) Bulk element superconducting microwave optomechanical device. (b) Illustration of heterodyne detection power spectrum. (c) Observed sideband asymmetry in the heterodyne power spectrum at $\bar{n}_b = 0.6$ (light grey) and $\bar{n}_b = 2.5$ (dark grey).

density of the mechanical oscillator itself. Heterodyne detection, therefore, allows the quantum asymmetry in the mechanical power spectrum to be directly probed.

Sideband asymmetry has been experimentally explored in a variety of cavity optomechanical systems, including, for example, phononic-photonic crystals [244], intracavity cold atom clouds [48], and superconducting microwave optomechanical systems [308]. Figure 3.4 shows the sideband asymmetry observed for a superconducting cavity optomechanical system cooled to millikelvin temperatures in a dilution refrigerator and further cooled using resolved sideband cooling (see Section 4.2.2) [308].

3.3.6.3 Characterisation of optomechanical parameters

Homodyne and heterodyne detection are valuable tools to characterise many of the important parameters in quantum optomechanical systems, the most important generally being the optomechanical cooperativity C and the mechanical occupancy $\bar{n}$. Appendix A discusses some useful methods to characterise these parameters that utilise the optical shot noise level and sideband asymmetry as a form of quantum ruler to achieve accurate calibration.

3.4 STANDARD QUANTUM LIMIT ON MECHANICAL POSITION MEASUREMENT

Now that we have determined the detected power spectral density of the output field of a cavity optomechanical system, including the radiation pressure induced measurement back-action on the mechanical oscillator, we are in a position to rigorously derive the standard quantum limit to optomechanical sensing which was first introduced for the case of a free mass in Section 3.1.

We being by renormalising the phase quadrature of the output field (Eq. (3.39)) into units of mechanical position:

$$\hat{Q}_{\text{det}}(\omega) \equiv \frac{\hat{X}_{\text{det}}^{\pi/2}(\omega)}{2\sqrt{\Gamma C_{\text{eff}}}} \tag{3.50}$$

$$= \underbrace{\hat{Q}^{(0)}(\omega)}_{\text{mechanics}} - \underbrace{\frac{1}{2\sqrt{\Gamma C_{\text{eff}}}} \left(\frac{\kappa/2 + i\omega}{\kappa/2 - i\omega} \right) \hat{Y}_{\text{in}}(\omega)}_{\text{measurement noise}} - \underbrace{2\sqrt{\Gamma C_{\text{eff}}} \chi(\omega) \hat{X}_{\text{in}}(\omega)}_{\text{back-action noise}}.$$

It is evident that, any estimate of the mechanical position from measurements on the output field will be contaminated by both the phase and amplitude quadratures of the input field. The phase quadrature contributes the usual measurement noise – or shot noise – familiar to many optical systems and sensors. The amplitude quadrature enters, in contrast, as a form of measurement back-action noise induced by the radiation pressure driving the mechanical motion (as we saw earlier in Section 3.2). Combined, these two noise sources are responsible for the standard quantum limit of measurement precision.

Again limiting ourselves to the case of coherent optical driving, where $S_{X_{\text{in}}X_{\text{in}}}(\omega) = S_{Y_{\text{in}}Y_{\text{in}}}(\omega) = 1/2$ and $\bar{S}_{X_{\text{in}}Y_{\text{in}}}(\omega) = 0$, the total homodyne detection noise power spectral density $\bar{S}_{\text{det}}(\omega)$ due to measurement imprecision and back-action can be shown from Eq. (3.50) to be

$$\bar{S}_{\text{det}}(\omega) = \frac{1}{8\eta\Gamma|C_{\text{eff}}|} + 2\Gamma |\chi(\omega)|^2 |C_{\text{eff}}|, \tag{3.51}$$

where, similar to our earlier treatment of homodyne and heterodyne detection, we have introduced the parameter η to account for the total detection efficiency of the measurement including both escape efficiency from the optical cavity and losses afterwards. Recognising that the mechanical position operator $\hat{q} = \sqrt{2}x_{zp}\hat{Q}$, this noise floor can be expressed in terms of the absolute position of the oscillator via $\bar{S}_{q_{\text{det}}q_{\text{det}}}(\omega) = 2x_{zp}^2\bar{S}_{\text{det}}(\omega)$, with the result having units of meters squared per hertz. Note also that, by limiting ourselves to both on-resonance optical driving ($\Delta = 0$), and the case where there are no correlations between the input quadratures of the bath, we are omitting important physics that provides a route to enhanced measurement precision. We consider these situations in detail in Section 5.4.

It is clear from Eq. (3.51) that, since the optical measurement noise decreases with increasing optomechanical cooperativity and the back-action noise increases with increasing cooperativity, an optimum cooperativity exists that minimises the total noise in the displacement measurement. This optimum is achieved at

$$|C_{\text{eff}}^{\text{opt}}| = \frac{1}{4\eta^{1/2}\Gamma|\chi(\omega)|}. \tag{3.52}$$

At the interaction strength given in Eq. (3.52) the measurement and back-action noise terms are exactly balanced, with the total detection uncertainty given by

$$\bar{S}_{\text{det}}^{\text{opt}}(\omega) = \frac{|\chi(\omega)|}{\eta^{1/2}}. \tag{3.53}$$

We see that, even with the optimum choice of interaction strength, probing the mechanical oscillator necessarily introduces additional noise. Since the mechanical susceptibility $\chi(\omega)$ is largest on resonance, this noise is maximised on resonance (see Fig. 3.5). The detection noise decreases with increasing measurement efficiency, and as $\eta \to 1$ reaches the *standard quantum limit*

$$\bar{S}_{\text{det}}^{\text{SQL}}(\omega) = |\chi(\omega)|, \tag{3.54}$$

which quantifies the best measurement precision that can be achieved for a given mechanical susceptibility in the absence of quantum correlations or back-action evading measurements.[6]

It is worth noting that in the free-mass limit where $\omega \gg \{\Omega, \Gamma\}$ relevant for example to gravitational wave interferometers (see Section 3.5), when normalised into absolute position units by multiplying through by $2x_{zp}^2$ the standard quantum limit of Eq. (3.54) is independent of both the mechanical resonance frequency and decay rate; only depending on the mechanical oscillator via its mass m.

Exercise 3.7 *Show this result.*

3.4.1 Incident power required to reach the standard quantum limit

Practically, it is useful to know the level of incident optical power required to reach the standard quantum limit at the mechanical resonance frequency, which we define as P^{SQL}. The optomechanical cooperativity which balances measurement and back-action noise can be re-expressed as an optimal intra-cavity optical coherent amplitude α^{opt} using Eq. (3.13), with the result

$$|C_{\text{eff}}^{\text{opt}}(\Omega)| = \frac{1}{4\eta^{1/2}} = \frac{4\alpha^{\text{opt}\,2}g_0^2}{\kappa\Gamma\left[1 + (2\Omega/\kappa)^2\right]}. \tag{3.55}$$

---

[6]Note that it is important to distinguish the *optomechanical* standard quantum limit discussed here from a different *metrological* standard quantum limit discussed widely in the quantum metrology literature. The former results from a balance between optical shot noise and radiation pressure noise. The latter is the result only of optical shot noise and is valid in the usual regime of photon flux low enough not to introduce significant radiation pressure noise. A rough formulation of the latter is that it is the optimal sensitivity possible in a linear measurement using coherent light in the absence of radiation pressure noise.

Using Eq. (2.43) the intracavity amplitude α can itself be related to the optical power P in watts incident on an optical cavity via

$$P = \hbar\Omega_L |\alpha_{\text{in}}|^2 \tag{3.56}$$

$$= \frac{\hbar\Omega_L\kappa}{4\eta_{\text{esc}}} \left[1 + \left(\frac{2\Delta}{\kappa}\right)^2\right] \alpha^2, \tag{3.57}$$

where the escape efficiency $\eta_{\text{esc}} = \kappa_{\text{in}}/\kappa$,[7] and as defined earlier, Ω_L is the laser frequency. Consequently, the incident optical power required to optimise the on-resonance ($\omega = \Omega$) measurement precision is

$$P^{\text{opt}} = \frac{\hbar\Omega_L\Gamma\kappa^2}{64g_0^2\,\eta^{1/2}\eta_{\text{esc}}} \left[1 + \left(\frac{2\Delta}{\kappa}\right)^2\right]\left[1 + \left(\frac{2\Omega}{\kappa}\right)^2\right], \tag{3.58}$$

where the terms in square brackets are corrections due to the cavity detuning and resolved-sideband suppression of the optomechanical interaction. The on-resonance standard quantum limit is reached when the detection is perfectly efficient ($\eta = \eta_{\text{esc}} = 1$), with $P^{\text{SQL}} \equiv P^{\text{opt}}(\eta = \eta_{\text{esc}} = 1)$. Taking parameters often found in cavity optomechanical systems of $\Omega_L/2\pi = 2\times 10^{14}$ Hz, $\kappa/2\pi = 10$ MHz, $\Delta = 0$, $\Omega/2\pi = 10$ MHz, $\Gamma/2\pi = 1$ kHz, and $g_0/2\pi = 100$ Hz, we find that in this particular case $P^{\text{SQL}} \approx 700$ nW.

3.4.2   Dependence of detection noise spectrum on incident power

The detection noise power spectral density $\bar{S}_{\text{det}}(\omega)$ is shown in Fig. 3.5 for a range of incident optical powers and perfect detection efficiency. It can be seen that for incident powers well below P^{SQL} the detection noise is spectrally flat. A Lorentzian peak appears near the mechanical resonance frequency as the incident power increases. When the incident power equals P^{SQL} the detection precision saturates the standard quantum limit on resonance but remains above it off resonance.

The back-action noise degrades the on-resonance detection precision once the incident power exceeds P^{SQL}. However, the off-resonance precision continues to improve, and, in the limit of perfect detection efficiency ($\eta = 1$), can reach the standard quantum limit. The frequencies at which this occurs can be found by equating the detection noise power spectral density in Eq. (3.51) to the standard quantum limit of Eq. (3.54) and solving for ω, with the result

$$\frac{\omega}{\Omega} = \left[1 \pm \frac{1}{Q}\sqrt{\left(\frac{P}{P^{\text{SQL}}}\right)^2 - 1}\right]^{1/2}. \tag{3.59}$$

Exercise 3.8 *Derive Eq. (3.59).*

---

[7]The total efficiency $\eta = \eta_{\text{esc}}\eta_{\text{det}}$ is the product of escape efficiency and the detection efficiency of photons that do escape from the cavity.

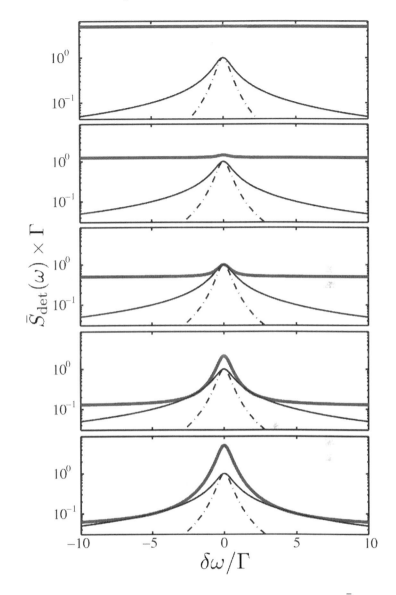

FIGURE 3.5 Total detection noise power spectral density $\bar{S}_{\mathrm{det}}(\omega)$ as a function of frequency, for various incident optical powers and perfect efficiency ($\eta = \eta_{\mathrm{esc}} = 1$). The frequency axis is centred on the mechanical resonance frequency with $\delta\omega = \omega - \Omega$. Thick solid curve: $\bar{S}_{\mathrm{det}}(\omega)$, thin solid curve: standard quantum limit $\bar{S}_{\mathrm{det}}^{\mathrm{SQL}}(\omega)$, thin dot-dashed curve: bath zero-point motion variance $\bar{S}_{\mathrm{det}}^{\mathrm{ZPL}}(\omega)$. For each trace the mechanical quality factor $Q = \Omega/\Gamma = 1000$, while from top to bottom $P/P^{\mathrm{SQL}} = \{0.1, 0.4, 1, 4, 10\}$.

The higher frequency solution is real as long as $P/P^{\mathrm{SQL}} \geq 1$, showing that once the incident power is sufficient to exceed the on-resonance standard quantum limit, in the limit of perfect efficiency there always exists a frequency above the mechanical resonance frequency at which the standard quantum limit is reached. If the additional criterion $Q^2 \geq (P/P^{\mathrm{SQL}})^2 - 1$ is satisfied, the standard quantum limit can also be reached at a second frequency beneath the mechanical resonance.

3.4.3 Optomechanical system as a probe of the mechanical bath

Notice that since the standard quantum limit of Eq. (3.54) is a function of the mechanical susceptibility, it is device dependent. That is to say, in a practical setting, if one wishes to enhance the precision of a measurement beyond the standard quantum limit of a particular device, this can be pursued not only by utilising quantum measurement techniques (as discussed in Chapter 5), but also by designing a device with smaller susceptibility at the frequencies of interest, or by altering the dynamics of the oscillator – for instance via dynamical optomechanical effects [43, 42, 155, 259, 296, 195, 197] – to directly modify the susceptibility. An alternative and more fundamental bound arises from thinking about optical measurements of the position of a mechanical oscillator as an indirect probe of the mechanical bath variable $\hat{P}_{\mathrm{in}}$. That is, the mechanical bath drives the mechanical oscillator and the optical field reads out the resulting motion. It is then natural to compare the precision of such a measurement to the zero-point uncertainty of the bath

$$\bar{S}_{P_{\mathrm{in}} P_{\mathrm{in}}}^{(\bar{n}=0)}(\omega) = \frac{\omega}{2\Omega}, \tag{3.60}$$

where here to retain validity well away from the mechanical resonance frequency we have chosen not make the usual quantum optics approximation, instead taking the zero temperature limit of the general bath power spectral densities in Eqs. (1.98).

When driven by the zero temperature bath of Eq. (3.60), the mechanical oscillator power spectral density in the absence of radiation pressure can be found from Eq. (3.41) to be

$$\bar{S}_{QQ}^{(\bar{n}=0)}(\omega) = \bar{S}_{\det}^{\mathrm{ZPL}}(\omega) = \frac{\omega}{Q}|\chi(\omega)|^2, \tag{3.61}$$

where, as usual, $Q = \Omega/\Gamma$ is the mechanical quality factor. We will see in Chapter 5 that this zero-point mechanical power spectral density presents a limit to the precision of continuous quantum measurements of a mechanical oscillator, even in the presence of quantum correlations. We term it the *zero-point limit*. A measurement that saturates this limit is just sufficient to resolve the zero-point uncertainty of the mechanical bath.

Comparing the zero-point limit to the standard quantum limit of

Eq. (3.54), we find that

$$\bar{S}_{\text{det}}^{\text{ZPL}}(\omega) \Big/ \bar{S}_{\text{det}}^{\text{SQL}}(\omega) = \frac{\omega}{Q} |\chi(\omega)|. \tag{3.62}$$

The two limits are equal at the mechanical resonance frequency where the mechanical susceptibility is at its maximum ($\chi(\Omega) = 1/\Gamma$), but diverge off-resonance where measurements at the standard quantum limit are incapable of resolving the mechanical zero-point fluctuations (see Fig. 3.5). As we will see in Chapter 5, quantum correlations in principle allow the zero-point limit to be reached at all frequencies; while back-action evading techniques allow it to be surpassed.

3.4.4   Standard quantum limit for displacement measurement

In many scenarios one is interested in measuring a classical signal encoded on the motion of a mechanical oscillator via, for example, an external force. In this case, the environmentally driven motion of the mechanical oscillator presents an additional source of noise. The symmetrised power spectral density of the detected mechanical position $\hat{Q}_{\text{det}}$ from Eq. (3.50) then determines the total noise floor of the measurement. This is given by

$$\begin{aligned}
\bar{S}_{Q_{\text{det}}Q_{\text{det}}}(\omega) &= \bar{S}_{QQ}^{(0)}(\omega) + \bar{S}_{\text{det}}(\omega) \tag{3.63} \\
&= \frac{\omega}{Q} |\chi(\omega)|^2 (2\bar{n}(\omega) + 1) + \bar{S}_{\text{det}}(\omega), \tag{3.64}
\end{aligned}$$

where we have again used the general bath power spectral densities in Eqs. (1.98), and the general frequency dependent bath occupation $\bar{n}(\omega)$ is defined in Eq. (1.85). Three interesting regimes are evident from inspection of this expression, in conjunction with Eq. (3.51) for $\bar{S}_{\text{det}}(\omega)$. At optical powers low enough that $|C_{\text{eff}}| < 8\eta\Gamma |\chi(\omega)|^2 (2\bar{n}(\omega) + 1)\,\omega/Q$, the precision of the measurement is insufficient to resolve even the thermal motion of the mechanical oscillator, and measurement noise dominates; when $8\eta\Gamma |\chi(\omega)|^2 (2\bar{n}(\omega) + 1)\,\omega/Q < |C_{\text{eff}}| < (\bar{n}(\omega) + 1/2)\,\omega/\Omega$, mechanical thermal noise dominates; and when $|C_{\text{eff}}| > (\bar{n}(\omega) + 1/2)\,\omega/\Omega$, back-action noise dominates.

Figure 3.6 shows the detected mechanical power spectral density $\bar{S}_{Q_{\text{det}}Q_{\text{det}}}(\omega)$ at the mechanical resonance frequency calculated from Eq. (3.64) as a function of incident optical power, phonon occupancy, and optical efficiency. As can be seen from Fig. 3.6a, with perfect efficiency the power spectral density is minimised at an incident power matching the on-resonance P^{SQL}. At lower powers optical measurement noise dominates, and at higher power back-action noise dominates, with a plateau present around P^{SQL} due to the thermal occupancy of the mechanical oscillator. Fig. 3.6b shows that the effect of measurement inefficiency is to increase the optical measurement noise, increasing the minimum achievable power spectral density, and pushing it to incident powers above P^{SQL}.

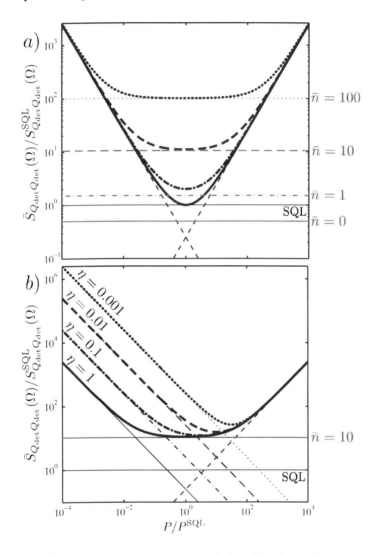

FIGURE 3.6 On-resonance power spectral density of continuous position measurement on a mechanical oscillator as a function of incident optical power, showing the standard quantum limit. (a) Power spectral density with varying thermal occupancy and $\eta = 1$, showing that the standard quantum limit (SQL) is reached when the mechanical oscillator is in its ground state and incident optical power equals P^{SQL}. (b) Power spectral density with varying optical measurement efficiency η and fixed on-resonance thermal occupancy $\bar{n}(\Omega) = 10$, showing that, as the efficiency is decreased, the optical power that attains the minimum power spectral density is increased and the overall measurement sensitivity is degraded. Dashed black lines with negative slope: measurement noise. Dashed black lines with positive slope: back-action noise.

In the limit that the thermal noise from the mechanical oscillator is negligible ($\bar{n} \ll 1$) and the optical measurement is performed with perfect efficiency ($\eta = 1$), the optimal measurement precision reaches the standard quantum limit for displacement measurement

$$\bar{S}^{SQL}_{Q_{det}Q_{det}}(\omega) = \frac{\omega}{Q}|\chi(\omega)|^2 + |\chi(\omega)|, \qquad (3.65)$$

at the optimal effective cooperativity of Eq. (3.52). At the mechanical resonance frequency the susceptibility $|\chi(\Omega)| = 1/\Gamma$. Remembering that the quality factor $Q \equiv \Omega/\Gamma$, we therefore see that a minimum consequence of continuous quantum measurement of the position of a mechanical oscillator is to introduce additional noise at the mechanical frequency with magnitude equal to the on-resonance mechanical zero-point fluctuations of the oscillator – i.e. to double the imprecision above that dictated by zero-point motion. As was illustrated in the simple example of a free mass in Section 3.1, this additional noise is required by quantum mechanics to ensure that the measurement does not violate the uncertainty principle.

Any realistic optomechanical system will have some mechanical thermal noise and optical inefficiencies, and therefore – without back-action evading measurements or non-classical correlations – will not achieve the standard quantum limit. Choosing the optimal effective cooperativity of Eq. (3.52), the optimal precision of a continuous displacement measurement in the presence of such imperfections can be found from Eq. (3.64) to be

$$\bar{S}^{opt}_{Q_{det}Q_{det}}(\omega) = \bar{S}^{SQL}_{Q_{det}Q_{det}}(\omega) + \bar{S}^{excess}_{Q_{det}Q_{det}}(\omega), \qquad (3.66)$$

where the excess noise above the standard quantum limit is

$$\bar{S}^{excess}_{Q_{det}Q_{det}}(\omega) = \frac{2\bar{n}(\omega)\,\omega}{Q}|\chi(\omega)|^2 + |\chi(\omega)|\left(\eta^{-1/2} - 1\right). \qquad (3.67)$$

It is natural to ask whether there are realistic situations in which the effects of thermal noise and inefficiencies are negligible, or in other words when is the criterion $\bar{S}^{excess}_{Q_{det}Q_{det}}(\omega) \ll \bar{S}^{SQL}_{Q_{det}Q_{det}}(\omega)$ satisfied? There are two particularly interesting regimes in which to analytically examine this question. The first is the on-resonance regime ($\omega = \Omega$), where the criterion becomes that

$$\bar{n} + \frac{\eta^{-1/2} - 1}{2} \ll 1. \qquad (3.68)$$

Therefore, the on-resonance standard quantum limit of displacement measurement can only be approached if both the mechanical bath is near its ground state, with thermal occupancy $\bar{n} \equiv \bar{n}(\Omega) \ll 1$, and the efficiency $\eta^{-1/2} - 1 \ll 2$. The efficiency requirement is not especially stringent, with the left- and right-hand sides of the inequality equal when $\eta = 1/9$. However, the requirement that the mechanical bath be close to its ground state precludes reaching the standard quantum limit in many scenarios.

The second interesting regime is the *free-mass regime*, where the frequencies of interest lie far above the mechanical resonance frequency, i.e., $\omega \gg \{\Omega, \Gamma\}$. In this regime, the mechanical susceptibility can be approximated as $|\chi(\omega)| = \Omega/\omega^2$, and both thermal noise and inefficiencies are negligible if

$$2\bar{n}(\omega)\frac{\Gamma}{\omega} + \eta^{-1/2} - 1 \ll 1. \tag{3.69}$$

That is, the two requirements $\bar{n}(\omega) \ll \omega/2\Gamma$ and $\eta^{-1/2} - 1 \ll 1$ must be satisfied. Quite strikingly, while, in the on-resonance case, cooling to near the ground state is necessary to approach the standard quantum limit, we see here that the thermal occupancy is always negligible sufficiently far into the free-mass regime. Therefore, it is possible for the standard quantum limit to impose the dominate constraint on precision even for a hot mechanical oscillator. The reason is that, as can be seen from inspection of Eqs. (3.67) and (3.54), the thermal noise decreases faster with frequency ω than the detection noise. The (much less significant) factor of two difference in the efficiency criterion compared to the on-resonance case arises similarly, because the mechanical zero-point motion decreases faster than the detection noise as ω increases (see Eq. (3.65)). Consequently, while the contributions to the standard quantum limit from zero-point motion and detection noise are exactly equal on-resonance, the zero-point motion contribution becomes negligible sufficiently far into the free-mass regime.

We can gain more intuition about the thermal occupancy requirement in the free-mass regime by making the high temperature approximation $\bar{n}(\omega) \approx k_B T / \hbar\omega$, in which case the condition becomes

$$\frac{\omega^2}{\Gamma} > \frac{2k_B T}{\hbar}. \tag{3.70}$$

This bears a close resemblance to the condition for quantum coherent oscillation of a mechanical oscillator given in Eq. (2.25). Here, however, the significance of the condition is that, for quantum noise to dominate, less than half a quanta of thermal heating must be introduced within one period of oscillation at frequency ω.

The detected mechanical oscillator position power spectral density is compared to the standard quantum limit of displacement measurement as a function of frequency in Fig. 3.7 (*left column*). The top subplot shows the power spectrum for incident power well below the standard quantum limit. In this regime the contribution from mechanical thermal motion may be resolvable above the measurement noise (as is the case here), but the contribution from mechanical zero-point fluctuations is not. The second subplot shows the power spectrum when the incident power $P = P^{\mathrm{SQL}}$. Here, the mechanical zero-point fluctuations are resolvable above the measurement noise, but remain obscured, in this case, by mechanical thermal fluctuations. As the incident power increases further, the contributions to the power spectrum from mechanical thermal and vacuum fluctuations are reduced relative to the optical

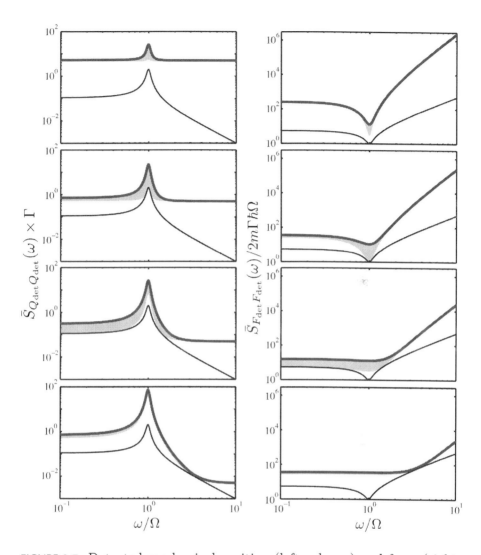

FIGURE 3.7 Detected mechanical position (left column) and force (right column) power spectral densities as a function of frequency. Bold line: power spectral density. Thin line: standard quantum limit. Shaded region: thermal contribution to power spectral density. From top to bottom, the vertically arrayed subplots have increasing incident powers of $P/P^{\mathrm{SQL}} = \{0.1, 1, 10, 100\}$. The mechanical oscillator thermal occupancy and quality factor are, respectively, $Q = 10$ and $\bar{n}(\Omega) = 10$, and the optical measurement efficiency is $\eta = 1$.

back-action contribution. It can be seen that – as expected given the $\bar{n} = 10$ thermal occupancy of the oscillator – the detected mechanical power spectral density never reaches the standard quantum limit at the mechanical resonance frequency. However, as we have observed earlier in this section, thermal occupancy does not preclude approaching the standard quantum limit away from the mechanical resonance. Indeed, this is possible, as shown in the bottom subplot.

3.5 STANDARD QUANTUM LIMIT FOR GRAVITATIONAL WAVE INTERFEROMETRY

Most relevant optomechanical measurements involve the optical transduction of a force applied to the mechanical oscillator, as will be discussed in Section 3.6. However, in some circumstances it is desired to measure direct modulations of the cavity path length rather than modulations that result from a force on the oscillator. The most notable example is gravitational wave interferometry, where an itinerant gravitational wave modulates the relative arm lengths of the interferometer directly. Here, for consistency, we will derive the standard quantum limit for path length measurement using an optical cavity. While gravitational wave detectors are interferometers, the essential effect is the same, and within a scaling factor accounting for the different geometry the results are identical. In an optical cavity, the net effect of a path length change is to alter the resonance frequency of the cavity, pushing it off resonance with the incident laser light. To model the standard quantum limit, we must therefore introduce an additional detuning term Δ_{sig} into the Hamiltonian. To do this, we begin with the nonlinearised Hamiltonian of Eq. (2.18), substituting in the total detuning $\Delta \to \Delta + \Delta_{\mathrm{sig}}$. The additional cavity detuning term is then $\hbar \Delta_{\mathrm{sig}} a^\dagger a$. We linearise this term in the usual way by making the substitution $a \to \alpha + a$, where α is the steady-state expectation value of the original operator, and neglecting the operator product term to find

$$\hbar \Delta_{\mathrm{sig}} a^\dagger a = \hbar \Delta_{\mathrm{sig}} \left(\alpha^2 + \sqrt{2} \alpha \hat{X} \right). \tag{3.71}$$

The first term in this expression is a constant in the steady-state and therefore has no effect on the steady-state dynamics and can be safely neglected. Setting the cavity detuning to zero when no signal is present, the linearised optomechanical Hamiltonian of Eq. (2.41) then becomes

$$\hat{H} = \frac{\hbar \Omega}{2} \left(\hat{Q}^2 + \hat{P}^2 \right) + \sqrt{2} \hbar \alpha \Delta_{\mathrm{sig}} \hat{X} + 2 \hbar g \hat{X} \hat{Q}. \tag{3.72}$$

Through geometric arguments the cavity detuning can be related to the change in path length l and strain h via

$$\frac{\Delta_{\mathrm{sig}}}{\Omega_c} = -\frac{l}{L} = -h, \tag{3.73}$$

where L and Ω_c are the cavity path length and resonance frequency, respectively. We then get

$$\hat{H} = \frac{\hbar\Omega}{2}\left(\hat{Q}^2 + \hat{P}^2\right) - \sqrt{2}\hbar\Omega_c\alpha h\hat{X} + 2\hbar g\hat{X}\hat{Q}. \tag{3.74}$$

The Langevin equations of motion derived from this Hamiltonian are the same as those from Eq. (2.41) except that an additional coherent displacement proportional to the strain h is introduced into Eq. (3.9b), which becomes

$$\dot{\hat{Y}} = -\frac{\kappa}{2}\hat{Y} + \sqrt{2\kappa}\hat{Y}_{\text{in}} + \sqrt{2}\Omega_c\alpha h - 2g\hat{Q}. \tag{3.75}$$

We can see immediately from this expression that the effect of the strain is identical to that of the position of the mechanical oscillator $\hat{Q}$ except scaled by a factor $-\Omega_c/\sqrt{2}g_0$ (remember that $g \equiv \alpha g_0$). Consequently, the strain sensitivity of a cavity optomechanical system can be related to sensitivity to displacements of the mechanical oscillator via

$$\bar{S}_{hh}(\omega) = 2\left(\frac{g_0}{\Omega_c}\right)^2 \bar{S}_{Q_{\text{det}}Q_{\text{det}}}(\omega). \tag{3.76}$$

This expression can be simplified to some degree in the special case of a Fabry–Pérot cavity. We showed in Section 2.3 that, for a Fabry–Pérot cavity, $g_0 = x_{zp}\Omega_c/L$. Defining the zero-point strain $h_{zp} \equiv x_{zp}/L$, we then have

$$\bar{S}_{hh}(\omega) = 2h_{zp}^2\bar{S}_{Q_{\text{det}}Q_{\text{det}}}(\omega). \tag{3.77}$$

Gravitational wave interferometers operate in the free-mass regime where $\omega \gg \{\Omega, \Gamma\}$, with their mirrors mounted on large low frequency suspension systems. In this limit the susceptibility $|\chi(\omega)| \approx \Omega/\omega^2$. Substituting this into Eq. (3.65) for the position standard quantum limit and using the relation in Eq. (3.77), we then arrive at the standard quantum limit of strain sensing

$$\bar{S}_{hh}^{\text{SQL}}(\omega) = \frac{2\Omega}{\omega^2}h_{zp}^2 \tag{3.78}$$

$$= \frac{\hbar}{mL^2\omega^2} \tag{3.79}$$

A similar result holds for strain sensing in a Michelson interferometer configuration, differing only by a factor of four [157]. We see that this limit scales favourably with the measurement frequency ω, the mass of the oscillator, and the cavity path length, while being independent of both the mechanical oscillator resonance frequency and mass. This motivates the use of long base-line interferometers with large mass end-mirrors in gravitational wave interferometry. For a cavity that has $L = 4$ km length, with one suspended mirror of mass $m = 1$ kg and a signal at $\omega/2\pi = 1$ kHz, Eq. (3.79) gives an impressive strain sensitivity of $\bar{S}_{hh}^{\text{SQL}}(\omega)^{1/2}/2\pi = 10^{-24}$ Hz$^{-1/2}$. As discussed in the previous section, the suppression of thermal noise with increasing ω above the mechanical resonance frequency makes it feasible to reach such sensitivities in future gravitational wave interferometers.

3.6 STANDARD QUANTUM LIMIT FOR FORCE MEASUREMENT

While displacement measurements such as those discussed in the previous section have the significant advantage of exhibiting a suppression of thermal noise off resonance, in typical applications of mechanical oscillators it is the forces applied to the mechanical oscillator, rather than the displacements they induce, that are of primary interest. Uncertainties in displacement measurement can be straightforwardly recast in terms of external forces using Eqs. (1.94) and (3.17). Equation (1.94) allows us to relate the power spectral density of the external force driving the oscillator to the power spectral density of the dimensionless bath momentum operator as $\bar{S}_{FF}(\omega) = 2\Gamma m\hbar\Omega\bar{S}_{P_{in}P_{in}}(\omega)$, while from Eq. (3.17) we have $\bar{S}_{QQ}^{(0)}(\omega) = 2\Gamma|\chi(\omega)|^2\bar{S}_{P_{in}P_{in}}(\omega)$. Consequently,

$$\bar{S}_{FF}(\omega) = \frac{\hbar m\Omega}{|\chi(\omega)|^2}\bar{S}_{QQ}^{(0)}(\omega). \tag{3.80}$$

Apart from establishing the coefficient converting a position power spectral density to a force power spectral density, this expression also makes clear that the precision of force measurements is, unsurprisingly, enhanced on resonance where the susceptibility $\chi(\omega)$ is at its maximum. Since the displacement and force power spectral densities are linearly related, Eq. (3.65) defines a standard quantum limit not only for displacement sensing, but also for force sensing. Cavity optomechanics experiments using the collective motion of a cold atom cloud cooled close to its ground state have recently demonstrated force sensing with precision within a factor of four of this limit [253].

Renormalising the displacement noise floor of Eq. (3.64) into newtons per root hertz by multiplying through by the coefficient $\hbar m\Omega/|\chi(\omega)|^2$ established in Eq. (3.80) yields the force noise floor of a cavity optomechanical force sensor

$$\bar{S}_{F_{det}F_{det}}(\omega) = 2m\Gamma\hbar\Omega\left[\frac{\omega}{\Omega}\left(\bar{n}(\omega) + \frac{1}{2}\right) + \frac{1}{16\eta\Gamma^2|\chi(\omega)|^2|C_{eff}|} + |C_{eff}|\right], \tag{3.81}$$

which is minimised for the same optimal effective cooperativity as given for displacement sensing in Eq. (3.52). The force power spectral density is shown as a function of frequency in the right-hand column of Fig. 3.7 for a range of cooperativity values, clearly displaying a minimum near the mechanical resonance frequency Ω and degrading away from resonance. This is in distinct contrast to the power spectral density of position sensing (Fig. 3.7 *left column*), which showed the reverse behaviour.

3.6.1 Bandwidth of optomechanical force sensing

The previous few sections show that measurement back-action introduces a fundamental quantum limit on the precision of mechanical oscillator based sensing. One might, therefore, think that there is no advantage to increasing the effective cooperativity above the optimum given in Eq. (3.52). In Chapter 5 we will see that, for special forms of measurement such as *back-action evading measurements*, the back-action penalty can, as the name suggests, be evaded. A second – and perhaps even more significant – advantage of increasing interaction strength is improved measurement bandwidth. Figure 3.7 (*right column*) shows that, where for low cooperativity the mechanical force power spectral density is sharply peaked around Ω, as the cooperativity is raised, the power spectral density broadens and eventually becomes constant at frequencies below Ω, increasing the sensing bandwidth.

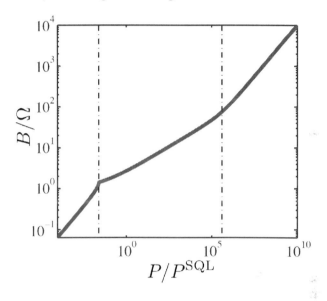

FIGURE 3.8 Bandwidth of cavity optomechanical force sensing as a function of incident power P in the limit where $\kappa \gg \omega$. Left-most vertical line: incident power at which the optical measurement noise is equal to the thermal noise at $\omega = 0$. Right-most vertical line: incident power at which back-action noise dominates thermal noise. The mechanical oscillator is taken to have a quality factor $Q = 100$ and thermal occupancy $\bar{n} = 10^5$, and the efficiency of the optical measurement is taken to be $\eta = 1$.

The force sensing bandwidth can reasonably be defined as the width of the region of frequency space over which the force noise power spectral density is within a factor of two of the minimum at $\omega = \Omega$. The frequencies $\omega_{3\,\text{dB}}^{\pm}$ at which the force noise power spectral density is exactly a factor of two above the minimum can be determined analytically from Eq. (3.81) in the regime where they lie well within the cavity linewidth ($\omega \ll \kappa$), so that $C_{\text{eff}}(\omega) \approx C$.

Making the quantum optics approximation that $(\bar{n}(\omega)+1/2)\omega/\Omega = \bar{n}(\Omega)+1/2$ they are found in this regime to be

$$\left(\frac{\omega_{3\,\mathrm{dB}}^{\pm}}{\Omega}\right)^2 = 1 - \frac{1}{2Q^2} \pm \frac{1}{Q}\left[\frac{1}{4Q^2} + 1 + 16\eta C\left(\bar{n} + \frac{1}{2} + C\right)\right]^{1/2}, \qquad (3.82)$$

where, as always when the argument is omitted, $\bar{n} \equiv \bar{n}(\Omega)$.

Exercise 3.9 *Show this result.*

Clearly, as the cooperativity increases, the separation between the low, and high, frequency solutions becomes larger, and therefore the bandwidth increases. The higher-frequency solution is always real. The lower-frequency solution, on the other hand, becomes imaginary once the cooperativity is sufficiently high that the contributions to the power spectral density from mechanical thermal noise and optical noise become comparable at zero frequency. This transition occurs at a cooperativity of $C = Q^2[8\eta(2\bar{n} + 1)]^{-1}$, at which point the sensing bandwidth extends all the way to $\omega = 0$.

Given the behaviour of the lower-frequency solution, the force sensing bandwidth B can be determined from Eq. (3.82) as

$$B = \omega_{3\,\mathrm{dB}}^{+} - \mathrm{real}\left\{\omega_{3\,\mathrm{dB}}^{-}\right\}. \qquad (3.83)$$

This bandwidth has several different regimes of behaviour as the cooperativity increases, as illustrated in Fig. 3.8. At relatively low cooperativity, where $[16\eta(\bar{n} + 1/2)]^{-1} \ll C \ll \{\bar{n} + 1/2, Q^2[16\eta(\bar{n} + 1/2)]^{-1}\}$ and the mechanical thermal noise is unresolved beneath the optical noise at zero frequency, the bandwidth can be shown to be

$$B = 4\Gamma\left[\eta C\left(\bar{n} + \frac{1}{2}\right)\right]^{1/2}, \qquad (3.84)$$

scaling as $C^{1/2}$. Once the mechanical thermal noise is resolved, in the limit $Q^2/[16\eta C(\bar{n} + 1/2)] \ll C \ll \bar{n} + 1/2$, the bandwidth becomes

$$B = 2\sqrt{\Gamma\Omega}\left[\eta C\left(\bar{n} + \frac{1}{2}\right)\right]^{1/4}, \qquad (3.85)$$

scaling as $C^{1/4}$; and when the measurement back-action dominates, in the limit $C \gg \{\bar{n} + 1/2, Q/4\eta^{1/2}\}$, it is

$$B = 2\sqrt{\Gamma\Omega C}\,\eta^{1/4}, \qquad (3.86)$$

scaling again as $C^{1/2}$. The reason for this improved bandwidth scaling in the back-action dominated regime is that, in additional to measurement noise that decreases as C^{-1} and allows the mechanical thermal noise to be resolved further from resonance, the back-action heating of the oscillator increases the total mechanical noise relative to the optical measurement noise. This second effect increases the bandwidth by *decreasing* the sensitivity on resonance, rather than increasing it in the wings of the resonance. Never-the-less, it can be seen that, while the scaling changes with cooperativity, the general trend of increasing bandwidth with C remains, even in the regime where the measurement is dominated by measurement back-action.

3.6.2 Sensitivity of classical force sensing

Micro- and nano-mechanical oscillators are commonly used as force and inertial sensors. Such sensors typically operate in a regime where the mechanical thermal noise dominates on resonance, and where the quantum back-action noise on the sensor is negligible. In this case, taking the relevant high temperature limit where $\bar{n}(\omega) + 1/2 \approx k_B T/\hbar\omega$ we obtain an optical shot-noise limited force noise floor of

$$S_{F_{\mathrm{det}}F_{\mathrm{det}}}^{\mathrm{class}}(\omega) = 2m\Gamma k_B T + \frac{m\hbar Q}{8\eta|\chi(\omega)|^2|C_{\mathrm{eff}}|}. \qquad (3.87)$$

The spectrally flat term on the left is the usual thermomechanical force sensing limit for micromechanical force sensors.[8] It can only be surpassed by engineering the bath coupled to the mechanical oscillator, for example, by cooling or squeezing it. The term on the right is the measurement noise floor, which has a minimum near the mechanical resonance frequency Ω and decreases as the optomechanical cooperativity increases. While the measurement noise appears to increase linearly with the mechanical quality factor Q, the mechanical susceptibility $\chi(\omega)$ is also a function of Q. In the limit that $\omega \gg \{\Omega, \Gamma\}$, for instance, $|\chi(\omega)|^2$ scales as Q^2 so that the measurement noise floor decreases as Q^{-1}. Substantial efforts have been made over many years in the nano- and micro-electromechanics community to both improve the measurement precision and thereby reach the thermomechanical noise dominated regime, and to reduce the noise floor in this regime by decreasing Γ and m [101].

One might think that the improved measurement precision provided, for example, by an optical cavity would provide no advantage in force sensing once the thermomechanical noise dominates the measurement. It is correct that improved measurement precision will not improve the force sensitivity at the peak of the mechanical resonance in this limit. However, as discussed in the previous section, improved precision will increase the frequency band over which the thermomechanical noise dominates measurement noise. This increases the bandwidth over which the maximum sensitivity can be achieved, and also allows improved sensitivity in measurements of broadband signals such as incoherent forces. This has allowed the demonstration of a

---

[8]Note that typically in the force sensing literature the decay rate is defined as the full-width-half-maximum of the mechanical resonance, and this results in a value equal to half the value used here. As a result, this classical force sensing limit is usually quoted as $S_{FF}^{\mathrm{class}}(\omega) = 4m\Gamma k_B T$.

range of high bandwidth, high sensitivity, cavity optomechanical sensors, including force sensors [115, 213, 136], inertial sensors [164], and magnetometers [107, 106].

CHAPTER 4

Coherent interaction between light and mechanics

CONTENTS

4.1	Strong coupling	....	94
4.2	Optical cooling of mechanical motion	....	96
	4.2.1	Effective temperature of the optical bath	97
		4.2.1.1	Coupling to both optical and mechanical baths	103
	4.2.2	Resolved sideband regime	105
		4.2.2.1	Weak coupling regime	107
		4.2.2.2	Strong coupling regime	109
		4.2.2.3	Mechanical occupancy achieved via resolved sideband cooling in the rotating wave approximation	111
		4.2.2.4	Approaching the ground state	111
		4.2.2.5	Thermodynamical understanding of optical cooling	112
4.3	Optomechanically induced transparency	....	114
	4.3.1	Hamiltonian	115
	4.3.2	Optomechanically induced absorption	116
		4.3.2.1	Output field	118
	4.3.3	Optomechanically induced transparency using a double-sided optical cavity	119
		4.3.3.1	Noise performance	121
		4.3.3.2	Experimental implementation	122
4.4	Optomechanical entanglement	....	122
	4.4.1	Hamiltonian and equations of motion	123
	4.4.2	Bogoliubov modes	124
	4.4.3	Optical and mechanical modes in the stable regime	126

4.4.4	Einstein–Podolsky–Rosen entanglement	127
4.4.5	Covariance matrix	128
4.4.6	Identifying and quantifying entanglement	128
4.4.7	Entanglement between intracavity field and mechanical oscillator	129
4.4.8	Entanglement of the mechanical oscillator with the external field	133
4.4.8.1	Wiener filtering	134
4.5	Mechanical squeezing of light	140
4.5.1	Basic concept	141
4.5.2	Understanding ponderomotive squeezing via the polaron transformation	142
4.5.2.1	Polaron transformation	142
4.5.2.2	Squeezing action in the polaron picture	143
4.5.2.3	Interaction with the coherent drive and the optical and mechanical baths	144
4.5.3	Squeezing spectra	145

In the previous chapter we examined the linearised radiation pressure interaction between the optical field and mechanical oscillator in a cavity optomechanical system in the case that the optical cavity is coherently driven on resonance. This chapter introduces radiation pressure-based coherent coupling between light and a mechanical oscillator that occurs when an optical detuning is introduced. We discuss effects such as resolved sideband cooling, optomechanically induced transparency, and the generation of optomechanical entanglement and squeezed states of light that manifest from this coherent interaction. Along the way we introduce the concepts of Wiener filtering to optimally estimate the mechanical position and momentum from the output optical field, methods to verify and quantify two-mode Gaussian entanglement, and the polaron transformation important for later chapters involving single photon optomechanics.

4.1 STRONG COUPLING

As a prelude to optomechanical cooling, it is illustrative to examine the dissipation-free dynamics generated by the linearised optomechanical Hamiltonian of Eq. (2.41) in the special case where the optical detuning equals the mechanical resonance frequency ($\Delta = \Omega$). The "X-X" form of the coupling term in this Hamiltonian brings to mind the simple problem of a pair of linearly coupled oscillators. Indeed, this is the essential physics described by the Hamiltonian. The choice of $\Delta = \Omega$, such that the driving laser field is detuned to the red side of the optical resonance by the mechanical resonance frequency, makes the effective resonance frequencies of the two oscillators degenerate,[1] such that they couple most effectively. Expressed in terms of annihilation operators, the optomechanical Hamiltonian is then

$$\hat{H} = \hbar \Omega a^\dagger a + \hbar \Omega b^\dagger b + \hbar g \left(a^\dagger + a \right) \left(b^\dagger + b \right), \tag{4.1}$$

where as usual g is the coherent amplitude boosted optomechanical coupling rate. This fully symmetric Hamiltonian may be straightforwardly diagonalised by transforming to the normal modes

$$c = \frac{1}{\sqrt{2}} (a + b) \tag{4.2}$$

$$d = \frac{1}{\sqrt{2}} (a - b), \tag{4.3}$$

where it is easy to show c and d preserve the Boson commutation properties of a and b (Eq. (1.2)). Substituting in for a and b in the Hamiltonian and expanding, after some work we find

$$\hat{H} = \hbar \left(\Omega + g \right) c^\dagger c + \hbar \left(\Omega - g \right) d^\dagger d. \tag{4.4}$$

Exercise 4.1 *Show this result.*

Several important facts can be gleaned from this expression. First, in this normal mode basis the Hamiltonian is indeed diagonalised, with the interaction between the modes removed and the dynamics of the system consequently substantially simplified. Second, c and d are a pair of independent quantum harmonic oscillators. Finally, the original coupling between modes a and b is replaced with opposing frequencies shifts on c and d, resulting in a frequency splitting of $2g$ between these modes (see experimental demonstration in Fig. 4.6).

In the time domain, c and d then exhibit simple harmonic oscillation with the well-known dynamics[2]

$$c(t) = c(0) e^{-i(\Omega + g)t} \tag{4.5a}$$
$$d(t) = d(0) e^{-i(\Omega - g)t}. \tag{4.5b}$$

Using $b = (c - d)/\sqrt{2}$ the dynamics of the mechanical oscillator can then be shown to be

$$b(t) = [b(0) \cos gt - ia(0) \sin gt] \, e^{-i\Omega t} \tag{4.6}$$
$$= b^{(0)}(t) \cos gt - ia^{(0)}(t) \sin gt, \tag{4.7}$$

where $a^{(0)}(t)$ and $b^{(0)}(t)$ are the annihilation operators that would describe the time evolution of the intracavity optical field and mechanical oscillator if

---

[1] Remember that, in Eq. (2.41), the optical field is in a rotating frame at the drive laser frequency.

[2] This can be shown, for example, using Eq. (1.20).

there was no optomechanical coupling ($g = 0$). A similar calculation for the optical field gives

$$a(t) = a^{(0)}(t) \cos gt - ib^{(0)}(t) \sin gt. \tag{4.8}$$

We see that, setting an optical detuning $\Delta = \Omega$, the optomechanical interaction allows the quantum state of the light and mechanical oscillator to be coherently and unitarily exchanged.

Of course, in the realistic scenario where dissipation is present, the exchange becomes imperfect. It is natural to then define two physical regimes, the *strong coupling* and *quantum-coherent coupling* regimes.[3] In the strong coupling regime $g > \{\kappa, \Gamma\}$, so that a full exchange between light and oscillator occurs within the mechanical and optical decay times and the non-degeneracy between normal modes is spectrally resolvable. In the quantum-coherent coupling regime $g > \{(2\bar{n}_L + 1)\kappa, (2\bar{n} + 1)\Gamma\}$ where $\bar{n}$ and $\bar{n}_L$ are the mechanical and optical bath mean occupancies, so that a full exchange between light and oscillator occurs within the quantum decoherence time of both the light and mechanical oscillator (see Section 2.5). The strong coupling and quantum-coherent coupling regimes were first experimentally realised in [130] and [295], respectively.

Notably, if $gt = \pi/2$ the optical and mechanical states are, in the dissipationless scenario of Eqs. (4.7) and (4.8), exactly exchanged. A continuous version of this state-swap is discussed in Section 8.3 as a method to realise quantum conversion between optical and microwave degrees of freedom. Even if the light is only in a relatively uninteresting coherent state,[4] a state-swap allows ground state cooling of the mechanical oscillator, with the typically high thermal occupancy mechanical state swapped out onto the optical field and the near zero thermal occupancy optical state swapped onto the mechanical oscillator. In the following section, we introduce a continuous version of this approach to mechanical cooling, including dissipative processes that act to limit the final mechanical occupancy.

4.2 OPTICAL COOLING OF MECHANICAL MOTION

In the previous chapter we introduced on-resonance ($\Delta = 0$) optical probing of mechanical motion, and showed that radiation pressure shot noise heats the mechanical oscillator. This back-action heating is a consequence of the fact that information about the mechanical motion is imprinted on the phase of the optical field, and is necessary to ensure that the Heisenberg uncertainty principle is not violated. However, as the simple example in the previous section

---

[3] Note that these regimes are different from the radiation pressure shot noise dominated regime introduced in Section 3.2, which requires that the optomechanical cooperativity $C \equiv 4g^2/\kappa\Gamma > \bar{n} + 1/2$. This radiation pressure shot noise dominated regime lies between the strong coupling and quantum-coherent coupling regimes.

[4] So that, in the displaced frame used in the linearised picture, it is in a vacuum state.

demonstrated, the presence of back-action heating does not necessarily preclude an overall optical cooling effect on the mechanical oscillator. Indeed as we saw briefly in Section 3.2.2, the mapping of the mechanical motion onto the intracavity field suggests a possible a cooling mechanism when $\Delta \neq 0$ based on dynamical back-action. The essential idea is that, when the optical cavity is detuned, the mechanical position is imprinted – at least in part – on the amplitude of the optical field which then back-acts through radiation pressure upon the mechanical oscillator. Since the optical cavity also induces a delay in the optical response, this dynamical back-action is retarded, with a component of the optical force being proportional to the velocity of the mechanical oscillator. Depending on the sign of this component, it either damps/cools or amplifies/heats the mechanical motion [37, 160].

An alternative and particularly powerful approach to understanding optomechanical cooling – and indeed coherent interactions between light and mechanical systems in general – is via an energy level diagram, as shown in Fig. 4.1. Here we observe that downwards going phonon number transitions are resonantly enhanced when the optical driving tone is red detuned, while upwards going transitions are enhanced by blue detuning. We will see in what follows that these two operations can be thought of, respectively, as beam splitting and parametric operations between the light and mechanical oscillator, with the former allowing cooling while the latter can be used to generate optomechanical entanglement.

4.2.1 Effective temperature of the optical bath

The optical field can be thought of as a thermal bath for the mechanical oscillator in a cavity optomechanical system, with radiation pressure shot noise introducing a random driving force (see, e.g., Section 3.3.1). We found in Section 1.2 that the temperature of a high-quality quantum oscillator that is linearly forced by a bath is governed by the ratio of bath power spectral densities at $\pm\Omega$. This relationship provides an elegant approach to determining the effect of the optical field on the temperature of the oscillator, as first observed in [191]. We follow the approach in that paper here.

To understand the effect of the optical field on the temperature of the mechanical oscillator, it is useful to consider first the case where the mechanical oscillator has zero intrinsic dissipation ($\Gamma \to 0$), or at least where the heating from the optical field dominates the heating from the mechanical bath.[5] In this case, the optical force makes the only significant contribution towards the force power spectral density experienced by the mechanical oscillator. In Section 3.3.1 we found the optical force $\hat{F}_L$ in a cavity optomechanical system in the linearised regime by taking the derivative of the optomechanical Hamiltonian with respect to $\hat{q}$. Here, we similarly use the nonlinearised Hamiltonian

---

[5]That is, in the regime where radiation pressure shot noise dominates mechanical thermal noise.

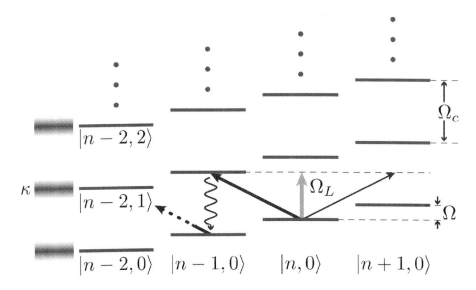

FIGURE 4.1 Energy level diagram for optical cooling of mechanical motion. $|n, m\rangle$: n-phonon, m-photon state. Ω_c: cavity resonance frequency, Ω_L: laser resonance frequency. Laser driving at frequency $\Omega_L < \Omega_c$ is closer to resonance for mechanical cooling transitions $|n, m\rangle \rightarrow |n - 1, m + 1\rangle$ than heating transitions $|n, m\rangle \rightarrow |n + 1, m - 1\rangle$, resulting in preferential cooling. In the resolved sideband limit where $\kappa \ll \Omega$, optimal cooling is achieved when $\Delta = \Omega_c - \Omega_L = \Omega$.

of Eq. (2.15) to obtain the force

$$\hat{F}_L(t) = \frac{\partial \hat{H}}{\partial \hat{q}} = \hbar G a^\dagger a. \tag{4.9}$$

The power spectral density of this force can be found from its autocorrelation function (see Eq. (1.43)), which is given by

$$\left\langle \hat{F}_L(t + \tau) \hat{F}_L(t) \right\rangle_{t=0} = \hbar^2 G^2 \left\langle a^\dagger(t + \tau) a(t + \tau) a^\dagger(t) a(t) \right\rangle_{t=0} \tag{4.10}$$

$$\approx \frac{\hbar^2 g^2}{x_{zp}^2} \left[\alpha^2 + 2\bar{n}_L + 2 \left\langle \hat{X}(t + \tau) \hat{X}(t) \right\rangle_{t=0} \right].$$

Here, we have made the usual substitution $a \rightarrow \alpha + a$ to displace away the coherent amplitude of the intracavity field, linearised the resulting expression by neglecting the term that does not contain the coherent amplitude α, substituted $\alpha G = \alpha g_0 / x_{zp} = g / x_{zp}$ (see Section 2.3), and used the relation

$\langle a^\dagger(t)a(t)\rangle = \bar{n}_L$ with $\bar{n}_L$ being the thermal occupancy of the displaced[6] incident optical field. As usual $\hat{X}$ is the amplitude quadrature of the intracavity field.

Exercise 4.2 *Derive this result for yourself.*

As we have discussed previously, for an optical field in thermal equilibrium with its environment, $\bar{n}_L$ is given by Eq. (1.7) and is essentially zero. Here we retain the optical occupancy explicitly, motivated both by the aim of clarifying the effect of a nonzero optical bath temperature and by the fact that, in realistic experiments, technical noise often raises the optical occupancy above its equilibrium value.

The optical force power spectral density can then be directly calculated using Eq. (1.43), with the result

$$S_{F_L F_L}(\omega) = \frac{\hbar^2 g^2}{x_{zp}^2} \left[\left(\alpha^2 + 2\bar{n}_L \right) \delta(\omega) + 2 S_{XX}(\omega) \right]. \qquad (4.11)$$

As might be expected, we see that this consists of a coherent mean force at $\omega = 0$ and broadband incoherent noise driving due to the amplitude quadrature power spectral density of the intracavity field. It is through this broadband incoherent forcing that the optical field acts like a thermal bath.

Using Eq. (1.43) the power spectral density of the intracavity field can be expressed in terms of frequency domain annihilation and creation operators as

$$S_{XX}(\omega) = \int_{-\infty}^{\infty} d\omega' \left\langle \hat{X}^\dagger(-\omega)\hat{X}(\omega') \right\rangle \qquad (4.12)$$

$$= \frac{1}{2} \int_{-\infty}^{\infty} d\omega' \left\langle \left(a(\omega) + a^\dagger(-\omega) \right) \left(a^\dagger(-\omega') + a(\omega') \right) \right\rangle. \qquad (4.13)$$

To find an analytical expression for this power spectral density, we must determine $a(\omega)$ and $a^\dagger(\omega)$. To do this we use the Hamiltonian of Eq. (2.18) in the rotating wave quantum Langevin equation (Eq. (1.112)). We simplify the problem by making one substantial approximation – that the intracavity optical field is not affected by the motion of the mechanical oscillator (i.e., we set $g_0 = 0$). This may seem like an unreasonable approximation on the face of it. However, it is appropriate as long as the optical cavity decay rate κ is sufficiently large to remove the fluctuations introduced to the optical field by the interaction with the mechanical oscillator. We will consider the case where this is not true in Section 4.2.2, where we find that the approximation is reasonable as long as κ is large enough that the optomechanical system is not operating within the strong coupling regime (i.e., $\kappa \gg g$).

Returning to the problem at hand, setting $g_0 = 0$ and taking the Fourier

---

[6]That is, the thermal occupancy if the coherent driving tone is displaced away.

transform of the equation of motion for the intracavity optical field, we find

$$a(\omega) = \frac{\sqrt{\kappa}}{\kappa/2 + i(\Delta - \omega)} a_{\text{in}}(\omega) = \chi_{\text{opt}}(\omega) a_{\text{in}}(\omega), \tag{4.14}$$

where, in the same spirit as the mechanical susceptibility $\chi(\omega)$, we have quantified the frequency response of the optical cavity via the *optical susceptibility*

$$\chi_{\text{opt}}(\omega) = \frac{\sqrt{\kappa}}{\kappa/2 + i(\Delta - \omega)}. \tag{4.15}$$

Substituting Eq. (4.14) into Eq. (4.13) and using the bath correlation relations properties in Eqs. (1.115a) and (1.115c), which are valid for our model of the optical bath since we have taken a rotating wave approximation, we then find that

$$S_{XX}(\omega) = \frac{1}{2}\left[\bar{n}_L \left|\chi_{\text{opt}}(-\omega)\right|^2 + (\bar{n}_L + 1)\left|\chi_{\text{opt}}(\omega)\right|^2\right]. \tag{4.16}$$

Exercise 4.3 *Show this result.*

It is notable that this optical amplitude quadrature spectral density is asymmetric in frequency. As discussed in Section 1.2.3, this is a key signature that the, in this case optical, bath is acting to heat or cool the mechanical oscillator. Neglecting the coherent driving term at $\omega = 0$, which acts only to statically displace the mechanical oscillator, we can now establish an analytical expression for the optical force power spectral density of Eq. (4.11):

$$S_{F_L F_L}(\omega) = \frac{\hbar^2 g^2}{x_{zp}^2}\left[\bar{n}_L \left|\chi_{\text{opt}}(-\omega)\right|^2 + (\bar{n}_L + 1)\left|\chi_{\text{opt}}(\omega)\right|^2\right]. \tag{4.17}$$

If the optical force is the only appreciable source of heating of the mechanical oscillator and the mechanical oscillator has sufficiently high quality that $S_{F_L F_L}(\omega)$ is flat across the mechanical resonance, the equilibrium phonon occupancy $\bar{n}_b$ and the optically induced mechanical decay rate Γ_{opt} can be determined by substitution of Eq. (4.17), respectively, into Eqs. (1.54b) and (1.62). Note that, since Eq. (1.54b) is a ratio of the power spectral densities at $\pm\Omega$, the prefactor in Eq. (4.17) plays no role in determining the equilibrium phonon occupancy, with the occupancy determined solely by the characteristics of the optical field. The results of these substitutions are

$$\Gamma_{\text{opt}} = g^2\left[\left|\chi_{\text{opt}}(\Omega)\right|^2 - \left|\chi_{\text{opt}}(-\Omega)\right|^2\right] \tag{4.18a}$$

$$\bar{n}_{b,\text{opt}} = \frac{g^2}{\Gamma_{\text{opt}}}\left[\bar{n}_L \left|\chi_{\text{opt}}(\Omega)\right|^2 + (\bar{n}_L + 1)\left|\chi_{\text{opt}}(-\Omega)\right|^2\right]$$

$$= \frac{\bar{n}_L \left|\chi_{\text{opt}}(\Omega)\right|^2 + (\bar{n}_L + 1)\left|\chi_{\text{opt}}(-\Omega)\right|^2}{\left|\chi_{\text{opt}}(\Omega)\right|^2 - \left|\chi_{\text{opt}}(-\Omega)\right|^2}. \tag{4.18b}$$

To understand these relations, it is worthwhile to consider three specific scenarios:

- If the optical driving field is on resonance ($\Delta = 0$), $|\chi_{\text{opt}}(\omega)| = |\chi_{\text{opt}}(-\omega)|$, so that the optically induced mechanical decay rate $\Gamma_{\text{opt}} = 0$ and $\bar{n}_{b,\text{opt}} = \infty$. Consistent with our observations from Section 3.2, in this regime the optical field causes heating and does not affect the mechanical damping rate. Since, in our current model the mechanical oscillator is not directly coupled to any other bath, there is no mechanical decay, so that the steady-state temperature is infinite.

- If the optical field is blue detuned ($\Delta < 0$), $|\chi_{\text{opt}}(-\omega)| > |\chi_{\text{opt}}(\omega)|$ so that $\{\Gamma_{\text{opt}}, \bar{n}_{b,\text{opt}}\} < 0$. In this case, each photon impinging on the optical cavity carries more energy than an intracavity photon. To enter the cavity, a scattering process must occur whereby the mechanical oscillator takes up some of the photons' energy (see Fig. 4.1). As a result, the optical field coherently adds energy to the mechanical oscillator, providing gain (or negative damping) to its motion. Since the mechanical oscillator has no other pathway to lose energy, its energy exponentially grows. The fact that $\bar{n}_{b,\text{opt}}$ is negative in this situation can be understood because the power spectral density of the optical bath that it is coupled to increases with frequency – the opposite behaviour to that of a system in thermal equilibrium. This results in an effective negative temperature within the Boltzmann factor (see Eq. (1.54b)).

- If the optical field is red detuned ($\Delta > 0$), $|\chi_{\text{opt}}(-\omega)| < |\chi_{\text{opt}}(\omega)|$ so that $\{\Gamma_{\text{opt}}, \bar{n}_{b,\text{opt}}\} > 0$. This is the reverse situation to that discussed above, with each incident optical photon carrying less energy than an intracavity photon. As a result, scattering processes cause a net optical damping of the mechanical oscillator, which then reaches a finite positive equilibrium phonon occupancy.

We examine the last of these three scenarios in detail in what follows.

Figure 4.2 shows the optically induced mechanical damping rate Γ_{opt} as a function of detuning for a range of resolved sideband parameters Ω/κ. It can be seen that, indeed, additional damping is introduced for $\Delta > 0$, and negative damping (or amplification) is introduced for $\Delta < 0$. The maximum damping/amplification occurs near detunings equal to $\Delta = \pm\Omega$,[7] with the range of frequencies that achieve effective damping/amplification narrowing to small regions around these detunings as the resolved sideband factor Ω/κ increases. This can be readily understood, because when $\Delta = \pm\Omega$ the scattering process that transfers energy between the optical field and the mechanical oscillator is resonant (see Fig. 4.1). Taking this special resonant case while

---

[7]It should be noted that, while the vertical axis of Fig. 4.2 is normalised to the optomechanical coupling rate g, in practise for a fixed incident optical power g is a function of the detuning Δ. As the detuning increases away from resonance for fixed incident power, the intracavity phonon number, and therefore g, is reduced. Taking this into account, one finds that, outside of the well-resolved sideband regime where $\Omega/\kappa \gg 1$, the detuning at which maximum optical damping/amplification occurs is shifted towards resonance (i.e., $|\Delta| < \Omega$).

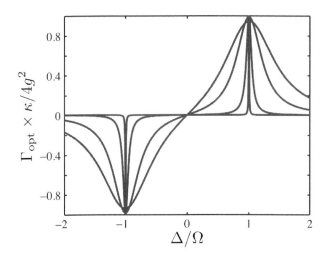

FIGURE 4.2 Optically induced mechanical damping rate Γ_{opt} as a function of cavity detuning Δ for resolved sideband ratios $\Omega/\kappa = \{1, 2, 10, 100\}$ in order of least to most sharply peaked curve.

driving on the red (cooling) side of the optical resonance $(\Delta = \Omega)$, we find

$$\Gamma_{\text{opt}}^{\Delta=\Omega} = \frac{4g^2}{\kappa}\left[1 + \left(\frac{\kappa}{4\Omega}\right)^2\right]^{-1} \tag{4.19a}$$

$$\bar{n}_{b,\text{opt}}^{\Delta=\Omega} = \bar{n}_L + \left(\frac{\kappa}{4\Omega}\right)^2(2\bar{n}_L + 1), \tag{4.19b}$$

This is exactly the scenario we considered in Section 4.1, where we showed that, with this detuning, the state of the optical field and mechanical oscillator swap at the rate $2g$. It is perhaps unsurprising, therefore, to find that this results in cooling of the mechanical oscillator. We observe from Eq. (4.19b) that, in the resolved sideband regime $(\kappa \ll \Omega)$, the mechanical oscillator equilibrates to the occupancy of the optical field $(\bar{n}_{b,\text{opt}}^{\Delta=\Omega} = \bar{n}_L)$, with additional deleterious heating introduced as the resolved sideband factor decreases. If light is shot-noise limited $(\bar{n}_L = 0)$, this additional heating introduces the fundamental limit [191, 314]

$$\bar{n}_{b,\text{opt}}^{\Delta=\Omega} = \left(\frac{\kappa}{4\Omega}\right)^2 \tag{4.20}$$

on the mechanical occupancy. Thus we see that, somewhat surprisingly, even when using an ideal minimum uncertainty coherent state to drive the optomechanical system, the presence of an optical cavity causes the optical field to act like a nonzero temperature bath for the mechanical oscillator.

4.2.1.1 Coupling to both optical and mechanical baths

In most scenarios, with the notable exception of laser trapped atoms, it is not realistic to neglect the coupling of the mechanical oscillator to its mechanical bath. In general, if the oscillator is coupled to two independent baths (labelled A and B here), the total force power spectral density that it experiences is simply

$$S_{FF}(\omega) = S_{FF}^A(\omega) + S_{FF}^B(\omega). \tag{4.21}$$

From Eqs. (1.54b) and (1.62) the equilibrium phonon occupancy of the oscillator is then just the weighted mean

$$\bar{n}_b = \left(\frac{x_{zp}}{\hbar}\right)^2 \left[\frac{S_{FF}^A(-\Omega) + S_{FF}^B(-\Omega)}{\Gamma_A + \Gamma_B}\right] \tag{4.22}$$

$$= \frac{\Gamma_A \bar{n}_b^A + \Gamma_B \bar{n}_b^B}{\Gamma_A + \Gamma_B}, \tag{4.23}$$

where $\bar{n}_b^A$ and $\bar{n}_b^B$, and Γ_A and Γ_B, are the phonon occupancies and damping rates, respectively, that would be obtained if the baths were individually coupled to the oscillator, as defined by Eqs. (4.18a) and (4.18b).

Exercise 4.4 *Show this result.*

Substituting the optical and mechanical parameters from earlier ($\{\bar{n}_A, \bar{n}_B\} \rightarrow \{\bar{n}_{b,\text{opt}}, \bar{n}\}$, $\{\Gamma_A, \Gamma_B\} \rightarrow \{\Gamma_{\text{opt}}, \Gamma\}$) into Eq. (4.23), it is then possible to analytically determine the equilibrium mechanical occupancy in the presence of both optical and mechanical baths, remembering that the result is only valid outside of the optomechanical strong coupling limit (specifically in the regime where $\kappa \ll g$).

Exercise 4.5 *Derive the analytic expression for yourself. Show that on-resonance optical driving ($\Delta = 0$) heats the mechanical oscillator without altering its decay rate and results in an equilibrium occupancy of*

$$\bar{n}_b = \bar{n} + |C_{\text{eff}}| (2\bar{n}_L + 1), \tag{4.24}$$

where C_{eff} is the effective cooperativity defined in Eq. (3.13), consistent with what we found in Section 3.2.

The equilibrium mechanical occupancy is plotted as a function of optical detuning in Fig. 4.3 for a range of optomechanical cooperativities C. As can be seen, when $|\Delta| \gg \Omega$ the mechanical occupancy asymptotes to $\bar{n}$ as expected. Net optical heating occurs for $\Delta < 0$, with the system exhibiting instability over the range of detunings for which $\Gamma_{\text{opt}} + \Gamma < 0$. Net cooling beneath the occupancy of the mechanical bath occurs for $\Delta > 0$, and as expected is strongly peaked near $\Delta = \Omega$. In this special resonant cooling regime ($\Delta = \Omega$), if the optical field is treated as a coherent state ($\bar{n}_L = 0$), we find that the equilibrium occupancy is

$$\bar{n}_b = \frac{\bar{n} + (\kappa/4\Omega)^2 (\bar{n} + C)}{1 + C + (\kappa/4\Omega)^2}. \tag{4.25}$$

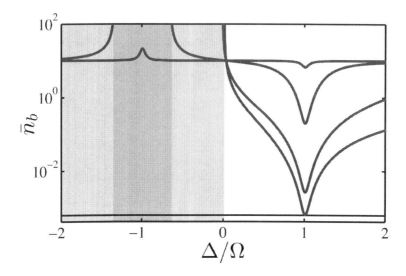

FIGURE 4.3 Optical cooling and amplification of the motion of a mechanical oscillator as a function of detuning Δ for various optomechanical cooperativities. Here the mechanical bath occupancy is $\bar{n} = 10$ and the resolved sideband parameter $\Omega/\kappa = 10$. The optomechanical cooperativity is $C = \{0.5, 50, 5 \times 10^3, 5 \times 10^5\}$ in order of the traces showing the least to most cooling at $\Delta = \Omega$. The thin horizontal line is the fundamental limit due to radiation pressure shot noise heating (Eq. (4.25)). The light shaded region is the region of instability for $C \geq 5000$ and the dark shaded region is the region of instability for $C = 50$.

In the limit where the resolved sideband factor $(\kappa/\Omega)^2 \ll \{\bar{n}/(\bar{n}+C), 1+C\}$, this can be approximated as

$$\bar{n}_b = \frac{\bar{n}}{1 + C}, \tag{4.26}$$

while in the alternative limit that the optomechanical cooperativity C dominates all other terms $(C \gg \{1 + (\kappa/4\Omega)^2, \bar{n}[(4\Omega/\kappa)^2 + 1]\})$ the mechanical occupancy reaches the fundamental limit set by optical radiation pressure heating given in Eq. (4.20).

In this textbook we will generally define being "close to the ground state" as meaning $\bar{n}_b < 1$. With this definition, we see that, in the first of the above limits, the ground state can be approached for $C > \bar{n} + 1$. Interestingly, in the limit that $\bar{n} \gg 1$, this condition approaches the requirement we found in Section 3.2 for the radiation pressure back-action heating of a mechanical oscillator to equal $\bar{n}$ in the nonresolved sideband limit with on-resonance optical driving.[8] The significance of the condition in Section 3.2 is that once $C > \bar{n}$, a measurement on the output field of the optical cavity is able to resolve the zero-point motion of the oscillator in a time shorter than the time one phonon enters from the bath.[9] Here, a similar conclusion may be drawn. When $C > \bar{n} + 1$, the coherent cooling provided by the optomechanical interaction is able to extract the full $\bar{n}$ of occupancy from the mechanical oscillator in a time that is short compared to the time within which one phonon enters from the bath.

4.2.2 Resolved sideband regime

In the previous section, we derived the final temperature of a mechanical oscillator in the presence of both optical and mechanical bath forcing, under the approximation that the optomechanical interaction introduces negligible heating to the intracavity optical field. As we discussed, this approximation is only valid if the optical decay rate κ is sufficiently high that the energy extracted from the mechanical oscillator does not remain in the optical cavity long enough to couple back into the oscillator, i.e., when the optomechanical system is outside of the strong coupling regime (see Section 4.1). In this section we use a quantum Langevin approach to include the effect of this recycling of energy back into the mechanical oscillator, while taking the ideal resolved sideband limit ($\kappa \ll \Omega$) where direct radiation pressure heating of the mechanical oscillator does not significantly raise its temperature (see Eqs. (4.20) and (4.26)). In the resolved sideband regime each intracavity photon on average remains in the cavity for longer than the mechanical period and therefore interacts more or less equally with all quadratures of mechanical motion. The interaction term in the Hamiltonian in Eq. (4.1) may then be simplified by making a rotating wave approximation which neglects the fast oscillating terms (ab and $a^\dagger b^\dagger$). An alternative way to justify this assumption is by considering the optomechanical energy level diagram in Fig. 4.1. It is clear that the scattering processes described by ab and $a^\dagger b^\dagger$ are off-resonance and therefore suppressed if $\kappa \ll \Omega$. With this approximation, the Hamiltonian becomes

$$\hat{H} = \hbar\Omega a^\dagger a + \hbar\Omega b^\dagger b + \hbar g \left(a^\dagger b + ab^\dagger \right). \tag{4.27}$$

This is a beam splitter Hamiltonian which acts to swap excitations between the mechanical oscillator and the optical field.

Exercise 4.6 *Use the quantum Langevin equation given in Eq. (1.112) to find*

---

[8]It should be noted that, while the conditions become identical in this limit, resolved sideband cooling remains much more difficult to achieve in practise. One reason for this is that radiation pressure heating can be achieved with an on-resonance optical drive, while optomechanical cooling requires that the optical field is detuned. As a result, a higher *incident* optical power is required to achieve the same intracavity optical power (and therefore the same C) in the case of optomechanical cooling.

[9]Therefore, requiring appreciable quantum back-action noise on the mechanical oscillator to sustain the Heisenberg uncertainty principle.

equations of motion for both a and b, valid within the rotating wave approximation. Then solve these equations for b in the frequency domain, ignoring transient boundary terms that arise due to the initial conditions, to obtain the result

$$b(\omega) = \chi_{bb}(\delta)\, b_{\text{in}}(\omega) + \chi_{ba}(\delta)\, a_{\text{in}}(\omega), \tag{4.28}$$

where $\delta = \Omega - \omega$ is the offset frequency from the mechanical resonance frequency, and

$$\chi_{bb}(\delta) = \sqrt{\Gamma}\left[\frac{\kappa/2 + i\delta}{(\Gamma/2 + i\delta)(\kappa/2 + i\delta) + g^2}\right] \tag{4.29a}$$

$$\chi_{ba}(\delta) = -\sqrt{\kappa}\left[\frac{ig}{(\Gamma/2 + i\delta)(\kappa/2 + i\delta)}\right] \tag{4.29b}$$

are the mechanical-bath-to-mechanical oscillator and optical-bath-to-mechanical oscillator susceptibilities, respectively.

Equation (4.28) matches our intuition from the unitary approach to strong optomechanical coupling considered in Section 4.1. b is a linear combination of b_{in} and a_{in}. The proportion arising from a_{in} is linearly proportional to g, and the proportion arising from b_{in} decreases with g. Hence, in the usual limit where the optical bath occupancy is much lower than the mechanical bath occupancy ($\bar{n}_L \ll \bar{n}$), the optomechanical coupling results in cooling.

For a stationary system, the mean occupancy of the mechanical mode can be calculated from the frequency spectrum of b using Parseval's theorem since

$$\bar{n}_b = \left\langle b^\dagger(t)b(t)\right\rangle \tag{4.30}$$

$$= \lim_{\tau \to \infty} \frac{1}{\tau} \int_{-\tau/2}^{\tau/2} \left\langle b^\dagger(t)b(t)\right\rangle dt \tag{4.31}$$

$$= \frac{1}{2\pi} \int_{-\infty}^{\infty} S_{bb}(\omega)\, d\omega \tag{4.32}$$

$$= \frac{1}{2\pi} \iint_{-\infty}^{\infty} \left\langle b^\dagger(\omega)b(\omega')\right\rangle d\omega\, d\omega', \tag{4.33}$$

where in the first step we have evoked stationarity, in the second used Parseval's theorem, and in the final step used the relation in Eq. (1.43). In the reasonable case that the mechanical and optical baths are uncorrelated, we then find that the mechanical occupancy is

$$\bar{n}_b = \frac{1}{2\pi}\left[\bar{n}\int_{-\infty}^{\infty}|\chi_{bb}(\delta)|^2\, d\delta + \bar{n}_L\int_{-\infty}^{\infty}|\chi_{ba}(\delta)|^2\, d\delta\right], \tag{4.34}$$

where we have used the frequency domain version of the bath correlation property in Eq. (1.115a), which is valid within the rotating wave approximation.

Exercise 4.7 *In the limit of no optomechanical coupling ($g = 0$) confirm that*

$|\chi_{bb}(\delta)|^2$ *is a Lorentzian centred around* $\delta = 0$ *(*$\omega = \Omega$*) with a width of* Γ*. Then show that, as should be expected, in this case the oscillator is equilibrated at the temperature of its environment with* $\bar{n}_b = \bar{n}$*.*

It should be clear from inspection of Eqs. (4.29) that, in general, $|\chi_{bb}(\delta)|^2$ and $|\chi_{ba}(\delta)|^2$ are not Lorentzian. This means that, in general, once an optomechanical interaction is introduced, the mechanical oscillator can no longer be thought of as an isolated mechanical oscillator in thermal equilibrium with its environment. It is still possible, of course, to determine its occupancy via Eq. (4.34). However, the integrals are not straightforward to solve analytically. To simplify matters, henceforth we take the experimentally relevant limit that the optical field is shot-noise limited so that $\bar{n}_L = 0$. In this case Eq. (4.34) reduces to

$$\bar{n}_b = \frac{\bar{n}}{2\pi} \int_{-\infty}^{\infty} |\chi_{bb}(\delta)|^2 \, d\delta. \tag{4.35}$$

To make progress in solving this integral, it is insightful to consider the characteristics of $|\chi_{bb}(\delta)|^2$ as a function of the optomechanical coupling strength. The thick lines in Fig. 4.4 show this. As can be seen, for low optomechanical coupling strengths the form of $|\chi_{bb}(\delta)|^2$ appears to be Lorenzian-like. The function broadens and decreases in amplitude as the optomechanical coupling strength increases, and eventually splits into two peaks separated by $2g$. As discussed in Section 4.1, this splitting is a characteristic signature of strong coupling between the intracavity optical field and the mechanical oscillator. These observations motivate us to search for separate approximate forms of $|\chi_{bb}(\delta)|^2$ that are accurate in the *strong coupling regime* where $g \gg \{\Gamma, \kappa\}$ and in the complimentary *weak coupling regime* where $\{g, \Gamma\} \ll \kappa$. We would note that, despite its name, the latter regime does not imply that radiation pressure shot noise may be neglected. It can be seen from Eq. (3.19) that, even in the weak coupling regime, radiation pressure shot noise can dominate the heating of the mechanical oscillator.

Exercise 4.8 *Convince yourself of this.*

4.2.2.1 Weak coupling regime

Considering – in generality – a mechanical oscillator that has a decay rate Γ and is coupled to an optical field with rate g, it can be anticipated that the energy of the oscillator will be predominantly confined to frequencies within $\Gamma + g$ of the mechanical resonance frequency. This presumption leads to the expectation that frequency components outside of the range $|\delta| \lesssim \Gamma + g$ should have little contribution to the integral in Eq. (4.35). In the weak coupling regime where $g \ll \kappa$, and assuming the usual situation where $\Gamma \ll \kappa$, this leads to the conclusion that the only frequencies that contribute significantly to the integral satisfy $|\delta| \ll \kappa$. It is then possible to approximate Eq. (4.29a)

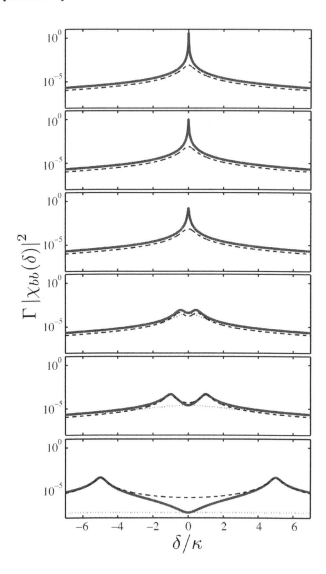

FIGURE 4.4 Modification of the absolute-squared mechanical susceptibility ($|\chi_{bb}(\delta)|^2$) due to resolved sideband cooling with varying optomechanical coupling strengths. Here $\Gamma/\kappa = 0.01$, and from top to bottom $g/\kappa = \{0, 0.05, 0.1, 0.5, 1, 5\}$, or expressed in terms of the optomechanical cooperativity $C = \{0, 1, 4, 100, 400, 1 \times 10^4\}$. The thick solid lines are predicted from the full model (Eq. (4.29a)), the dotted lines are predicted using the weak-coupling approximation (Eq. (4.37)), and the dashed lines are using the strong coupling approximation (Eq. (4.40)).

as

$$\chi_{bb}(\delta) \approx \frac{\sqrt{\Gamma}}{\Gamma/2 + 2g^2/\kappa + i\delta} \tag{4.36}$$

$$= \frac{\sqrt{\Gamma}}{\Gamma/2 \times (1 + C) + i\delta}, \tag{4.37}$$

where C is the usual optomechanical cooperativity of Eq. (3.14). This is, indeed, a Lorenzian with an increased width of

$$\Gamma' = (1 + C)\,\Gamma, \tag{4.38}$$

and on-resonance peak height $\chi_{bb}(0)$ reduced compared to the bare mechanical oscillator by a factor of $1 + C$.

Exercise 4.9 *Show that the optical contribution to Γ' in Eq. (4.38) agrees with Eq. (4.19a) in the very good cavity limit ($\kappa \ll \Omega$).*

Equation (4.37) is illustrated by the dotted lines in Fig. 4.4. In the weak coupling regime the agreement with the full mechanical susceptibility is near perfect, with major discrepancies becoming evident as the strong coupling regime is approached.

Applying this result for $\chi_{bb}(\delta)$ into Eq. (4.35), we find that the mean mechanical occupancy for resolved sideband cooling in the weak coupling limit is

$$\bar{n}_b = \frac{\bar{n}}{1 + C}. \tag{4.39}$$

This expression is identical to the result we found earlier (Eq. (4.26)) when neglecting both mechanical heating of the optical field and radiation pressure heating of the mechanical oscillator.

4.2.2.2 Strong coupling regime

As discussed earlier, as the strong coupling regime is approached the mechanical phonon number spectrum characterised by $\chi_{bb}(\delta)$ starts to take on the appearance of a double-peaked Lorenzian (see Figs. 4.4 and 4.6). Each of the peaks represents one of the pair of hybridised optomechanical modes described in Section 4.1. We know from Section 4.1 that the separation of the peaks should equal $2g$. In the weak coupling regime the decay of the cavity field is fast compared to the optomechanical coupling rate ($g \ll \kappa$) such that the heat transfer from the mechanical oscillator into the optical environment is rate-limited by g. Here, on the other hand, since $g \gg \kappa$, the exact opposite is true, with the rate of heat transfer limit by κ. As a result, one should expect the linewidths of the hybrid modes to be determined by κ and Γ rather than g. This leads to the conclusion that frequency components in the mechanical power spectrum that are outside of the ranges $|\delta \pm g| \lesssim \Gamma + \kappa$ should not contribute appreciably to the mechanical occupancy. Using the strong coupling criterion $g \gg \{\kappa, \Gamma\}$, this can be rewritten in the relaxed form $|\delta \pm g| \ll g$.

Returning to the general form of the susceptibility in Eq. (4.29a), and utilising the approximations outlined in the previous paragraph, we find after some work that in the strong coupling regime $|\chi_{bb}(\delta)|^2$ is well approximated by

$$|\chi_{bb}(\delta)|^2 = \frac{\Gamma/4}{\left(\frac{\Gamma+\kappa}{4}\right)^2 + (\delta+g)^2} + \frac{\Gamma/4}{\left(\frac{\Gamma+\kappa}{4}\right)^2 + (\delta-g)^2}. \qquad (4.40)$$

Exercise 4.10 *Show this for yourself.*

This function is represented by the dashed lines in Fig. 4.4, showing good agreement with the full absolute-squared mechanical susceptibility in the strong coupling regime close to the hybrid modes resonance frequencies, and, as should be expected, poor agreement in the weak coupling regime. Unlike the weak measurement regime, here both the height and shape of the Lorenzian peaks are independent of g. Furthermore, since the absolute-squared suscepti- bility, and therefore the mechanical power spectrum, is not a single Lorenzian, the mechanical oscillator cannot be thought of simply as a single oscillator cou- pled to a bath in thermal equilibrium. Each hybrid optomechanical mode can, however, be thought of in this way.

As can be seen from Eq. (4.40), the width of each Lorenzian is

$$\Gamma' = \frac{\Gamma + \kappa}{2}. \qquad (4.41)$$

This is exactly the usual result for a pair of strongly coupled oscillators: the de- cay rate of the hybrid modes is the average of the uncoupled decay rates since the energy of each hybrid mode is shared equally between the two oscillators.

Compared with the peak of the bare amplitude-squared mechanical sus- ceptibility when no optomechanical coupling is present, the peak of each of the two Lorenzians in Eq. (4.40) is reduced by the fraction

$$\frac{|\chi_{bb}(\delta \pm g)|^2}{|\chi_{bb}(0)|^2_{g=0}} \approx \left(\frac{\Gamma}{\Gamma + \kappa}\right)^2. \qquad (4.42)$$

Evidently, this ratio, and consequently the level of mechanical cooling, criti- cally depends on the ratio of mechanical to optical decay rates.

Similarly to the weak coupling regime, integrating Eq. (4.42) over δ yields the occupancy of the mechanical mode in the strong coupling limit, with the result

$$\bar{n}_b = \bar{n}\left(\frac{\Gamma}{\Gamma + \kappa}\right) \approx \bar{n}\left(\frac{\Gamma}{\kappa}\right), \qquad (4.43)$$

where the approximate solution is valid in the realistic scenario that the optical cavity decays much faster than the mechanical oscillator ($\kappa \gg \Gamma$). We see that, within the strong coupling regime, the final equilibrium occupancy is simply equal to the mechanical bath occupancy suppressed by the ratio of mechanical to optical decay rates.

4.2.2.3 Mechanical occupancy achieved via resolved sideband cooling in the rotating wave approximation

Figure 4.5 provides an example of the reduction in phonon occupancy achieved by resolved sideband cooling in the rotating wave approximation and compares the full numerical integration of Eq. (4.35) with the weak and strong coupling approximations derived in the previous two subsections. As can be seen, good agreement is achieved in the regimes of validity of each model.

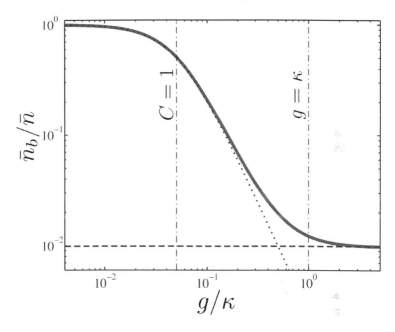

FIGURE 4.5 Final mechanical occupancy achieved with resolved sideband cooling in the rotating wave approximation. Here we set $\Gamma/\kappa = 0.01$. The dotted and dashed lines are the predictions when using the additional weak (Eq. (4.39)) and strong (Eq. (4.43)) coupling approximations, respectively.

4.2.2.4 Approaching the ground state

Optomechanical cooling is a commonly used technique in atom and ion trapping experiments, where operation in the resolved sideband regime routinely enables cooling to the ground state as an initialisation step for experiments in quantum computing and quantum information science [86, 202, 176]. Doppler cooling, on the other hand, is analogous to cavity optomechanical cooling in the nonresolved sideband regime [154].

Optomechanical cooling of a macroscopic mechanical oscillator was first reported in the microwave frequency domain by Braginskii, Manukin, and Tikhonov in 1970, using a radio frequency waveguide resonator with one end consisting of a metal plate suspended on a quartz fibre [38]. The resolved sideband regime was achieved in 1995 using microwave driving and readout of a superconducting high-Q niobium resonant mass gravitational wave antenna [31, 83].

In the optical domain, passive radiation pressure cooling was observed in 2006 [16, 120, 250] in three different microcavity architectures,[10] and the resolved sideband regime was achieved in 2008 using both microtoroidal and microsphere optomechanical systems [251, 218]. Figure 4.6 shows a resolved sideband cooling experiment performed using a microtoroidal optomechanical system [295], which clearly shows the hybridisation of the optical and mechanical modes that occurs within the strong coupling regime.

From Eqs. (4.20), (4.39), and (4.43), one can see that there are three requirements to approach the ground state via resolved sideband cooling

$$\Gamma\bar{n} \ll \kappa \ll \left\{\Omega, \frac{g^2}{\Gamma\bar{n}}\right\}. \tag{4.44}$$

The first requirement arises because the optical cavity decay rate presents a bottleneck on how fast heat can leave the system, the second comes from heating due to the off-resonant fast rotating terms in the Hamiltonian, and the third is a condition on the required strength of optomechanical coupling. Resolved sideband cooling experiments that approached the ground state ($\bar{n}_b < 1$) were first achieved in 2011, both using a photonic-phononic crystal architecture [66] and using a superconducting lumped element electromechanical system [280].

4.2.2.5 Thermodynamical understanding of optical cooling

In Section 4.2.1.1 we found that the steady-state phonon occupancy of an oscillator coupled to two baths equilibrates to a mean of the occupancies of the baths, weighted by the respective coupling rates to each bath (see Eq. (4.23)). This provides an alternative way to think about – and to derive – the optomechanical cooling results from Sections 4.2.2.1 and 4.2.2.2, requiring only the appropriate choice of bath occupancies and system-bath coupling rates. The mechanical oscillator is, of course, coupled to both a hot bath (its environment) with phonon occupancy $\bar{n}$ and a coupling rate Γ, and a cold bath (the optical field). We showed in Section 4.2.1 that, if the optical driving field is in a coherent state, the effective phonon occupancy of the cold bath approaches zero (see Eq. (4.20)) in the well-resolved sideband regime ($\kappa \ll \Omega$). All that remains

---

[10]Note that we use the term "passive radiation pressure cooling" here to distinguish this work from feedback cooling where the position of the mechanical oscillator is measured and active feedback forces are then applied to the mechanical oscillator. Feedback cooling is introduced in Chapter 5.

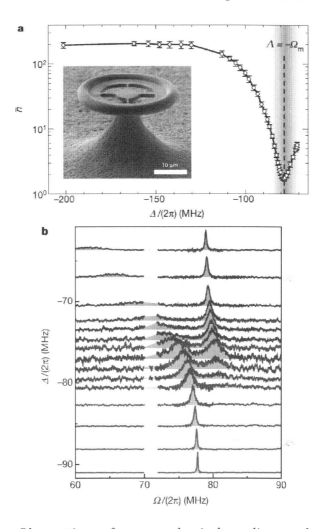

FIGURE 4.6 Observation of optomechanical cooling and quantum-coherent coupling. Adapted by permission from Macmillan Publishers Ltd: *Nature* [295], copyright 2012. Note that here Ω_m and Ω respectively denote the mechanical resonance and optical sideband frequencies. (a) Mechanical occupancy (termed $\bar{n}$ here) as a function of detuning Δ (our $-\Delta$). Inset: microtoroidal optomechanical system. (b) Mechanical power spectral density observed on output optical field showing the avoided crossing characteristic of strong coupling. The dip at exactly $\Delta = -\Omega$ shown in Fig. 4.4 for strong hybridisation is not observed here because the optical field probes the mechanical position spectral density, rather than the phonon number power spectral density.

to determine the occupancy of the mechanical oscillator in the well-resolved sideband regime is to determine the coupling rate between the mechanical oscillator and the optical bath. As we showed in Sections 4.2.2.1 and 4.2.2.2, this coupling rate is different in the weak and strong coupling regimes, with the optical cavity decay rate introducing a bottle-neck in the latter case.

Let us first consider the weak coupling regime, where the heat introduced to the intracavity optical field from the mechanical oscillator decays out of the cavity sufficiently fast to prevent the possibility of reheating the mechanics. In this regime, we can determine the coupling rate of the mechanical excitation into the optical cold bath by inspection of $\chi_{bb}(\delta)$. We found in Section 4.2.2.1 that, in the weak coupling regime, this is well approximated by the Lorenzian given in Eq. (4.36). We can immediately see from this expression that the effect of the optical field is to introduce a second decay channel with decay rate

$$\Gamma_{\text{opt}} = \frac{4g^2}{\kappa}. \tag{4.45}$$

It is worth commenting that this rate is identical to the rate $\mu = 4g^2/\kappa$ that information about the mechanical position is encoded on the output optical field in the nonresolved sideband limit which we identified in Section 3.3. Using this rate, along with the other parameters defined above, in Eq. (4.23) we exactly reproduce Eq. (4.39) for $\bar{n}_b$ from Section 4.2.2.1 in the well sideband resolved weak coupling regime.

In Section 4.2.2.2 we found that, in the strong coupling regime, the coupling rate between the mechanical oscillator and the optical bath is

$$\Gamma_{\text{opt}} = \kappa. \tag{4.46}$$

This can be understood since, in this regime, energy is exchanged back and forth between the cavity field and the mechanical oscillator until it eventually decays out of the cavity into the optical environment at rate κ. Using this rate in Eq. (4.23), we immediately retrieve the same steady-state occupancy as we derived in Section 4.2.2.2 (Eq. (4.43)).

4.3 OPTOMECHANICALLY INDUCED TRANSPARENCY

We have now seen how red detuning of the optical drive field to an optomechanical system can enable mechanical cooling. More complicated protocols are possible by using multiple optical drive fields. Optomechanically induced transparency is one prominent example [4, 309, 246]. Optomechanically induced transparency is analogous to electromagnetically induced transparency (EIT), which allows the absorption spectrum and dispersion of ensembles of atoms to be engineered and is an important tool for quantum memories and repeaters [104]. Just as in EIT, in optomechanically induced transparency two laser fields are injected into the optical cavity, a weak *probe field* at a frequency near the optical resonance frequency, and a strong *control field*. The control field is red detuned by the mechanical frequency in the same manner as the

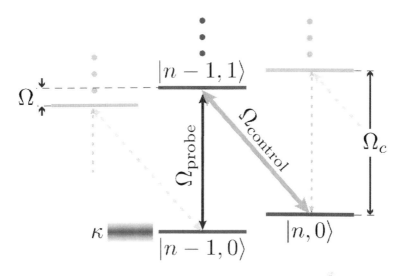

FIGURE 4.7 Energy level diagram for optomechanically induced transparency showing the control and probe optical tones at frequencies Ω_{control} and Ω_{probe}, respectively. Similarly to Fig. 4.1 for optomechanical cooling, $|n, m\rangle$ represents an n-phonon, m-photon state.

cooling field in resolved sideband cooling. This pump-probe scheme is illustrated in the energy level diagram shown in Fig. 4.7 and allows a similar level of control over the absorption and dispersion of optical fields as EIT.

4.3.1 Hamiltonian

It is most convenient to solve for the dynamics of optomechanically induced transparency in the interaction picture at the cavity resonance frequency, such that $\Delta = 0$. Starting with the standard general optomechanical Hamiltonian of Eq. (2.18) within a rotating frame for the optical field, we then have

$$\hat{H} = \hbar\Omega b^\dagger b + \hbar g_0 a^\dagger a \left(b^\dagger + b \right). \tag{4.47}$$

Treating the control field as classical and much brighter than the probe field, this Hamiltonian may be linearised using a similar approach to that which we followed in Section 2.7.

We begin by reexpressing the annihilation operator a as

$$a \to a + \alpha e^{i\Omega t}, \tag{4.48}$$

where the first term constitutes the quantum fluctuations and any coherent contribution from the probe field,[11] while the second term represents the co-

---

[11] Which must be small compared with α.

herent control tone which, in the same manner as the drive tone in resolved
sideband cooling treated earlier, is red detuned from the cavity by the mechanical resonance frequency Ω. The term $a^\dagger a$ in the Hamiltonian (Eq. (4.47))
can then be expanded as

$$a^\dagger a \;\rightarrow\; \alpha^2 + a^\dagger a + \alpha \left[a^\dagger e^{i\Omega t} + a e^{-i\Omega t} \right] \tag{4.49}$$

$$\approx\; \alpha \left[a^\dagger e^{i\Omega t} + a e^{-i\Omega t} \right]. \tag{4.50}$$

Here we have neglected the constant α^2 term, which, as we saw in Section 2.7,
introduces a constant term in the Hamiltonian as well as a static displacement
to the mechanical oscillator that has no effect on the dynamics of the system.
We have also neglected the $a^\dagger a$ term in the usual linearisation approximation,
with the probe field and vacuum optomechanical coupling rate g_0 assumed to
be sufficiently weak so that this term does not appreciably affect the system
dynamics. Substituting this expression into the Hamiltonian, we arrive at

$$\hat{H} \;=\; \hbar\Omega b^\dagger b + \hbar g \left[a^\dagger e^{i\Omega t} + a e^{-i\Omega t} \right] \left(b^\dagger + b \right) \tag{4.51}$$

$$\approx\; \hbar\Omega b^\dagger b + \hbar g \left[a b^\dagger e^{-i\Omega t} + a^\dagger b e^{i\Omega t} \right], \tag{4.52}$$

where, similar to our previous treatment of resolved sideband cooling, in the
approximation we have neglected terms that are not resonant and are therefore suppressed in the resolved sideband limit (see Fig. 4.7). Apart from the
explicit time dependence, this Hamiltonian is very similar to the Hamiltonian
for resolved sideband cooling given in Eq. (4.27). The oscillation in time is
crucial for the operation of optomechanically induced transparency and can
be thought of experimentally as a beating between the probe and control fields
that causes near DC fluctuations in the probe to be mixed up to the mechanical resonance frequency and thereby to strongly interact with the mechanical
oscillator. In optomechanically induced transparency probe excitations are
converted to mechanical oscillations and then back into the probe field again.
The essential idea is to set up a perfect destructive interference between the
intracavity probe field and the fluctuations it drives onto the mechanics when
they return to the cavity. Thereby, the probe field cannot exist in the cavity,
and the cavity becomes transparent.

4.3.2 Optomechanically induced absorption

We begin our treatment of optomechanically induced transparency by considering the open system dynamics of a single-sided optomechanical system – that
is, one that has only a single optical input/output channel (see Fig. 4.8 (left)).
This turns out to enable optomechanically induced *absorption* but not transparency. In Section 4.3.3, we will extend the treatment to a double-sided optomechanical system which does exhibit transparency. Similar to our approach
to resolved sideband cooling in Section 4.2.2, within the regime of validity of
the rotating wave approximation Eq. (1.112) may be used to obtain the equa-

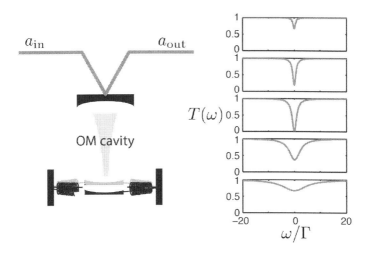

FIGURE 4.8 Optomechanically induced absorption in a single-sided optomechanical system. *Left:* Experimental schematic. *Right:* Modelled transmissivity $T(\omega)$ as a function of frequency ω for optomechanical cooperativities of $C = \{0.1, 0.4, 1, 4, 10\}$ from top to bottom, and $\kappa = 10\Gamma$.

tions of motion

$$\dot{a} = -\frac{\kappa}{2}a - igbe^{i\Omega t} + \sqrt{\kappa}a_{\text{in}} \qquad (4.53)$$

$$\dot{b} = -\frac{\Gamma}{2}b - i\Omega b - igae^{-i\Omega t} + \sqrt{\Gamma}b_{\text{in}} \qquad (4.54)$$

for the intracavity optical field and mechanical oscillator.

Exercise 4.11 *Solve these equations of motion in the frequency domain to show that in the steady state*

$$a(\omega) = \chi_{aa}(\omega)a_{\text{in}}(\omega) + \chi_{ab}(\omega)b_{\text{in}}(\omega + \Omega), \qquad (4.55)$$

where $\chi_{aa}(\omega)$ and $\chi_{ab}(\omega)$ are the light-to-light and mechanical oscillator-to-light susceptibilities

$$\chi_{aa}(\omega) = \sqrt{\kappa}\left[\frac{\kappa}{2} - i\omega + \frac{g^2}{\Gamma/2 - i\omega}\right]^{-1} \qquad (4.56a)$$

$$\chi_{ab}(\omega) = -\sqrt{\Gamma}\left(\frac{ig}{\Gamma/2 - i\omega}\right)\left[\frac{\kappa}{2} - i\omega + \frac{g^2}{\Gamma/2 - i\omega}\right]^{-1}. \qquad (4.56b)$$

As can be seen, the light-to-light susceptibility is modified from the usual Lorentzian describing a bare optical resonance, whilst the presence of the mechanical oscillator introduces a second input noise term in the equation, driving the intracavity optical field.

4.3.2.1 Output field

The output field from the system can be found using the general input-output relation of Eq. (1.125) with the result

$$a_{\text{out}} = t(\omega)a_{\text{in}}(\omega) + l(\omega)b_{\text{in}}(\omega + \Omega), \tag{4.57}$$

where

$$
\begin{align}
t(\omega) &= 1 - \sqrt{\kappa}\,\chi_{aa}(\omega) \tag{4.58a} \\
l(\omega) &= -\sqrt{\kappa}\,\chi_{ab}(\omega). \tag{4.58b}
\end{align}
$$

Here $t(\omega)$ is a complex frequency-dependent transmission coefficient that quantifies the fraction of the incident field that remains in the output field of the optomechanical system, while $l(\omega)$ quantifies the fraction of the mechanical bath fluctuations that are imprinted onto the output field. As we found in Section 4.2.2, in the resolved sideband regime with a strong drive field that is red detuned to $\Delta = \Omega$, the optomechanical interaction is of the form of a beam splitter, or equivalently a two-mode linear scattering interaction. From this perspective $t(\omega)$ and $l(\omega)$ can be thought of as scattering amplitudes, with $l(\omega)$ not only quantifying the level of mechanical fluctuations imprinted on the output field, but also the loss (or absorption) of the optical field by the mechanical bath – i.e., it is a frequency-dependent loss coefficient.

Exercise 4.12 *Show that the optomechanical scattering process is energy conserving, with*

$$T(\omega) + L(\omega) = 1, \tag{4.59}$$

where the transmissivity $T(\omega)$ and absorptivity $L(\omega)$ are defined as $T(\omega) \equiv |t(\omega)|^2$ and $L(\omega) \equiv |l(\omega)|^2$. Show that, in the limit $\omega \ll \kappa$, the absorptivity is given by the Lorentzian

$$L(\omega) = \frac{C\Gamma^2}{(1+C)^2(\Gamma/2)^2 + \omega^2}, \tag{4.60}$$

where, as usual, $C = 4g^2/\kappa\Gamma$ is the optomechanical cooperativity.

As the above exercise demonstrates, in this configuration the optomechanical system is acting as an optical absorber, with peak absorption at sideband frequencies that are close to Ω higher than the coherent drive frequency (i.e., near $\omega = 0$). In the previous section we examined the effect of the optical field on the mechanical oscillator when a red-detuned coherent drive tone is applied, showing that cooling occurs due to (cold) optical fluctuations being transferred onto the oscillator. Here we see the complimentary effect on the optical field from the same process, with the mechanical oscillator absorbing optical energy. The absorptive feature is Lorentzian with a width $(1 + C)\Gamma$ matching the optomechanically broadened mechanical linewidth of resolved sideband cooling in the weak coupling regime (Eq. (4.38)), and exhibits a peak absorption of $4C/(1+C)^2$ when $\omega = 0$.

The transmissivity $T(\omega)$ of this optomechanical absorber is plotted as a function of frequency in Fig. 4.8 (*right*) for a range of optomechanical cooperativities. One observation that can be made from this figure is that, in the special case of $C = 1$, the optical field is perfectly absorbed by the mechanical oscillator at $\omega = 0$. Note that this does not imply that this choice of cooperativity allows the mechanical oscillator to be cooled to its ground state. As we saw in the previous section, significantly more stringent requirements must be satisfied to achieve this (see Eq. 4.44).

4.3.3 Optomechanically induced transparency using a double-sided optical cavity

In the case of a single-sided cavity considered in the previous section, when there is no optomechanical interaction ($g = C = 0$) energy conservation requires that the transmissivity $T(\omega) = 1$ for all sideband frequencies.

Exercise 4.13 *Confirm that this is the case from Eq. (4.58a).*

Consequently, in this single-sided configuration it is clearly not possible for the optomechanical interaction to enhance the transmissivity as required for optomechanically induced transparency. This motivates us to consider the case of a double-sided cavity as depicted in Fig. 4.9 (*left*). We envision that the cavity decay occurs through two optical ports with decay rates of κ_1 and κ_2, respectively, though one of these ports may in fact arise due to optical losses and therefore not be readily accessible to the experimenter. The total cavity decay rate is then $\kappa = \kappa_1 + \kappa_2$. While the broad features of optomechanically induced transparency are evident for any choice of decay rates, we choose to consider a balanced two-sided cavity here, with $\kappa_1 = \kappa_2 = \kappa/2$. This has the advantage that, without optomechanical interaction, the optical field is fully impedance matched into the cavity through one input/output port, and out through the other input/output port.

For a balanced two-sided cavity, the input optical field of Eq. (4.57) has equal contributions from both optical ports so that

$$a_{\text{in}} = \frac{a_{\text{in},1} + a_{\text{in},2}}{\sqrt{2}}. \tag{4.61}$$

Exercise 4.14 *Substituting this into Eq. (4.55) and using the input-output relations of Eq. (1.125), show that the two output fields are*

$$a_{1,\text{out}} = t(\omega)a_{1,\text{in}}(\omega) + r(\omega)a_{2,\text{in}}(\omega) + l(\omega)b_{\text{in}}(\omega + \Omega) \tag{4.62a}$$

$$a_{2,\text{out}} = t(\omega)a_{2,\text{in}}(\omega) + r(\omega)a_{1,\text{in}}(\omega) + l(\omega)b_{\text{in}}(\omega + \Omega), \tag{4.62b}$$

where now

$$t(\omega) = 1 - \frac{\sqrt{\kappa}}{2}\chi_{aa}(\omega) \tag{4.63a}$$

$$r(\omega) = -\frac{\sqrt{\kappa}}{2}\chi_{aa}(\omega) \tag{4.63b}$$

$$l(\omega) = -\sqrt{\frac{\kappa}{2}}\chi_{ab}(\omega), \tag{4.63c}$$

with $r(\omega)$ being a complex frequency-dependent reflection coefficient describing the coupling from one optical field to the other.

Confirm that, analogously to the case of a single-sided optical cavity, energy is conserved with

$$T(\omega) + R(\omega) + L(\omega) = 1, \tag{4.64}$$

where the reflectivity $R(\omega)$ is defined as $R(\omega) \equiv |r(\omega)|^2$.

In the limit $\omega \ll \kappa$ the transmissivity $T(\omega)$ of this two-sided optomechanical system is well approximated by the Lorenzian

$$T(\omega) = \frac{C^2(\Gamma/2)^2}{(1+C)^2(\Gamma/2)^2 + \omega^2}, \tag{4.65}$$

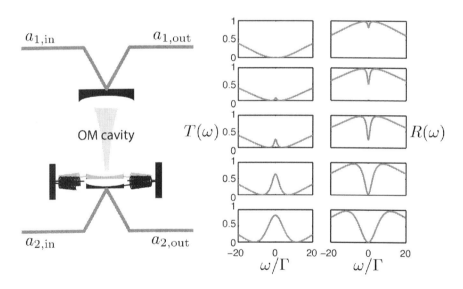

FIGURE 4.9 Optomechanically induced transparency in a double-sided optomechanical system. *Left:* Experimental schematic. *Right:* Modelled transmissivity $T(\omega)$ (left column) and reflectivity $R(\omega)$ (right column) as a function of frequency ω for optomechanical cooperativities of $C = \{0.1, 0.4, 1, 4, 10\}$ from top to bottom, and $\kappa = 50\Gamma$.

which is quite similar in form to the absorption spectrum in the single-sided case (Eq. (4.60)). We see that, when there is no optomechanical interaction ($C = 0$), the optomechanical system has no on-resonance ($\omega = 0$) transmission. As the interaction strength increases, a transparency window now appears with peak transmissivity of $T(0) = C^2/(1 + C)^2$ and width equal to the resolved sideband cooled mechanical linewidth $(1 + C)\Gamma$. Asymptotically, at $C \to \infty$ the on-resonance transmissivity approaches unity; while at $C = 1$ the optomechanical system acts as a balanced but lossy beam splitter with $T(0) = R(0) = 1/4$ and $L(0) = 1/2$. This behaviour is shown in Fig. 4.9 (*right*).

4.3.3.1 Noise performance

From the above observations it should now be clear that, in the linearised regime with coherent driving on the red sideband, a cavity optomechanical system acts as a tuneable beam splitter between the incident optical cavity fields and the mechanical bath fluctuations, with a sharp spectral response dependent on the mechanical decay rate and optomechanical cooperativity. It is interesting to ask what conditions are required for the fluctuations introduced by the mechanical oscillator to be negligible, such that the oscillator acts only as a sort of controllable noise-free valve connecting the two fields. Assuming that the input optical fields are both in coherent states and the mechanical bath is, as usual, in a thermal state with occupancy $\bar{n}$, the power spectral density of an arbitrary quadrature $\hat{X}^\theta_{1,\text{out}}$ of output field 1 can be easily shown using Eqs. (1.43), (1.118a), and (4.62) to be

$$S_{X^\theta_{1,\text{out}} X^\theta_{1,\text{out}}}(\omega) = \frac{T(\omega)}{2} + \frac{R(\omega)}{2} + L(\omega)(\bar{n} + 1/2), \qquad (4.66)$$

with an identical result for output field 2, where we have neglected the coherent peak from each optical field at $\omega = 0$. Making use of Eq. (4.64) we find the condition on the absorptivity

$$L(\omega) < \frac{1/2}{\bar{n} + 1} \qquad (4.67)$$

for the first two (optical) terms to dominate the last (mechanical) term. This regime can be achieved in both the high and low cooperativity limits. In the more interesting high cooperativity limit ($C \gg 1$), for instance, the on-resonance absorptivity is $L(0) = 2/C$ and the condition becomes $C > 4(\bar{n}+1)$, which is a similar criterion as that to achieve ground state cooling in the weak coupling sideband-resolved regime (second inequality in Eq. (4.44)). If this condition is satisfied, the effect of the optomechanical interaction is to modify the optical susceptibility, creating a sharp transparency feature that – at least on-resonance – switches the output optical fields of the system without adding any appreciable fluctuations from the mechanical bath.

While the treatment in this section is only valid in the linearised regime, we introduce single-photon optomechanical beam splitters which operate in the nonlinearisable single-photon strong coupling regime in Sections 6.3 and 6.4.

4.3.3.2 Experimental implementation

Optomechanically induced transparency has been experimentally demonstrated in a range of different architectures. It was first achieved in a microtoroidal optomechanical system [309]. Figure 4.10 shows the experimental apparatus, the overused transparency window, and resulting broadening and deepening of the transparency as a function of optomechanical cooperativity. One particularly remarkable feature of optomechanically induced transparency is demonstrated here. While a typical microcavity might have a resonance linewidth of megahertz or gigahertz, a typical micromechanical oscillator has a linewidth in the range of millihertz to kilahertz. The optomechanical transparency allows these ultra narrow linewidths to be mapped onto an optical field (see the exceptionally sharp transparency feature in Fig. 4.10b). One proposed application of such a sharp resonance is to apply a strongly frequency dependent rotation to squeezed light so that it can be used to exceed the standard quantum limit of the measurement of mechanical motion over a broad frequency band [185, 233] (we discuss this concept further in Section 5.4.2). It has proved practically difficult to achieve such rotations through other means.

4.4 OPTOMECHANICAL ENTANGLEMENT

In the previous two sections we considered the scenario where the optical field was red detuned from the optical resonance frequency, showing that this results in cooling of the mechanical oscillator and can be used to facilitate an optomechanically induced transparency for the optical field. It is interesting to consider the opposite regime where the optical field is blue detuned. Here, each incident photon carries more energy than an intracavity photon. When photons enter the cavity, the additional energy is taken up by the mechanical oscillator. One might initially imagine that this process would be undesirable, acting only to heat the mechanical oscillator. However, this is not the case. In fact it acts to correlate the intracavity field and the mechanical oscillator and ultimately generate entanglement between them, which is the topic of this section.

A wide range of theoretical studies have been performed on optomechanical entanglement, with some of the earlier seminal works including [35, 188, 333, 299, 301, 116]. It was first experimentally demonstrated for a lumped element superconducting microwave optomechanical system in 2013 [217] (see Fig. 4.16).

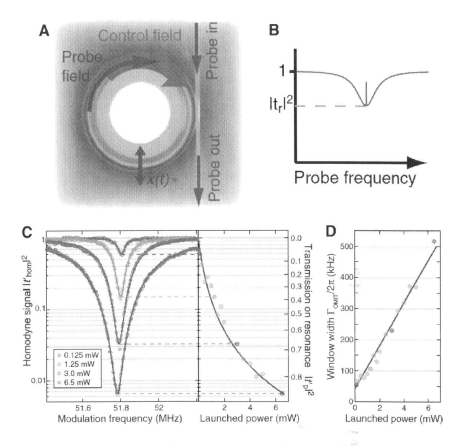

FIGURE 4.10 Experimental demonstration of optomechanically induced transparency. From [309]. Reprinted with permission from AAAS. (a) Microtoroidal optomechanical system showing probe and control fields. (b) Theoretical model of probe transmission as a function of frequency. (c) Observed homodyne signal showing a sharp dip near the mechanical resonance frequency, which in this configuration is evidence of an increased probe transmission at that frequency. (d) Width of transparency window as a function of laser power.

4.4.1 Hamiltonian and equations of motion

Similarly to our treatment of resolved sideband cooling in Section 4.2.2, for simplicity we consider the resolved sideband limit where $\kappa \ll \Omega$. In Section 4.2.2, this limit allowed us to apply a rotating wave approximation and neglect the off-resonant ab and $a^\dagger b^\dagger$ terms in the Hamiltonian of Eq. (4.1), resulting in a beam splitter-like Hamiltonian between a and b. Here we choose a blue detuning of $\Delta = -\Omega$ rather than red detuning, so that these terms are resonant (consider the energy level diagram in Fig. 4.1), and the $ab^\dagger$ and $a^\dagger b$ terms are off-resonant. Neglecting the off-resonant terms, in an interaction picture with the optical field and mechanical oscillator rotating at the cavity and mechanical resonance frequencies, respectively, the optomechanical Hamiltonian of Eq. (4.1) then becomes

$$\hat{H} = \hbar g \left(ab + a^\dagger b^\dagger \right). \tag{4.68}$$

The effect of this Hamiltonian is immediately apparent – it is the Hamiltonian of a parametric interaction that generates (or annihilates) correlated photon-phonon pairs. It may appear that pair production of this kind would violate energy conservation. Indeed, unlike the case of a beam splitter, this Hamiltonian is not energy conserving in its own right, with the energy required to generate photon-phonon pairs derived from the coherent optical drive field. Parametric processes are the fundamental processes behind all forms of bipartite Gaussian entanglement. The ability to tune between beam splitter and parametric interactions via detuning is a particularly useful tool to control the physics of optomechanical systems.

The rotating wave quantum Langevin equation of Eq. (1.112) can be used, as usual, to arrive at equations of motion for the evolution of the intracavity field and mechanical oscillator:

$$\dot{a} = igb^\dagger - \frac{\kappa}{2}a + \sqrt{\kappa}a_{\text{in}} \tag{4.69a}$$

$$\dot{b} = iga^\dagger - \frac{\Gamma}{2}b + \sqrt{\Gamma}b_{\text{in}}. \tag{4.69b}$$

4.4.2   Bogoliubov modes

While the equations of motion in Eq. (4.69) can be straightforwardly solved by taking the Fourier transform and solving simultaneously, it is instructive to instead first diagonalise them.

Exercise 4.15 *Show that substitution of a and b in terms of*

$$c^- = \frac{1}{\sqrt{2}} \left(ira + \frac{1}{r}b^\dagger \right) \tag{4.70a}$$

$$c^+ = \frac{1}{\sqrt{2}} \left(\frac{1}{r}a + irb^\dagger \right) \tag{4.70b}$$

diagonalises Eqs. (4.69), with the resulting uncoupled equations of motion given by

$$\dot{c}^- = -\frac{\gamma^-}{2}c^- + ir\sqrt{\frac{\kappa}{2}}a_{\text{in}} + \frac{1}{r}\sqrt{\frac{\Gamma}{2}}b_{\text{in}}^\dagger \tag{4.71a}$$

$$\dot{c}^+ = -\frac{\gamma^+}{2}c^+ + \frac{1}{r}\sqrt{\frac{\kappa}{2}}a_{\text{in}} + ir\sqrt{\frac{\Gamma}{2}}b_{\text{in}}^\dagger, \tag{4.71b}$$

with

$$r = \frac{1}{\sqrt{2}g} \left[\frac{\kappa - \Gamma}{2} + \sqrt{\left(\frac{\kappa - \Gamma}{2} \right)^2 + 4g^2} \right]^{1/2} \tag{4.72a}$$

$$\gamma^\pm = \frac{\kappa + \Gamma}{2} \mp \sqrt{\left(\frac{\kappa - \Gamma}{2} \right)^2 + 4g^2}. \tag{4.72b}$$

This transformation to modes $c^\pm$ is a form of two-mode *Bogoliubov transformation*, used by Nickolai Bogoliubov to study the physics of superfluity and superconductivity [34]. Note that here $c^\pm$ are not bosonic modes. From Eqs. (4.70) we find that $c^\pm$ and $c^\mp$ do not commute with each other:

$$\left[c^\pm, c^{\mp \dagger} \right] = \pm i, \tag{4.73}$$

while

$$\left[c^\pm, c^{\pm \dagger} \right] = \pm \frac{1}{2} \left(\frac{1}{r^2} - r^2 \right). \tag{4.74}$$

It is apparent from this second commutation relation that each operator commutes with its own conjugate for the special case of $r = 1$, so that, for any arbitrary functions f and h, $[f(c^\pm, c^{\pm \dagger}), h(c^\pm, c^{\pm \dagger})] = 0$. We therefore see that, unlike the usual boson annihilation and creation operators, for $r = 1$ there exists no Heisenberg uncertainty principle between the quadratures

$$\hat{X}_c^- \equiv \frac{1}{\sqrt{2}} \left(c^{-\dagger} + c^- \right) \overset{r=1}{\equiv} \frac{1}{\sqrt{2}} \left(\hat{X}_M - \hat{Y}_L \right) \tag{4.75a}$$

$$\hat{Y}_c^- \equiv \frac{i}{\sqrt{2}} \left(c^{-\dagger} - c^- \right) \overset{r=1}{\equiv} \frac{1}{\sqrt{2}} \left(\hat{X}_L - \hat{Y}_M \right), \tag{4.75b}$$

where $\hat{X}_M$ and $\hat{Y}_M$ are mechanical quadrature operators (see Eqs. (1.135)) rotating at the mechanical resonance frequency Ω, which we term the *mechanical position quadrature* and the *mechanical momentum quadrature*, respectively,[12] and we have introduced the subscript L to clearly distinguish the quadratures of the intracavity optical field from those of the mechanical oscillator.

Exercise 4.16 *Determine the equivalent of Eqs. (4.75) for the quadratures $\hat{X}_c^+ \equiv (c^{+\dagger} + c^+)/\sqrt{2}$ and $\hat{Y}_c^+ \equiv i(c^{+\dagger} - c^+)/\sqrt{2}$.*

We see from Eqs. (4.75) and the commutation relation $[\hat{X}_c^-, \hat{Y}_c^-] \overset{r=1}{=} 0$ that $\hat{X}_M$ can in principle be perfectly correlated to $\hat{Y}_L$, while simultaneous perfect correlations exist between $\hat{X}_L$ and $\hat{Y}_M$. This is the essential feature of two-mode Gaussian entanglement [237] and demonstrates the role that Bogoliubov modes play in understanding such entanglement.

---

[12]Note that the appearance of mechanical quadrature operators here, rather than the dimensionless position $\hat{Q}$ and momentum $\hat{P}$, arises because – unlike our approach in previous parts of the textbook – in the Hamiltonian of Eq. (4.68) we have moved into an interaction picture for the mechanical oscillator rotating at Ω.

From Eq. (4.72b) it is apparent that each of the modes $c^\pm$ decay at different rates that depend on κ, Γ, and g, with c^- (c^+) decaying faster (slower) than the mean of the optical and mechanical decay rates. Some straightforward algebra shows that γ^+ is negative for $C > 1$ where, as usual, C is the optomechanical cooperativity. $C = 1$ therefore constitutes a threshold for instability, above which c^+ grows exponentially as a result of the optomechanical interaction, never reaching a steady state. This is an example of parametric instability [159, 61]. While this exponential growth can, in principle, be stabilised – for instance, using feedback [134] or a secondary optical cooling tone [231] – or mitigated by using short optical pulses [217], here we restrict our analysis to the intrinsically stable regime where $C < 1$.

4.4.3 Optical and mechanical modes in the stable regime

Taking the Fourier transform of Eqs. (4.71), we arrive at uncoupled equations of motion for $c^\pm(\omega)$ at steady state in the stable regime

$$c^-(\omega) = \left(\frac{1}{\gamma^-/2 - i\omega}\right)\left[ir\sqrt{\frac{\kappa}{2}}\hat{a}_{\rm in}(\omega) + \frac{1}{r}\sqrt{\frac{\Gamma}{2}}\hat{b}^\dagger_{\rm in}(-\omega)\right] \quad (4.76a)$$

$$c^+(\omega) = \left(\frac{1}{\gamma^+/2 - i\omega}\right)\left[\frac{1}{r}\sqrt{\frac{\kappa}{2}}\hat{a}_{\rm in}(\omega) + ir\sqrt{\frac{\Gamma}{2}}\hat{b}^\dagger_{\rm in}(-\omega)\right]. \quad (4.76b)$$

Using Eqs. (4.70), the frequency domain annihilation operators for the intracavity optical field and mechanical oscillator can then be directly obtained. Expressed in terms of quadrature operators, they are

$$\hat{X}_L(\omega) = \chi_{aa}(\omega)\hat{X}_{L,\rm in}(\omega) + \chi_{ab}(\omega)\hat{Y}_{M,\rm in}(\omega) \quad (4.77a)$$

$$\hat{Y}_L(\omega) = \chi_{aa}(\omega)\hat{Y}_{L,\rm in}(\omega) + \chi_{ab}(\omega)\hat{X}_{M,\rm in}(\omega) \quad (4.77b)$$

$$\hat{X}_M(\omega) = \chi_{bb}(\omega)\hat{X}_{M,\rm in}(\omega) + \chi_{ba}(\omega)\hat{Y}_{L,\rm in}(\omega) \quad (4.77c)$$

$$\hat{Y}_M(\omega) = \chi_{bb}(\omega)\hat{Y}_{M,\rm in}(\omega) + \chi_{ba}(\omega)\hat{X}_{L,\rm in}(\omega), \quad (4.77d)$$

where the susceptibilities $\chi_{ij}(\omega)$ are

$$\chi_{aa}(\omega) = \frac{\sqrt{\kappa}\,(\Gamma/2 - i\omega)}{(\kappa/2 - i\omega)(\Gamma/2 - i\omega) - g^2} \quad (4.78a)$$

$$\chi_{bb}(\omega) = \frac{\sqrt{\Gamma}\,(\kappa/2 - i\omega)}{(\kappa/2 - i\omega)(\Gamma/2 - i\omega) - g^2} \quad (4.78b)$$

$$\chi_{ab}(\omega) = \frac{g\sqrt{\Gamma}}{(\kappa/2 - i\omega)(\Gamma/2 - i\omega) - g^2} \quad (4.78c)$$

$$\chi_{ba}(\omega) = \frac{g\sqrt{\kappa}}{(\kappa/2 - i\omega)(\Gamma/2 - i\omega) - g^2}. \quad (4.78d)$$

Exercise 4.17 *Derive these expressions.*

4.4.4 Einstein–Polodsky–Rosen entanglement

Gaussian two-mode entanglement, as produced by the linearised optomechanical interaction, was first considered by Einstein, Podolsky, and Rosen in 1935 [100]. They considered two quantum particles A and B described by the wave function

$$\psi_{AB}(x_A, x_B) = \int e^{ip(x_A - x_B)/\hbar} dp. \tag{4.79}$$

Since it is not possible to write this wave function in a product form $\psi_A(x_A)\,\psi_B(x_B)$, the state is inseparable and subsystems A and B are entangled. Einstein, Podolsky, and Rosen were particularly interested in the correlations exhibited between the particles with this wave function and the implications of these correlations on our understanding of quantum mechanics. They recognised that a perfect measurement of $\hat{x}_A$ with result x_A will collapse the state of particle B into the position eigenstate $\psi_{B|A}(x_B) = \delta(x_A)$; while similarly, a perfect momentum measurement collapses particle B into the momentum eigenstate. Since subsystems A and B can in general be space-like separated,[13] the ability to predict both the position and momentum of particle B with perfect precision using different measurements on particle A introduces a conflict between quantum mechanics and local realism – either wave function collapse must occur faster than the speed of light or the position and momentum of particle B must exist with less indeterminism than required by the Heisenberg uncertainty principle. This apparent paradox is now referred to as the *Einstein–Podolsky–Rosen paradox*.

The perfect Einstein–Podolsky–Rosen state exhibits perfectly correlated position and momentum quadratures, such that

$$\left\langle \hat{X}_A(t) - \hat{Y}_B(t) \right\rangle = 0 \tag{4.80a}$$

$$\left\langle \hat{Y}_A(t) - \hat{X}_B(t) \right\rangle = 0, \tag{4.80b}$$

where we have rotated the correlations compared to the previous paragraph such that they occur between position and momentum quadratures since, as we found in the previous section, this is the form of correlation generated by the optomechanical interaction. We see that, for a perfect Einstein–Podolsky–Rosen state, a noise-free measurement of the momentum quadrature of B collapses A into a position eigenstate, while similarly a noise-free measurement of the position quadrature of B collapses A into a momentum eigenstate.

By inspection of Eqs. (4.77), we can immediately observe that the optomechanical interaction produces correlations of a form similar to what would be expected for an ideal Einstein–Podolsky–Rosen state, but that the correla-

---

[13]While in the wave function of Eq. (4.79) the two particles are co-located, a separation s can be introduced straightforwardly via the modification $\psi_{AB}(x_A, x_B) = \int e^{ip(x_A - x_B + s)/\hbar} dp$.

tions are imperfect. In the following two subsections, we introduce the standard methods to quantify two-mode Gaussian entanglement in the presence of such imperfections.

4.4.5 Covariance matrix

In the linearised regime considered here, the entanglement generated via the optomechanical interaction between light and a mechanical oscillator is a form of two-mode Gaussian entanglement. Two-mode Gaussian states are fully characterised by their quadrature expectation values and the covariance matrix

$$\mathbf{M} = \begin{bmatrix} V_{X_A X_A} & V_{X_A Y_A} & V_{X_A X_B} & V_{X_A Y_B} \\ V_{Y_A X_A} & V_{Y_A Y_A} & V_{Y_A X_B} & V_{Y_A Y_B} \\ V_{X_B X_A} & V_{X_B Y_A} & V_{X_B X_B} & V_{X_B Y_B} \\ V_{Y_B X_A} & V_{Y_B Y_A} & V_{Y_B X_B} & V_{Y_B Y_B} \end{bmatrix} = \left[\begin{array}{c|c} \mathbf{A} & \mathbf{C} \\ \hline \mathbf{C}^T & \mathbf{B} \end{array} \right] \tag{4.81}$$

where

$$V_{\mathcal{A}\mathcal{B}} \equiv \frac{\left\langle \hat{\mathcal{A}}(t)\hat{\mathcal{B}}(t) \right\rangle + \left\langle \hat{\mathcal{B}}(t)\hat{\mathcal{A}}(t) \right\rangle}{2} - \left\langle \hat{\mathcal{A}}(t) \right\rangle \left\langle \hat{\mathcal{B}}(t) \right\rangle \tag{4.82}$$

is the covariance between operators $\hat{\mathcal{A}}$ and $\hat{\mathcal{B}}$, the 2×2 matrices $\mathbf{A}$ and $\mathbf{B}$ quantify the variances of the optical and mechanical quadratures and the internal correlations within each subsystem, while $\mathbf{C}$ quantifies the correlations between the mechanical oscillator and the optical field.

4.4.6 Identifying and quantifying entanglement

Diagonalisation of the covariance matrix of Eq. (4.81) yields a pair of *symplectic eigenvalues* $\nu_\pm$ [256]. As first recognised simultaneously by Duan and Simon [97, 261], the condition

$$\nu_- < \frac{1}{2} \tag{4.83}$$

is necessary and sufficient for two-mode Gaussian entanglement, where ν_- is the smaller of the two eigenvalues. For a general covariance matrix, the symplectic eigenvalues are given by

$$\nu_\pm = \left[\frac{1}{2} \left(\tilde{\Delta} \pm \sqrt{\tilde{\Delta}^2 - 4\det(\mathbf{M})} \right) \right]^{1/2}, \tag{4.84}$$

where $\det(...)$ is the determinant and

$$\tilde{\Delta} = \det(\mathbf{A}) + \det(\mathbf{B}) - 2\det(\mathbf{C}). \tag{4.85}$$

The significance of the smaller symplectic eigenvalue with regards to quantum correlations may be understood in the following way. Consider a pair of collective observables of the subsystems A and B

$$\hat{u} = \hat{X}_A + c\hat{X}_B \tag{4.86a}$$
$$\hat{v} = \hat{Y}_A - c\hat{Y}_B, \tag{4.86b}$$

where c is a real constant. If subsystems A and B share no correlations, or indeed share only classical correlations introduced via local operations and classical communication, then the minimum possible product of standard deviations of $\hat{u}$ and $\hat{v}$ is easily shown to be

$$\min\left\{\sigma(\hat{u})\sigma(\hat{v})\right\}_{\text{uncor}} = \frac{1+c^2}{2}, \tag{4.87}$$

and is achieved when both $\hat{u}$ and $\hat{v}$ are symmetric minimum uncertainty states, i.e., vacuum or coherent states.

Exercise 4.18 *Convince yourself of this.*

The criterion of Duan and Simon can be interpreted as stating that subsystems A and B are inseparable if-and-only-if there exists a choice of c for which arbitrary local operations on each individual subsystem are able to bring the product of standard deviations $\sigma(\hat{u})\sigma(\hat{v})$ beneath this value – i.e., if the joint uncertainty in $\hat{u}$ and $\hat{v}$ can be made smaller than the smallest possible uncertainty in the absence of quantum correlations. For a given inseparable state, the smaller symplectic eigenvalue ν_- quantifies exactly this ratio. Specifically,

$$\nu_- = \frac{1}{2}\min\left\{\frac{\sigma(\hat{u})\sigma(\hat{v})}{\min\left\{\sigma(\hat{u})\sigma(\hat{v})\right\}_{\text{uncor}}}\right\} = \min\left\{\frac{\sigma(\hat{u})\sigma(\hat{v})}{1+c^2}\right\}, \tag{4.88}$$

where, here, the minimisation is taken over c and all possible local operations.[14]

The *logarithmic negativity* is a convenient and commonly used parameter to quantify the strength of a given entanglement resource and has the attractive properties of both being additive for multiple independent entangled states and quantifying the maximum distillable entanglement [229]. For two-mode Gaussian entangled states it is given by

$$L = \min\left\{0, -\log_2\left(2\nu_-\right)\right\}. \tag{4.89}$$

4.4.7 Entanglement between intracavity field and mechanical oscillator

Given the optomechanical covariance matrix, Eqs. (4.89) and (4.84) allow quantification of the entanglement between the intracavity optical field and

---

[14]Note that here we express the criterion of Duan in a product form as introduced in [36], rather than the more familiar sum form. The product form is in the spirit of the Heisenberg uncertainty relation or the criteria for the Einstein–Podolsky–Rosen paradox [237], and also achieves the minimum ν_- for a wider class of covariance matrices.

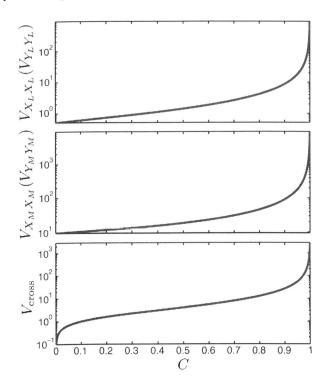

FIGURE 4.11 Covariance matrix elements of intracavity optomechanical entanglement as a function of optomechanical cooperativity C. Model parameters: $\bar{n} = 9$, $\bar{n}_L = 0.009$, $\kappa = 10\,\Gamma$.

the mechanical oscillator. By inspection of Eqs. (4.77) we can immediately recognise that, as long as there are no correlations in the optical and mechanical baths, there will be no correlations between the amplitude and phase quadratures of the optical field, or indeed between the position and momentum quadratures of the mechanical oscillator,[15] so that

$$V_{X_L Y_L} = V_{Y_L X_L} = V_{X_M Y_M} = V_{Y_M X_M} = 0. \qquad (4.90)$$

Furthermore, the correlations between light and mechanics are all cross-quadrature with

$$V_{X_L X_M} = V_{X_M X_L} = V_{Y_L Y_M} = V_{Y_M Y_L} = 0. \qquad (4.91)$$

Therefore, the only covariance matrix elements that must be determined are the variances $V_{X_L X_L}$, $V_{Y_L Y_L}$, $V_{X_M X_M}$, and $V_{Y_M Y_M}$, and the covariances

---

[15]This should be clear because the bath terms that appear in the equations for $\hat{X}_L(\omega)$ and $\hat{Y}_M(\omega)$ are different from those that appear in the equations for $\hat{Y}_L(\omega)$ and $\hat{X}_M(\omega)$.

$V_{X_L Y_M}$, $V_{Y_M X_L}$, $V_{Y_L X_M}$, and $V_{X_M Y_L}$. For each of these elements the operators $\mathcal{A}$ and $\mathcal{B}$ in $V_{\mathcal{AB}}$ commute. Furthermore, since the optomechanical Hamiltonian is linearised, $\langle \hat{X}_L \rangle = \langle \hat{Y}_L \rangle = \langle \hat{X}_M \rangle = \langle \hat{Y}_M \rangle = 0$ (see Section 2.7). These two properties allow the definition of the covariance matrix elements in Eq. (4.82) to be reexpressed more simply as

$$V_{\mathcal{AB}} = \left\langle \hat{\mathcal{A}}(t)\hat{\mathcal{B}}(t) \right\rangle \tag{4.92}$$

$$= \frac{1}{2\pi} \int_{-\infty}^{\infty} d\omega \, S_{\mathcal{AB}}(\omega) \tag{4.93}$$

$$= \frac{1}{2\pi} \iint_{-\infty}^{\infty} d\omega \, d\omega' \left\langle \hat{\mathcal{A}}^\dagger(-\omega)\hat{\mathcal{B}}(\omega') \right\rangle, \tag{4.94}$$

where to arrive at this result we have, as usual, used Parseval's theorem and the definition of the power spectral density in Eq. (1.43).

Exercise 4.19 *Using Eqs. (4.77) and (4.94) show that the nonzero covariance matrix elements are*

$$V_{X_L X_L} = V_{Y_L Y_L}$$

$$= \bar{n}_L + \frac{1}{2} + \left(\frac{C}{1-C}\right)\left(\frac{1}{1+\kappa/\Gamma}\right)(\bar{n}_L + \bar{n} + 1) \tag{4.95a}$$

$$V_{X_M X_M} = V_{Y_M Y_M}$$

$$= \bar{n} + \frac{1}{2} + \left(\frac{C}{1-C}\right)\left(\frac{1}{1+\Gamma/\kappa}\right)(\bar{n}_L + \bar{n} + 1) \tag{4.95b}$$

$$V_{\text{cross}} = \left(\frac{\sqrt{C}}{1-C}\right)\left(\frac{1}{\sqrt{\kappa/\Gamma} + \sqrt{\Gamma/\kappa}}\right)(\bar{n}_L + \bar{n} + 1) \tag{4.95c}$$

where $V_{\text{cross}} = V_{X_L Y_M} = V_{Y_M X_L} = V_{X_M Y_L} = V_{Y_L X_M}$.

We see from Eqs. (4.95) that the variances of the optical field and mechanical oscillator are each equal to their respective bath variances plus a modification that depends on the optomechanical cooperativity C, the ratio of optical and mechanical decay rates, and the sum of the mechanical and optical bath variances; with cross-correlations growing from zero as the optomechanical cooperativity increases. The functional dependence of the variances and covariances on cooperativity are shown in Fig. 4.11. The equal magnitude of the two optical quadrature variances, as well as the two mechanical quadrature variances, and all of the cross-variances, is a result of limiting our treatment to high-quality oscillators for which it is possible to make a rotating wave approximation. In this regime any fast fluctuations that can cause differences between the variances are averaged out. It can also be observed that, as the system approaches instability ($C \to 1$), all three correlation matrix elements approach infinity.

With the covariance matrix elements determined, the logarithmic negativity can be straightforwardly calculated using Eqs. (4.84) and (4.89). The

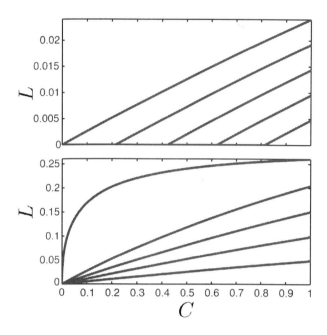

FIGURE 4.12 Logarithmic negativity of intracavity optomechanical entanglement as a function of optomechanical cooperativity C, with $\kappa = 10\,\Gamma$. *Top:* Mechanical bath occupancy held fixed at $\bar{n} = 9$, with optical bath occupancy increasing over the range $\bar{n}_L = \{0, 0.002, 0.004, 0.006, 0.008\}$ from the top to bottom trace. *Bottom:* Optical bath occupancy held fixed at $\bar{n}_L = 0$, with mechanical bath occupancy increasing over the range $\bar{n} = \{0, 2, 4, 6, 8\}$ from the top to bottom trace.

resulting expressions are not, in general, particularly illuminating. Instead of reproducing them here, the logarithmic negativity is plotted as a function of optomechanical cooperativity for a range of different optical $\bar{n}_L$ and mechanical $\bar{n}$ bath occupancies and $\kappa/\Gamma = 10$ in Fig. 4.12. We see that, in general, the entanglement improves as the cooperativity increases. From Fig. 4.12(*top*) it is clear that when the mechanical bath has nonzero occupancy, there is a threshold cooperativity beneath which no entanglement is present, with the threshold increasing as the optical bath occupancy increases. By contrast, Fig. 4.12(*bottom*) shows that, in the realistic scenario where the optical bath occupancy approaches zero, no such threshold is evident, with entanglement existing for any mechanical bath occupancy and any nonzero cooperativity. Indeed, this result can be shown to be true in general, so long as the optical decay is fast compared to the mechanical decoherence rate, specifically in circumstances where $\kappa > \bar{n}\Gamma$.

As the optomechanical cooperativity approaches the point of instability $(C \to 1)$, entanglement is present as long as

$$\left(\frac{\kappa}{\Gamma}\right)\bar{n}_L + \left(\frac{\Gamma}{\kappa}\right)\bar{n} < 1, \qquad (4.96)$$

which restricts the optical and mechanical decoherence rates to be smaller than the mechanical and optical decay rates, respectively.

4.4.8 Entanglement of the mechanical oscillator with the external field

In the usual regime that $\kappa \gg \Gamma$, the interaction of the optomechanical system with the external optical field occurs much more rapidly than with the mechanical bath. While this interaction degrades the entanglement present between the intracavity optical field and the mechanical oscillator, since the output optical field is generally accessible experimentally, it is natural to ask how strongly it is entangled to the mechanical oscillator. Using the input-output relations of Eq. (1.125) and the intracavity field quadratures of Eqs. (4.77a) and (4.77b), the output field quadratures from the optomechanical system can be expressed in the frequency domain as

$$\hat{X}_{L,\text{out}}(\omega) = \left[1 - \sqrt{\kappa}\chi_{aa}(\omega)\right]\hat{X}_{L,\text{in}}(\omega) - \sqrt{\kappa}\chi_{ab}(\omega)\hat{Y}_{M,\text{in}}(\omega) \qquad (4.97a)$$

$$\hat{Y}_{L,\text{out}}(\omega) = \left[1 - \sqrt{\kappa}\chi_{aa}(\omega)\right]\hat{Y}_{L,\text{in}}(\omega) - \sqrt{\kappa}\chi_{ab}(\omega)\hat{X}_{M,\text{in}}(\omega). \qquad (4.97b)$$

As we saw in Section 1.4, in the Markov and rotating wave approximations the input and output fields of an optical cavity can be thought of as a train of infinitesimally separated infinitely narrow optical pulses. At any moment in time t the mechanical oscillator will exhibit entanglement with some ensemble of the output pulses, with the correlations decaying over a time scale that can be expected to depend in some way on the optical cavity and mechanical linewidths κ and Γ. Given this decay of correlations, there exists an optimal temporal mode for the output field that displays maximal correlations (and therefore entanglement) with the oscillator at time t. We would like to identify this optimal mode.

To approach this problem, we define a temporal mode of the output field with modeshape $u(t)$, and annihilation operator

$$a_u(t) = u(t) * a_{\text{out}}(t), \qquad (4.98)$$

where $u(t)$ is normalised such that

$$\int_{-\infty}^{\infty} |u(t)|^2 dt = 1. \qquad (4.99)$$

Exercise 4.20 *Show that this normalisation of $u(t)$ ensures that a_u obeys the usual Boson commutation relation*

$$[a_u(t), a_u^\dagger(t)] = 1. \qquad (4.100)$$

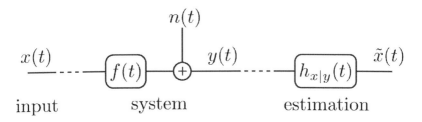

$$n(t)$$

$$x(t) \quad\quad \boxed{f(t)} \quad \oplus \quad y(t) \quad\quad \boxed{h_{x|y}(t)} \quad \tilde{x}(t)$$

input   system   estimation

FIGURE 4.13 Schematic diagram of the use of classical Wiener filtering to estimate a continuous-in-time signal $x(t)$ after it has been processed by a system that filters it ($f(t)$) and adds noise ($n(t)$). $\tilde{x}(t)$ is the final estimate.

The entanglement between this output temporal mode and the mechanical oscillator can be quantified in a manner similar to the case for the intracavity field dealt with in the previous section. However, as discussed already, before doing this we would like to determine the mode that exhibits maximal entanglement with the mechanical oscillator. This is the topic of the next section.

4.4.8.1 Wiener filtering

Estimation of a signal $x(t)$ from a measurement $y(t)$ is a common problem in classical control systems and information processing. For stationary processes and additive noise $n(t)$, the measurement $y(t)$ is related to the signal via $y(t) = f(t) * x(t) + n(t)$, where $f(t)$ is some filter function. In this scenario, the optimal estimation strategy is to apply a *Wiener filter* $h_{x|y}(t)$ to $y(t)$, retrieving an estimate of the signal $\tilde{x}(t) = h_{x|y}(t) * y(t)$ (see Fig. 4.13) [311].

Since linear quantum systems such as that described by the linearised optomechanical Hamiltonian of Eq. (2.18) are unable to generate Wigner function negativity, their statistics may be fully explained through an equivalent classical process. As such, results from classical information processing can be readily applied [90]. In our specific case, Wiener filtering can be used to determine the optimal filter to apply to a measurement on the output field from a cavity optomechanical system to estimate the position quadrature (or momentum quadrature) of the mechanical oscillator at time t. As well as providing key information to practically perform measurement-based conditioning protocols such as feedback cooling as discussed in Chapter 5, this process identifies the temporal modes of the output field that are maximally correlated to each of the position and momentum quadratures of the mechanical oscillator at time t.

Both causal and noncausal Wiener filters exist [53]. Noncausal filters are generally used in information processing where one has access to the full measurement record, while causal filters are appropriate for real-time control ap-

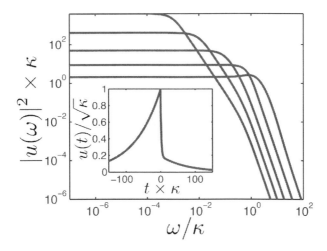

FIGURE 4.14 Wiener filter for optimal entanglement between a mechanical oscillator and an external optical field, with optomechanical cooperativity $C = 0.9$ and bath occupancies $\bar{n} = \bar{n}_L = 0$ and for $\kappa/\Gamma = \{0.1, 1, 10, 100, 1000\}$ from the bottom to the top trace. Inset, corresponding time-domain wave-packet for $\kappa = 100\,\Gamma$. Note that these filters are noncausal.

plications where only the measurement record prior to the time t is available. In assessing optomechanical entanglement we should, in principle, use a causal filter since at time t the mechanical oscillator can only be entangled with the external optical field that has already interacted with it and has left the system before that time. However, for simplicity here we instead use a noncausal filter.[16] Once we have obtained solutions for the profile of the optimal noncausal temporal mode, we will be able to assess how close to causal they are. The exact noncausal Wiener filter is given in the frequency domain by [53]

$$h_{x|y}(\omega) = \frac{S_{yx}(\omega)}{S_{yy}(\omega)}, \tag{4.101}$$

where $S_{yy}(\omega)$ and $S_{yx}(\omega)$ are the usual self- and cross-power spectral densities.

As can be observed in Eqs. (4.97), only the mechanical momentum quadrature appears on the output optical amplitude quadrature; similarly, only the mechanical position quadrature appears on the output optical phase quadrature. The Wiener filters that allow optimal estimation of each mechanical

---

[16]The reader is encouraged to approach the same problem using causal filtering.

quadrature are then given by

$$h_{X_M|Y_{L,\text{out}}}(\omega) = \frac{S_{Y_{L,\text{out}} X_M}(\omega)}{S_{Y_{L,\text{out}} Y_{L,\text{out}}}(\omega)} \qquad (4.102a)$$

$$h_{Y_M|X_{L,\text{out}}}(\omega) = \frac{S_{X_{L,\text{out}} Y_M}(\omega)}{S_{X_{L,\text{out}} X_{L,\text{out}}}(\omega)}. \qquad (4.102b)$$

These filter functions are not identical, in general, which leads to some ambiguity as to the temporal mode that is maximally entangled to the mechanical oscillator. Here, however, since our analysis is limited to the regime in which the rotating wave approximation is valid, the filters coincide. Using Eqs. (4.97), they can be found to be

$$h(\omega) = \frac{\left(1 - \sqrt{\kappa}\chi_{aa}(\omega)\right) \chi_{ba}^*(\omega) \left(\bar{n}_L + 1/2\right) - \sqrt{\kappa}\chi_{ab}(\omega)\chi_{bb}^*(\omega) \left(\bar{n} + 1/2\right)}{\left|1 - \sqrt{\kappa}\chi_{aa}(\omega)\right|^2 \left(\bar{n}_L + 1/2\right) + \kappa \left|\chi_{ab}(\omega)\right|^2 \left(\bar{n} + 1/2\right)}.$$
$$(4.103)$$

The temporal modeshape that is optimally entangled to the mechanical oscillator is then given in the frequency domain by

$$u(\omega) = Nh(\omega), \qquad (4.104)$$

where the normalisation constant N is

$$N = \left[\int_{-\infty}^{\infty} |h(t)|^2 \, dt\right]^{-1/2} = \sqrt{2\pi} \left[\int_{-\infty}^{\infty} |h(\omega)|^2 \, d\omega\right]^{-1/2}. \qquad (4.105)$$

The modeshape defined in Eq. (4.104) is shown in Fig. 4.14 for the specific case of $\bar{n}_L = \bar{n} = 0$ and an optomechanical cooperativity of $C = 0.9$. The main figure plots the modeshape in the frequency domain as a function of the ratio κ/Γ. It can be seen that, when $\Gamma \ll \kappa$, the filter is essentially a causal low-pass filter with a cut-off frequency of approximately Γ – that is, as might be expected, the temporal width of the optimal mode-shape is determined by the mechanical decay time. As Γ approaches and eventually exceeds κ, the modeshape is modified subtly due to the approach of the strong coupling regime, evidenced by the slight resonance feature in the bottom trace of the figure, and becomes increasingly noncausal.

The figure inset shows the optimal modeshape in the temporal domain for $\kappa/\Gamma = 100$. As can be seen, the modeshape can be approximately described as a causal single-sided exponential with a decay rate of Γ. However, for this ratio of optical to mechanical decay rates, there still exists some noncausal contribution in the form of a forwards-in-time exponential of reduced amplitude. The causal contribution is larger because at time t the mechanical oscillator is correlated via the past radiation pressure to the optical field that entered the cavity prior to t, but is clearly not correlated to the optical field that enters the cavity subsequently. As a result, the optical field at times earlier than t carries a greater amount of information about the mechanical oscillator and is weighted more strongly by the filter. It is this effect that, for sufficiently high optomechanical cooperativities and optical decay rates, results in an approximately causal filter.

From Eq. (4.98) we see that, in the frequency domain, the quadratures of the temporal mode $u(t)$ are simply $\hat{X}_u(\omega) = u(\omega)\hat{X}_{L,\text{out}}(\omega)$ and $\hat{Y}_u(\omega) = u(\omega)\hat{Y}_{L,\text{out}}(\omega)$. The covariance matrix elements between these quadratures and the mechanical oscillator can then be calculated in a similar manner as the intracavity case of Section 4.4.7, allowing quantification of the level of entanglement present between the two systems. The resulting logarithmic negativity is shown for $\kappa/\Gamma = 1000$ and various bath occupancies in Fig. 4.15. In the top figure, the mechanical bath occupancy is held fixed at $\bar{n} = 0$ while the optical bath occupancy is increased from zero to, ultimately, infinity (dashed trace), showing that, for $\bar{n} = 0$, the entanglement between the external cavity field and the mechanical oscillator is reduced as $\bar{n}_L$ increases but is always present. In the bottom trace the optical occupancy is held fixed at $\bar{n}_L = 0$. Again, we see that the entanglement is degraded as the mechanical bath occupancy increases. However, here, a threshold optomechanical cooperativity is introduced below which no entanglement is present. Since the system is unstable for $C > 1$, this ultimately introduces a maximum mechanical bath occupancy beyond which entanglement is only possible if some additional technique is introduced to stabilise the system.

It may seem surprising that, if the mechanical bath is in its ground state, the output optical field is always entangled to the mechanical oscillator independent of the optical occupancy $\bar{n}_L$. However, this can be understood in the following way. Consider an attempt to estimate the mechanical position quadrature $\hat{X}_M$ from a measurement of the field exiting the optomechanical system. As we have already discussed, the optimal estimate is given by $\hat{X}_M^{\text{est}}(t) = h_{X_M|Y_{L,\text{out}}}(t) * \hat{Y}_{L,\text{out}}(t)$, with an uncertainty of $\langle (\hat{X}_M(t) - \hat{X}_M^{\text{est}}(t))^2 \rangle$. When the mechanical bath is in its ground state, the uncertainty of this estimate is smaller than the mechanical zero-point uncertainty for any choice of $\bar{n}_L$.

Exercise 4.21 Exercise. *Show this result numerically, or otherwise.*

As such, an optical phase quadrature measurement conditionally prepares a mechanical state with squeezed position quadrature (see Section 5.3.2 for further discussion of mechanical squeezed state preparation via measurement). Similarly, a measurement of the output optical amplitude quadrature conditionally prepares a momentum quadrature squeezed mechanical state. [17] The observables $\hat{u} = \hat{X}_M(t) - \hat{X}_M^{\text{est}}(t)$ and $\hat{v} = \hat{Y}_M(t) - \hat{Y}_M^{\text{est}}(t)$ then clearly exhibit joint uncertainty beneath the minimum uncertainty that is possible when only classical correlations are present (see Eq. (4.87) in Section 4.4.6), which, as we discussed in Section 4.4.6, is a sufficient criterion for entanglement.

---

[17] Note that, since $\hat{X}_L$ and $\hat{Y}_L$ cannot be measured simultaneously without a noise penalty, this does not allow the Heisenberg uncertainty principle to be violated.

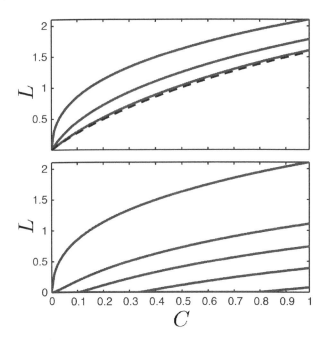

FIGURE 4.15 Logarithmic negativity of entanglement between mechanical oscillator and output optical field as a function of optomechanical co-operativity C, with $\kappa = 1000\,\Gamma$. *Top:* Mechanical bath occupancy held fixed at $\bar{n} = 0$, with optical bath occupancy increasing over the range $\bar{n}_L = \{0, 1, 10\}$ from the top to bottom trace. Dashed line: logarithmic negativity as $\bar{n}_L \to \infty$. *Bottom:* Optical bath occupancy held fixed at $\bar{n}_L = 0$, with mechanical bath occupancy increasing over the range $\bar{n} = \{0, 1, 2, 4, 8\}$ from the top to bottom trace.

As mentioned above, optomechanical entanglement was first demonstrated in 2013 [217] using a lumped element superconducting microwave optomechanical system. In this implementation a pulsed optomechanics protocol was used [292], with an initial blue-detuned pulse generating entanglement and a second red-detuned pulse transferring the mechanical state out onto the microwave field. Ultimately, time-delayed entangled was observed between two microwave pulses. The experimental protocol and final covariance matrix are shown in Fig. 4.16.

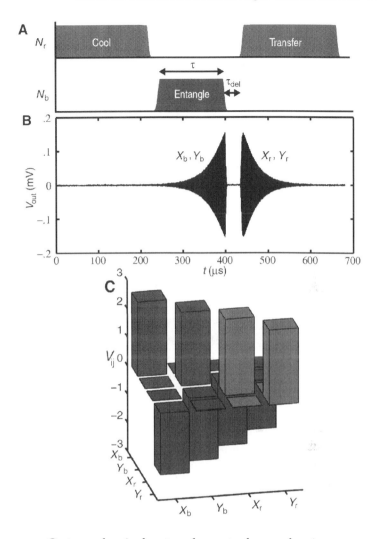

FIGURE 4.16 Optomechanical entanglement observed using a superconducting optomechanical system of the form shown in Fig. 2.2 (*top left*).
From [217]. Reprinted with permission from AAAS. (A) Pulsed protocol to cool the mechanical oscillator, entangle it with a microwave field,
and transfer the mechanical state out onto a second microwave field.
(B) Detected microwave field as a function of time, showing the exponentially amplified initial entangled field leaving the microwave circuit
followed by a second field as the mechanical motion is later transferred
onto the intracavity field and decays out of the circuit. (C) Measured
covariance matrix with off-diagonal elements evidencing entanglement.
Here X_b and Y_b are the mechanical, and X_r and Y_r are the optical (or
strictly speaking, microwave) quadrature operators.

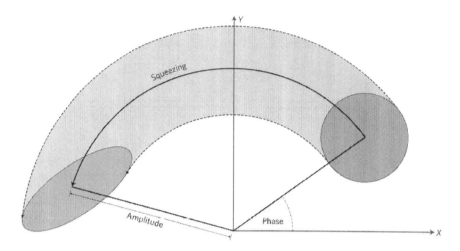

FIGURE 4.17 Ball and stick diagram of the concept of Kerr squeezing. Reprinted by permission from Macmillan Publishers Ltd: *Nature Photonics* [49], copyright 2013. The Kerr effect generates an intensity-dependent phase shift which skews the quantum noise distribution of light and can result in squeezing of the noise on one quadrature beneath the vacuum noise level.

4.5 MECHANICAL SQUEEZING OF LIGHT

From the perspective of the optical field, the optomechanical interaction imparts an intensity-dependent phase shift. The intensity of the optical field imparts momentum on the mechanical oscillator. This shifts its position, altering the phase of the optical field. Intensity-dependent phase shifts, or *Kerr nonlinearities*, are a common method to produce optical squeezing [263] (see Fig. 4.17) and can even generate optical states with Wigner negativity [329].[18] It is therefore unsurprising that the optomechanical interaction is capable of producing squeezed light. However, there are some characteristic differences between this *ponderomotive squeezing* produced by interaction with a mechanical oscillator and squeezing produced by other nonlinear interactions that generate an intensity-dependent phase shift, such as the Kerr effect in optical fibres [260]. Most importantly, the mechanical resonance introduces strong dispersion to the squeezing spectra and restricts the bandwidth of squeezing, while coupling to the mechanical bath can introduce significant degradation on the levels of achievable squeezing. This section will quantitatively introduce mechanical squeezing of light and examine these effects in some detail.

---

[18]Though exceptionally strong nonlinearities are required to achieve this.

4.5.1 Basic concept

While ponderomotive squeezing is generally thought of as a continuous steady-state process, the basic idea may be understood by considering a pair of unitary interactions between the optical field and the mechanical oscillator. The essential goal is to introduce correlations between the optical amplitude and phase quadratures mediated by the mechanical oscillator. In the first unitary interaction the optical phase quadrature is displaced by a factor proportional to the mechanical position and the mechanical momentum is displaced by a factor proportional to the optical amplitude, i.e.,

$$\hat{Y}' = \hat{Y} + \lambda\hat{Q} \tag{4.106a}$$
$$\hat{P}' = \hat{P} + \lambda\hat{X}, \tag{4.106b}$$

where here λ is a proportionality constant that quantifies the strength of the interaction, and – since we treat ponderomotive squeezing without moving into a rotating frame for the mechanical oscillator – we have returned to our usual notation, with $\hat{X}$ and $\hat{Y}$ representing the optical amplitude and phase quadratures, and $\hat{Q}$ and $\hat{P}$ representing the mechanical position and momentum. The optical amplitude quadrature and mechanical position are left unaffected by the interaction.

If the mechanical oscillator is then allowed to evolve for a quarter of a cycle, the perturbed momentum rotates into a position, $\hat{Q}'' = -\hat{P}'$. A second interaction of the same form as Eq. (4.106) then results in the final optical quadratures

$$\hat{X}'' = \hat{X} \tag{4.107a}$$
$$\hat{Y}'' = \hat{Y} + \lambda\left(\hat{Q} - \hat{P}\right) - \lambda^2\hat{X}. \tag{4.107b}$$

We see that this series of interactions imprints the optical amplitude quadrature onto the optical phase quadrature, introducing a correlation, and also imprints the original mechanical position and momentum operators on the light. It is introducing exactly the intensity (or amplitude) dependent phase shift we expect of an optical Kerr nonlinearity. This shears the distribution of quantum noise on the optical field in a similar way to the illustration in Fig. 4.17.

Given the correlations between $\hat{X}''$ and $\hat{Y}''$, there will exist some rotated quadrature of the optical field $\hat{Y}''^\theta$ where $\hat{X}$ is eliminated.

Exercise 4.22 *Using the definition of the rotated operator in Eq. (1.17b), which applies equally well to the phase quadrature operator as to the dimensionless momentum operator, show that $\hat{X}$ is eliminated on the $\hat{Y}''$ quadrature when* $\tan\theta = -\lambda^2$, *and that choosing this quadrature angle,*

$$\hat{Y}''^\theta = \frac{\hat{Y} + \lambda\left(\hat{Q} - \hat{P}\right)}{\sqrt{\lambda^4 + 1}}. \tag{4.108}$$

It is evident from this expression that, even including the mechanical position and momentum fluctuations, as the interaction strength $\lambda \to \infty$, the variance of $\hat{Y}''^\theta$ will approach zero, in principle allowing perfect optical squeezing.

4.5.2 Understanding ponderomotive squeezing via the polaron transformation

The intuition that a continuous optomechanical interaction can be thought of, from the perspective of the light, as a Kerr nonlinearity can be put on a solid foundation by making use of the polaron transformation to diagonalise the optomechanical Hamiltonian $\hat{H}$. This approach is common in condensed matter physics where it is useful to treat problems involving linear coupling between electrons and phonons (see, for example, [187]). The essential idea is to apply a displacement to the mechanical oscillator that corrects for the shift in the mechanical oscillator equilibrium position due to the interaction with the optical field.

The magnitude of the displacement due to the optomechanical interaction may be found, for example, by completing the square on the full (nonlinearised) Hamiltonian of Eq. (2.18).

Exercise 4.23 *Show that, to third order in the operators,*[19] *Eq. (2.18) may be re-expressed as*

$$\hat{H} = \hbar \Delta a^\dagger a + \frac{\hbar \Omega}{2} \left[\hat{P}^2 + \left(\hat{Q} + \frac{\sqrt{2} g_0}{\Omega} a^\dagger a \right)^2 \right]. \tag{4.109}$$

This expression makes clear that the radiation pressure interaction displaces the dimensionless mechanical position by $-\sqrt{2} g_0 a^\dagger a / \Omega$, consistent with our mean field observations from Chapter 2.

4.5.2.1 Polaron transformation

In the polaron transformation, one cancels this displacement by applying an opposite but equivalent displacement via the unitary operator

$$\hat{S} = \exp \left[i \frac{\sqrt{2} g_0}{\Omega} a^\dagger a \hat{P} \right]. \tag{4.110}$$

The Hamiltonian of Eq. (2.18) is then transformed to

$$\hat{\tilde{H}} \equiv \hat{S}^\dagger \hat{H} \hat{S} \tag{4.111}$$

$$= \hbar \Delta a^\dagger a - \hbar \chi_0 \left(a^\dagger a \right)^2 + \frac{\hbar \Omega}{2} \left(\hat{Q}^2 + \hat{P}^2 \right), \tag{4.112}$$

---

[19] By this we mean neglecting any terms that involve the product of more than three operators.

where we have defined the *single-photon optical frequency shift*

$$\chi_0 \equiv \frac{g_0^2}{\Omega}, \tag{4.113}$$

and identify operators in the polaron frame via the bar accent. As we will see in Chapter 6, this is an important parameter for single-photon optomechanics.

Exercise 4.24 *Show that*

$$\hat{\bar{Q}} \equiv \hat{S}^\dagger \hat{Q} \hat{S} = \hat{Q} - \frac{\sqrt{2}g_0}{\Omega} a^\dagger a, \tag{4.114}$$

making use of the Hadamard lemma

$$e^{\hat{A}} \hat{B} e^{-\hat{A}} = \hat{B} + [\hat{A}, \hat{B}] + \frac{1}{2!}[\hat{A}, [\hat{A}, \hat{B}]] + \frac{1}{3!}[\hat{A}, [\hat{A}, [\hat{A}, \hat{B}]]] + \dots$$

Then derive the polaron transformed Hamiltonian of Eq. (4.112) from Eq. (2.18), using the general property of a unitary operator $\hat{U}$ that $\hat{U}^\dagger \hat{A} \hat{B} \hat{U} = \hat{U}^\dagger \hat{A} \mathbb{1} \hat{B} \hat{U} = \hat{U}^\dagger \hat{A} \hat{U} \hat{U}^\dagger \hat{B} \hat{U}$, where $\mathbb{1}$ is the identity operator.

This new Hamiltonian makes a number of things transparent about the optomechanical interaction. Most particularly, we see that the dynamics of the mechanical oscillator and intracavity field are now independent. The optomechanical interaction term in Eq. (2.18) has been replaced with the photon-photon interaction term $\chi_0 \left(a^\dagger a\right)^2$ characteristic of a pure optical Kerr nonlinearity with no dependence on the dynamics of the mechanical oscillator. Notice that this term is proportional to g_0 squared, since the effective photon-photon interaction involves two interactions with the mechanical oscillator, one to drive the motion of the oscillator and the other to transduce the motion back onto the optical field.[20] This conversion of an optomechancal nonlinearity to a purely optical one was the primary purpose of the polaron transformation. In the absence of other nonlinear terms in the Hamiltonian, it is well known that the optical Kerr nonlinearity is capable of generating squeezed states of light [328, 263, 260] and even, in principle, Schrödinger's cat-like states [329].

4.5.2.2 Squeezing action in the polaron picture

We can understand why a term proportional to $(a^\dagger a)^2$ will generate squeezing of the optical field by expanding and linearising this term, replacing $a \to \alpha + a$ and neglecting terms that are lower than second order in α:

$$\left(a^\dagger a\right)^2 \to \alpha^4 + \alpha^2 + 2\alpha^3(a^\dagger + a) + 4\alpha^2 a^\dagger a + \alpha^2 \left(a^{\dagger 2} + a^2\right). \tag{4.115}$$

The first two terms in this expansion are static and have no effect on the dynamics of a, the third is a displacement, and the fourth a frequency shift.

---

[20] As we saw in our toy example in Section 4.5.1.

The last term can be seen to produce correlated pairs of photons within the cavity. It is this term that is responsible for squeezing the intracavity field. Using Eqs. (1.17)[21] we can reexpress

$$a^{\dagger 2} + a^2 = -4\hat{X}^{\pi/4}\hat{Y}^{\pi/4} \tag{4.116}$$

within a constant, where $\hat{X}^{\pi/4}$ and $\hat{Y}^{\pi/4}$ are quadrature operators that are rotated by $\pi/4$ from the amplitude and phase quadratures. Setting the overall cavity detuning to zero, including the shift from Eq. (4.115), and neglecting the static and displacement terms from Eq. (4.115), the linearised Hamiltonian describing the intracavity field is then simply

$$\hat{\bar{H}} = 4\hbar\chi\hat{X}^{\pi/4}\hat{Y}^{\pi/4}, \tag{4.117}$$

where $\chi \equiv \chi_0\alpha$ is the coherent amplitude boosted optical frequency shift. This is the classical Hamiltonian for parametric squeezing.

Exercise 4.25 *Using the quantum Langevin equation of Eq. (1.112) show that this Hamiltonian acts to amplify the $\hat{X}^{\pi/4}$ quadrature while deamplifying (and therefore squeezing) the $\hat{Y}^{\pi/4}$ quadrature.*

4.5.2.3 Interaction with the coherent drive and the optical and mechanical baths

While the above discussion appears to suggest that, by performing a polaron transformation, we have found an exceptionally simple method to model ponderomotive squeezing, unfortunately this is not the case. While the polaron transform diagonalises the system Hamiltonian, it introduces coupling between the mechanical oscillator and the optical field through its effect on the drive and system-bath interaction Hamiltonians. This can be seen straightforwardly by applying the polaron transformation to the drive term in the Hamiltonian of Eq. (2.35).

Exercise 4.26 *Using the same approach as in Exercise 4.24 show that*

$$\bar{a} \equiv \hat{S}^{\dagger}a\hat{S} = a\exp\left[i\frac{\sqrt{2}g_0}{\Omega}\hat{P}\right], \tag{4.118}$$

and therefore that, in the polaron frame, the drive term in Eq. (2.35) is dependent on the mechanical momentum quadrature $\hat{P}$.

Similarly, applying the polaron transformation to the system-bath terms in Eqs. (1.69) and (1.111) one finds that, in the polaron frame, the system-bath coupling terms introduce a new optomechanical interaction, with the coupling

---

[21]These equations can be applied to the quadrature operators as well as the dimensionless position and momentum, with $\hat{Q} \to \hat{X}$ and $\hat{P} \to \hat{Y}$.

rate between the optical field and its bath dependent on the mechanical position. This is a form of dissipative optomechanical coupling, as discussed in Section 2.8. A linear coupling is also introduced between the intracavity optical field and the mechanical bath. This coupling is direct – i.e., it is not mediated by the mechanical oscillator.

Overall, these effects mean that, while the polaron transformation is useful for understanding the unitary dynamics of ponderomotive squeezing, it does not yield significant benefits for modelling the nonunitary dynamics. We return to the polaron transformation in Chapter 6, where it proves to be highly useful for studying quantum optomechanics at the single-photon level. Henceforth in this section, we use the usual optomechanical Hamiltonian.

4.5.3 Squeezing spectra

In Sections 3.2 and 3.3 we examined radiation pressure shot noise heating of a mechanical oscillator and the standard quantum limit to measurements of mechanical motion using the linearised optomechanical Hamiltonian of Eq. (2.41) in the zero detuning limit. However, in those sections we examined only the effect of the quantum noise of the light on the temperature of the mechanical oscillator and the information contained on the phase quadrature of the output optical field about the mechanical motion. We did not look at correlations induced by radiation pressure between the amplitude and phase of the output optical field.

While it is possible to generate ponderomotive squeezing in the general case where $\Delta \neq 0$, here, for simplicity, we again restrict ourselves to the zero detuning case.[22] To examine correlations in the output field, we must express the output optical phase quadrature $\hat{Y}_{\text{out}}$ in terms of the optical and mechanical input fluctuations. This is straightforward to do using Eqs. (3.29b) and (3.12), with the result

$$\hat{Y}_{\text{out}}(\omega) = -\left(\frac{\kappa/2 + i\omega}{\kappa/2 - i\omega}\right) \hat{Y}_{\text{in}}(\omega) + 2\Gamma\chi(\omega)\left[\sqrt{2C_{\text{eff}}}\hat{P}_{\text{in}}(\omega) - 2C_{\text{eff}}\hat{X}_{\text{in}}(\omega)\right],$$

(4.119)

where we have returned to our usual definitions of the optical amplitude and phase quadratures ($\hat{X}$ and $\hat{Y}$, respectively) and dimensionless position and momentum operators ($\hat{Q}$ and $\hat{P}$, respectively). We see that, through the optomechanical interaction, the input fluctuations of the optical amplitude quadrature are imprinted on the output optical phase quadrature, just as we found earlier for the simple model of two discrete and temporally separated interactions (Section 4.5.1). This induces correlations that are at the heart of ponderomotive squeezing. It should be noted that the correlations are enhanced close to the mechanical resonance due to the mechanical susceptibility pre-factor $\chi(\omega)$. Furthermore, as the optomechanical cooperativity C_{eff} in-

---

[22]The case of nonzero detuning displays the same qualitative behaviour.

creases, the contribution from the optical amplitude increases at a faster rate than that from mechanical input fluctuations.

Using Eq. (4.119) in combination with the output optical amplitude given in Eq. (3.29a), an arbitrary output quadrature at phase angle θ may be determined via Eq. (1.17a), with the result

$$\hat{X}_{\text{out}}^{\theta}(\omega) = -\left[\left(\frac{\kappa/2 + i\omega}{\kappa/2 - i\omega}\right)\cos\theta + 4\Gamma C_{\text{eff}}\chi(\omega)\sin\theta\right]\hat{X}_{\text{in}}(\omega) \quad (4.120)$$
$$-\left(\frac{\kappa/2 + i\omega}{\kappa/2 - i\omega}\right)\sin\theta\,\hat{Y}_{\text{in}}(\omega) + 2\Gamma\sqrt{2C_{\text{eff}}}\chi(\omega)\sin\theta\,\hat{P}_{\text{in}}(\omega).$$

The symmetrised power spectral density that would be measured via perfect homodyne detection can then be calculated using Eqs. (1.65) and (1.99) and is given by

$$\bar{S}_{X_{\text{out}}^{\theta} X_{\text{out}}^{\theta}}(\omega) = \frac{1}{2} + 8\Gamma^2|\chi(\omega)|^2\,|C_{\text{eff}}|\left(\bar{n} + |C_{\text{eff}}| + \frac{1}{2}\right)\sin^2\theta$$
$$+\Gamma\,|C_{\text{eff}}|\left(\chi(\omega) + \chi^*(\omega)\right)\sin 2\theta, \quad (4.121)$$

where for simplicity we have taken the output field to be shot noise limited ($\bar{n}_L = 0$). The first term is the original quantum noise on the quadrature in the absence of any optomechanical interaction. The second term is a form of mechanical heating proportional to the variance of the mechanical position in the presence of radiation pressure driving and is always positive. The third and final term is the correlation term responsible for ponderomotive squeezing and gives the power spectral density a Fano-like shape. For the output quadrature to exhibit quantum squeezing below the shot noise level ($1/2$), this term must be negative and have a magnitude larger than the second term. This leads to the necessary and sufficient condition for quantum squeezing ($\bar{S}_{X_{\text{out}}^{\theta} X_{\text{out}}^{\theta}}(\omega) < 1/2$),

$$\bar{n} + |C_{\text{eff}}| + \frac{1}{2} < \frac{\Omega^2 - \omega^2}{2\Gamma\Omega\tan\theta}, \quad (4.122)$$

where we have made use of the definition of the mechanical susceptibility given in Eq. (1.102). Thus we see that quantum squeezing cannot occur exactly on the mechanical resonance ($\omega = \Omega$). By contrast, at all other frequencies quantum squeezing is always present for some range of phase angles θ, since for any frequency $\omega \neq \Omega$ the right-hand side of Eq. (4.122) goes to infinity as $\theta \to 0$ from either above or below. We can further observe that, for a given phase angle θ, squeezing will only exist on one side of the mechanical resonance frequency, with the side showing squeezing dictated by the sign of $\tan\theta$. The ponderomotive squeezing predicted from Eq. (4.121) is plotted for a range of parameters in Fig. 4.18a and b, showing this asymmetric frequency response around the mechanical resonance.

Minimising Eq. (4.121) over θ yields the optimal angle θ_{opt} to achieve

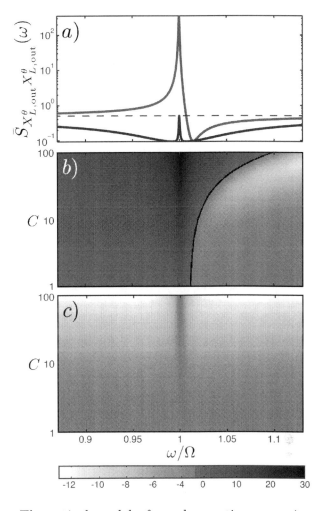

FIGURE 4.18 Theoretical model of ponderomotive squeezing, using the parameters $\kappa/\Omega = 10$, $\bar{n} = 10$, and $Q = 1,000$. (a) Squeezing as a function of frequency ω for an optomechanical cooperativity of $C = 50$. Top trace: power spectral density of the output quadrature $\bar{S}_{X^\theta_{L,\mathrm{out}}X^\theta_{L,\mathrm{out}}}(\omega)$ with phase angle $\theta = \pi/25$. Bottom trace: power spectral density choosing the optimal phase angle $\theta_{\mathrm{opt}}(\omega)$ for each frequency. Dashed line: shot noise level. (b) $10\log_{10}\{\bar{S}_{X^\theta_{L,\mathrm{out}}X^\theta_{L,\mathrm{out}}}(\omega)\}$ as a function of frequency and optomechanical cooperativity for $\theta = \pi/25$. Solid trace: contour of $\bar{S}_{X^\theta_{L,\mathrm{out}}X^\theta_{L,\mathrm{out}}}(\omega){=}1/2$. (c) $10\log_{10}\{\bar{S}_{X^\theta_{L,\mathrm{out}}X^\theta_{L,\mathrm{out}}}(\omega)\}$ as a function of frequency and optomechanical cooperativity choosing the optimal phase angle $\theta_{\mathrm{opt}}(\omega)$.

maximum squeezing, given by

$$\tan 2\theta_{\text{opt}} = \frac{\omega^2 - \Omega^2}{2\Gamma\Omega\left(n + |C_{\text{eff}}| + 1/2\right)}. \tag{4.123}$$

We see that the squeezing angle rotates as a function of frequency, crossing zero at the mechanical resonance frequency. The optimal squeezing at each frequency is shown for a range of parameters in Fig. 4.18a and c. We observe that, indeed, no squeezing is possible on the mechanical resonance, while broadband squeezing is achieved, in principle, over all other frequencies, with the level of squeezing increasing with effective cooperativity. The ponderomotive squeezing is resonantly enhanced near the mechanical resonance frequency, such that substantial levels of squeezing typically only exist in a relatively narrow band of frequencies. This contrasts techniques to generate squeezed light using standard nonlinear optical materials which are able to routinely generate broadband squeezing.

As we will see in Section 5.4.3, the ability to generate squeezing via the optomechanical interaction provides one path to overcome the standard quantum limit introduced in Section 3.4. However, as will be discussed in that section, the squeezing angle rotation evident in Eq. (4.123) complicates this approach to sub-standard quantum limited measurement, resulting in noise suppression only in a small band of frequencies unless specialised measurement techniques referred to as *variation measurements* are employed (see Fig. 5.9).

Ponderomotive squeezing was first proposed in 1994 [103, 189] and first demonstrated in 2012 [52]. It has been observed in several optomechanical architectures including intracavity cold atom ensembles [52], silicon nitride membrane optomechanical systems [232], and optomechanical zipper cavities [245]. Figure 4.19 shows an optomechanical zipper cavity and the observed optical power spectrum for a fixed homodyne phase angle. It can be observed that the power spectrum exhibits the expected Fano-like feature, with a small region of quantum squeezing observed at frequencies just below the mechanical resonance frequency.

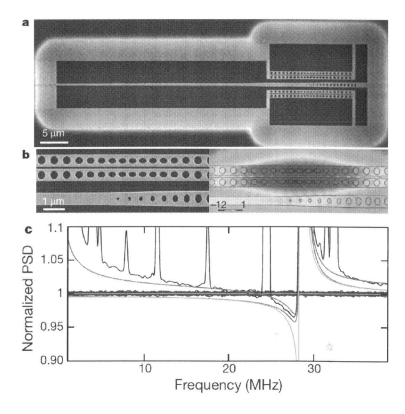

FIGURE 4.19 Ponderomotive squeezing observed from an optomechanical zipper cavity. Adapted by permission from Macmillan Publishers Ltd: *Nature* [245], copyright 2013. (a) and (b) Scanning electron micrscope images and optical modelling of the zipper cavity structure. (c) Optical power spectral density observed for one homodyne phase angle and normalised to the shot noise level. PSD: power spectral density of output field.

CHAPTER 5

Linear quantum control of mechanical motion

CONTENTS

5.1   Stochastic master equation including dissipation 152

     5.1.1   Measured photocurrent 153

     5.1.2   Itō calculus 154

     5.1.3   Evolution of the density matrix 155

         5.1.3.1   Evolution of operator expectation values 155

         5.1.3.2   Interaction with the environment 155

         5.1.3.3   Simultaneous measurement of multiple observables 156

5.2   Feedback cooling 156

     5.2.1   Quadrature photocurrents 157

     5.2.2   Stochastic master equation 159

     5.2.3   Evolution of an oscillator under continuous position measurement 159

         5.2.3.1   Evolution of the quadrature expectation values 159

         5.2.3.2   Evolution of the variances and covariance 160

         5.2.3.3   Steady-state solution 161

     5.2.4   Obtaining estimates for the quadratures 163

     5.2.5   Applying feedback to cool the oscillator 164

5.3   Back-action evading measurement 166

     5.3.1   Momentum measurement on a free mass 167

     5.3.2   General back-action evading measurement 167

     5.3.3   Two-tone driving 169

         5.3.3.1   Hamiltonian 169

         5.3.3.2   Evolution of mechanical quadratures ... 171

         5.3.3.3   Mechanical quadrature measurement rates and conditional squeezing 172

    5.3.4   Detuned parametric amplification 173
          5.3.4.1   Parametric amplification 175
          5.3.4.2   Back-action evasion 176
          5.3.4.3   Conditional squeezing 178
5.4   Surpassing the standard quantum limit using squeezed
    light 179
    5.4.1   Sub-standard quantum limit position measurement
          on a free mass 179
    5.4.2   Squeezed input field 183
    5.4.3   Ponderomotive squeezing 189

Building on Chapters 3 and 4, this chapter introduces the concepts of measurement-based quantum control of mechanical systems in the linearised approximation for the optomechanical interaction, including techniques such as feedback cooling and back-action evading measurements using both stroboscopic optical driving and parametric mechanical driving. To model these techniques we introduce the stochastic master equation approach. We then discuss the application of squeezed light fields and ponderomotive optical squeezing to overcome the standard quantum limit introduced in Chapter 3. These methods are important both to improve the precision of quantum sensors and for quantum state tomography of mechanical systems.

5.1   STOCHASTIC MASTER EQUATION INCLUDING DISSIPATION

In Chapters 3 and 4 our analysis was performed primarily in the Heisenberg picture, using quantum Langevin equations. While it is possible to model feedback cooling and back-action evading techniques this way (for example, see [81]), an alternative and more versatile approach to model continuous measurements on quantum systems is to use stochastic master equations in the Schrödinger picture. Stochastic master equation methods have the substantial benefit of straightforward extension to scenarios where the Hamiltonian or measurement is nonlinear. Consequently, as well as applying them here to feedback cooling and back-action evasion, we use them to model nonlinear optomechanical processes such as phonon counting and synchronisation in later chapters.

It is of course well known that measurements on a quantum system project the system into a new quantum state which depends on the outcome of the measurement. The result of a noise-free, or *von Neumann*, measurement is to project the state onto one of the eigenstates of the observable. However, in practice no measurement is entirely noise free. Realistic measurements act to reduce – rather than eliminate entirely – the uncertainty of the observable being measured. This wider, indeed complete, class of measurements are referred to as *positive operator-valued measures* (or POVM).

To derive a general stochastic master equation for the evolution of a quantum system under continuous Gaussian measurement, one considers a sequence of POVMs performed on the system and takes the limit of asymptotically weak measurements separated by an infinitesimally small time. This approach has been well treated in several textbooks and review papers (we recommend [143, 318]), so we state only the results here.

5.1.1 Measured photocurrent

We consider continuous measurement of a system observable $\hat{c}$ with Gaussian-distributed noise. The output classical photocurrent $i(t)$ from the measurement can then be modelled as a Wiener process such as is commonly used to model Brownian motion:

$$i(t) = \frac{dq}{dt}, \qquad (5.1)$$

where

$$dq = \sqrt{\eta}\,\langle\hat{c}\rangle(t)\,dt + dW(t), \qquad (5.2)$$

and here q is the electronic charge flowing through a circuit, not to be confused with $\hat{q}$ the mechanical position operator. $\langle\hat{c}\rangle(t)$ is the expectation value of $\hat{c}$ at time t given the known unitary dynamics of the system and the prior measurement record, and $dW(t)$, termed the *infinitesimal Wiener increment* or *innovation*, quantifies the deviation of $i(t)$ from this expected value over a time interval dt [110, 92]. The Wiener increment may be defined in terms of an ideal unit Gaussian white noise process $\xi(t)$ via

$$dW(t) \equiv W(t + dt) - W(t) = \xi(t)dt, \qquad (5.3)$$

where $\langle\langle\xi(t)\xi(t')\rangle\rangle = \delta(t - t')$, and the double angle brackets signify the ensemble average over all realisation of the Wiener process. Note that, in this definition, since $\xi(t)$ has infinite bandwidth, the derivative of $W(t)$ is singular.[1] In reality, of course, the finite bandwidth of detection processes removes this singularity.

Wiener increments have the following properties:

$$\langle\langle dW(t)\rangle\rangle = 0 \qquad (5.4a)$$

$$\langle\langle dW^2(t)\rangle\rangle = dW^2(t) = dt \qquad (5.4b)$$

$$\langle\langle dW(t)dW(t')\rangle\rangle = dW(t)dW(t') = 0 \text{ for } t' \neq t. \qquad (5.4c)$$

It may seem strange that in Eqs. (5.4b) and (5.4c) we equate a deterministic ensemble average with the square of a classical stochastic variable. Indeed, this is not particularly precise. More strictly, Wiener increments are only ever evaluated within integrals, and within an integral one may always make the substitutions in Eqs. (5.4b) and (5.4c).

---

[1] This is why we do not write $\frac{dW}{dt}$ here.

Exercise 5.1 *Consider the following Itō stochastic differential equation describing a simple Ornstein–Uhlenbeck process,*

$$dx = -\gamma dt + B dW(t) \qquad (5.5)$$

where dW is the Wiener increment described above and B is a constant.

i. Show that the solution to this equation is

$$x(t) = x(0)e^{-\gamma t} + B \int_0^t e^{-\gamma(t-t')} dW(t'). \qquad (5.6)$$

ii. Using the Itō calculus integration rule

$$\int_0^{t_1} \int_0^{t_2} \langle\!\langle dW(t) dW(t') \rangle\!\rangle f(t) g(t') = \int_0^{min(t_1,t_2)} dt\, f(t) g(t) \qquad (5.7)$$

show that the stationary two-time correlation function is given by

$$\lim_{t\to\infty} \langle\!\langle x(t+\tau)x(t) \rangle\!\rangle = \frac{B^2}{2\gamma} e^{-\gamma\tau} \qquad (5.8)$$

with $\tau \geq 0$.

5.1.2 Itō calculus

We may already gather from Eqs. (5.4) that stochastic calculus will obey different rules than regular deterministic calculus. Many of these rules were formulated by Kiyoshi Itō, including the identification $dW^2 = dt$ in Eq. (5.4b), which is termed the *Itō rule*. Stochastic calculus is often therefore referred to as *Itō calculus*. In *Itō calculus* integrals can be defined analogously to regular calculus. However, some ambiguity enters in their exact form, leading to two common versions named after Itō and Ruslan Stratonovich, respectively. This ambiguity arises from the fact that, while in regular calculus one neglects terms higher than first order in the increment dt, in Itō calculus second-order terms in the increment dW must be retained since – as can be seen from the Itō rule – they are first order in dt. Here, we use Itō integrals, which require the chain rule of calculus to be modified.[2] Stratonovich integrals, on the other hand, retain the usual chain rule and therefore the usual rules of calculus, but introduce correlations between the integrand and the increment. The reader is referred to [110] for a good introduction to Itō calculus, and a detailed discussion of Itō and Stratonovich integrals.

---

[2] The new chain rule is termed *Itō's lemma*.

5.1.3 Evolution of the density matrix

The evolution of the density matrix of a system interacting with its environment and undergoing continuous measurement of the observable $\hat{c}$ can be described by the stochastic master equation [318, 143]

$$d\rho = \frac{1}{i\hbar}\left[\hat{H},\rho\right]dt + \mathcal{L}_{\mathrm{env}}\rho\, dt + \mathcal{D}[\hat{c}]\rho\, dt + \sqrt{\eta}\,\mathcal{H}[\hat{c}]\rho\, dW \qquad (5.9)$$

where the Liouvillian superoperator $\mathcal{L}_{\mathrm{env}}$ describes the system-environment interaction, η is the detection efficiency, and $\mathcal{D}$ and $\mathcal{H}$ are the Lindblad and measurement superoperators defined by

$$\mathcal{D}[\hat{c}]\rho \equiv \hat{c}\rho\hat{c}^\dagger - \frac{1}{2}\left(\hat{c}^\dagger\hat{c}\rho + \rho\hat{c}^\dagger\hat{c}\right) \qquad (5.10a)$$

$$\mathcal{H}[\hat{c}]\rho \equiv \hat{c}\rho + \rho\hat{c}^\dagger - \left\langle\hat{c}+\hat{c}^\dagger\right\rangle\rho. \qquad (5.10b)$$

This master equation was first introduced, albeit without the environmental coupling term, by Belavkin [29]. Here $\rho(t)$ describes a *quantum trajectory* that the system follows in one experimental iteration with measurement record $dW(t)$. The first term in Eq. (5.9) describes the Hamiltonian evolution of the system, the second describes the influence of the environment, the third describes the back-action noise that is introduced as a result of the measurement, and the final term describes the conditioning of the system based on information gained through the measurement. Combined, the effect of measurement back-action and conditioning is to attempt to drive the state of the system into an eigenstate of the measurement operator $\hat{c}$. It should be recognised that, when referring to the "state of the system" here, we really mean the conditional state given the knowledge the observer has gained during the course of the measurement.

5.1.3.1 Evolution of operator expectation values

Using Eq. (5.9) the evolution of the expectation value of an arbitrary system operator $\hat{A}$ can be determined in the usual way via

$$d\langle\hat{A}\rangle = \mathrm{Tr}\left[\hat{A}\,d\rho\right], \qquad (5.11)$$

where $\mathrm{Tr}[\hat{A}\,d\rho]$ is the trace.

5.1.3.2 Interaction with the environment

If the environment of the quantum system is a thermal equilibrium bath that is linearly coupled to the system, such as was considered in Section 1.3.1, one can follow a similar approach to Section 1.3.1 to determine the environmental coupling term ($\mathcal{L}_{\mathrm{env}}\rho$) in Eq. (5.9) (see, e.g., [214]). In the Markov and rotating wave approximations we find

$$\mathcal{L}_{\mathrm{env}}\rho = \Gamma(\bar{n}+1)\mathcal{D}[b]\rho + \Gamma\bar{n}\mathcal{D}[b^\dagger]\rho. \qquad (5.12)$$

5.1.3.3 Simultaneous measurement of multiple observables

It is straightforward to model scenarios involving simultaneous measurements of multiple observables $\hat{c}_j$ that have uncorrelated noise ($\langle\langle dW_j dW_k\rangle\rangle = dW_j dW_k = 0$, $j \neq k$) by including additional the measurement and conditioning terms to Eq. (5.9). Specifically, one makes the substitutions

$$\mathcal{D}[\hat{c}]\rho\,dt \quad \rightarrow \quad \sum_j \mathcal{D}[\hat{c}_j]\rho\,dt \tag{5.13a}$$

$$\sqrt{\eta}\,\mathcal{H}[\hat{c}]\rho\,dW \quad \rightarrow \quad \sum_j \sqrt{\eta_j}\,\mathcal{H}[\hat{c}_j]\rho\,dW_j, \tag{5.13b}$$

where η_j is the efficiency of measurement j and dW_j is its associated Wiener increment [143]. While we do not consider correlations between the measurements or with the environment here, we note that a general stochastic master equation has also been derived in [317] that applies in such circumstances.

5.2 FEEDBACK COOLING

Here we apply the stochastic master equation approach introduced in the previous section to feedback cooling, as a first straightforward example of its use in quantum optomechanics. In feedback cooling, a continuous measurement is made of the position of the mechanical oscillator. This measurement allows an estimate of the velocity of the oscillator. Applying a force that opposes this velocity viscously damps and cools the oscillator. To achieve optimal cooling, it is important to understand the best approach to estimate the velocity of a mechanical oscillator and the appropriate filter to use when applying this estimate as a force back upon the mechanical oscillator. The stochastic master equation approach provides this information as well as the final occupancy that may be achieved by the protocol. For a comprehensive early treatment of feedback cooling via the stochastic master equation approach, we recommend [91], and for similar treatments using the quantum Langevin approach we refer the reader to [81, 300]. Here, we follow the treatment given in [92].

Feedback cooling of cavity optomechanical systems was proposed by Stefano Mancini and colleagues in 1998 [190]. It was pioneered experimentally by Antoine Heidmann, Michel Pinard, and co-workers, who demonstrated cooling of a mechanical compressional mode of one end-mirror of a high finesse Fabry–Pérot cavity in the optical domain as early as 1999 [75, 226]. These techniques have since been utilised in a range of other optomechanical architectures, including cryogenic Fabry–Pérot cavities [230], intracavity atomic force microcopy cantilevers [161], whispering-gallery mode optomechanical systems [180, 174, 134, 115, 136], levitated trapped particles [178, 119], the end-mirror of interferometric gravitational wave detectors [2], and cavities with gram-scale mirrors [80, 204].

The ideas of feedback cooling are also important for the related technique of photothermal cooling, demonstrated first in 2004 [199]. In photothermal cooling, a cavity detuning acts to encode mechanical position information onto the amplitude quadrature of the intracavity field. Absorption of some component of this field then heats the bulk material making up the mechanical oscillator, with thermal expansion applying a position-dependent feedback force. With sufficient delay in the thermal response, this can result in a cooling effect [238]. This effect can be exactly modelled as feedback cooling, with the thermal frequency response of the system acting as the filter which determines the position estimate that is fed back onto the mechanical oscillator.

5.2.1 Quadrature photocurrents

To model feedback cooling via a continuous position measurement using the stochastic master equation of Eq. (5.9), we begin by choosing the measurement operator

$$\hat{c} = \sqrt{2\mu}\hat{Q} \tag{5.14}$$

which, from Eq. (5.1), results in the measured continuous photocurrent

$$i(t)\,dt = \sqrt{2\mu\eta}\langle\hat{Q}\rangle(t)\,dt + dW(t). \tag{5.15}$$

In [91] an analytic model for feedback cooling was obtained using Eqs. (5.9), (5.14), and (5.15). Here, however, we wish to simplify the analysis by limiting ourselves to the scenario where the mechanical oscillator is sufficiently high quality and the measurement rate (to be determined later) is sufficiently weak that a continuous position measurement can be viewed as a form of heterodyne measurement of two orthogonal mechanical quadratures of motion $\hat{X}_M$ and $\hat{Y}_M$ that rotate with the resonance frequency of the mechanical oscillator [92]. As we will see, this is a form of rotating wave approximation which is particularly convenient for later sections on back-action evading measurements.

The quantum trajectories approach to heterodyne measurement has been well treated in [255, 318] and extended to the case of continuous position measurement of an oscillator in [92]. We follow those references closely here. From Eqs. (1.17) and (1.135) the dimensionless position operator can be expressed as $\hat{Q} = \cos\Omega t\,\hat{X}_M - \sin\Omega t\,\hat{Y}_M$. The photocurrent of Eq. (5.15) is then

$$i(t)\,dt = \sqrt{2\mu\eta}\left[\cos\Omega t\,\langle\hat{X}_M\rangle(t) - \sin\Omega t\,\langle\hat{Y}_M\rangle(t)\right]dt + dW(t). \tag{5.16}$$

The experimenter can perform filtering operations on this photocurrent to extract – at least in the regime in which the rotating wave approximation is valid – independent estimates for $\hat{X}_M$ and $\hat{Y}_M$. It is obvious from the form of the photocurrent that the correct choice of filter for each quadrature will involve a part oscillating at Ω, with the oscillation phase allowing the two quadratures to be distinguished. We therefore define the quadrature photocurrents as

$$i_X(t) = h(t) * [\cos\Omega t\,i(t)] \tag{5.17a}$$

$$i_Y(t) = -h(t) * [\sin\Omega t\,i(t)], \tag{5.17b}$$

where $*$ is the usual convolution function and $h(t)$ is a causal filter function. Considering

$$\cos \Omega t \, i(t) \, dt = \sqrt{\frac{\mu \eta}{2}} \left[(1 + \cos 2\Omega t) \, \langle \hat{X}_M \rangle (t) + \sin 2\Omega t \, \langle \hat{Y}_M \rangle (t) \right] dt + \cos \Omega t \, dW,$$

$$(5.18)$$

we see that Eq. (5.17a) contains a component near zero frequency that is proportional to $\langle \hat{X}_M \rangle$, and components oscillating at twice the mechanical resonance frequency proportional to $\langle \hat{X}_M \rangle$ and $\langle \hat{Y}_M \rangle$. Within the regime of validity of the rotating wave approximation, the bandwidth of the fluctuations of $\langle \hat{X}_M \rangle$ and $\langle \hat{Y}_M \rangle$ is small compared with the mechanical frequency. It is then possible to choose a filter function $h(t)$ which only passes the low-frequency term and therefore provides a linear estimate of $\langle \hat{X}_M \rangle$.

Let us envision a measurement of duration τ. If τ is simultaneously long compared to the mechanical oscillation period and short enough that $\langle \hat{X}_M \rangle$ and $\langle \hat{Y}_M \rangle$ are essentially stationary over the measurement duration, the optimal filter function is a top-hat function of width τ. The quadrature photocurrents can then be well approximated as [92, 318]

$$i_X(t) \, dt \approx \sqrt{\mu \eta} \langle \hat{X}_M \rangle \, dt + dW_X(t) \qquad (5.19a)$$

$$i_Y(t) \, dt \approx \sqrt{\mu \eta} \langle \hat{Y}_M \rangle \, dt + dW_Y(t), \qquad (5.19b)$$

where dW_X and dW_Y are uncorrelated Wiener increments, i.e., $dW_X(t) \, dW_Y(t') = 0$, and we have chosen the normalisation $\int_{-\infty}^{\infty} h(t) dt = \sqrt{2}$ for convenience later. We see, therefore, that, in the regime of validity of the rotating wave approximation, where the mechanical frequency is much larger than all other rates of the system, a continuous measurement of mechanical position is equivalent to performing simultaneous independent measurements of the quadratures $\hat{X}_M$ and $\hat{Y}_M$, each with efficiency reduced by a factor of two compared to the measurement of position $\hat{Q}$ (compare Eqs. (5.19) with Eq. (5.15)).

As we will see in Section 5.2.4, to optimally estimate the mechanical quadratures from the full measurement record of the quadrature photocurrents in Eqs. (5.19), a second filter $v(t)$ must be applied to each photocurrent, e.g., $\langle \hat{X}_M \rangle = v(t) * i_X(t)$. These filters generate weighted averages of the quadrature measurement records, with weightings chosen to maximise the precision of the estimate. It is useful to note that, while the optimal short-time filter function (a top-hat function) must be selected for $h(t)$ to arrive at the correct weightings for W_X and W_Y in Eqs. (5.19), as long as $v(t)$ varies slowly over the characteristic time scales of $h(t)$, the exact form of $h(t)$ selected by an experimenter[3] is unimportant to the final optimal estimate. This can be shown straightforwardly using the associativity property of the convolution.

---

[3] Or, possibility more accurately, determined by their detector bandwidth and data acquisition system.

5.2.2 Stochastic master equation

Given the discussion in the previous section, we see that, within the rotating wave approximation, it is possible to model feedback cooling via the stochastic master equation of Eq. (5.9), including simultaneous, equal efficiency, measurements of the mechanical X and Y quadratures quantified by the measurement operators $\hat{c}_X = \sqrt{\mu}\,\hat{X}_M$ and $\hat{c}_Y = \sqrt{\mu}\,\hat{Y}_M$. Using Eqs. (5.13), this results in the interaction picture stochastic master equation

$$d\rho = \frac{1}{i\hbar}\left[\hat{H}_I, \rho\right]dt + \Gamma(\bar{n}+1)\mathcal{D}[b]\rho\, dt + \Gamma\bar{n}\mathcal{D}[b^\dagger]\rho\, dt \qquad (5.20)$$
$$+\sqrt{\mu}\left[\mathcal{D}[\hat{X}_M]\rho\, dt + \sqrt{\eta}\,\mathcal{H}[\hat{X}_M]\rho\, dW_X + \mathcal{D}[\hat{Y}_M]\rho\, dt + \sqrt{\eta}\,\mathcal{H}[\hat{Y}_M]\rho\, dW_Y\right],$$

where $\hat{H}_I$ is the Hamiltonian in an interaction picture rotating at the mechanical resonance frequency. In general, only numerical solutions are possible for this stochastic master equation. However, for the specific case of continuous measurement on a harmonic oscillator that we are interested in – and in general for Hamiltonians that are quadratic in position and momentum [142] – as long as the initial state of the system is Gaussian, the statistics remain Gaussian under continuous linear measurement for all time. In these circumstance, the evolution of the mechanical oscillator is fully determined by the means of the position and momentum quadratures and the covariance matrix (see Section 4.4.5), with analytical solutions possible for each term. The assumption of a Gaussian initial state is generally not particularly restrictive [143]. For instance, a mechanical oscillator in thermal equilibrium with its environment will naturally be in such a state. Furthermore, after measuring for a sufficiently long period, the initial state of the system becomes insignificant in determining its future dynamics.

5.2.3 Evolution of an oscillator under continuous position measurement

5.2.3.1 Evolution of the quadrature expectation values

The evolution of the means of the X and Y quadratures can be obtained by applying Eq. (5.11) to the stochastic master equations of Eq. (5.20), with $\hat{A} = \hat{X}_M$ and $\hat{Y}_M$, respectively. Assuming that the Hamiltonian describing the unitary evolution is just that of a harmonic oscillator, so that in the interaction picture $\hat{H}_I = 0$, the results are

$$d\langle\hat{X}_M\rangle = -\frac{\Gamma}{2}\langle\hat{X}_M\rangle dt + 2\sqrt{\eta\mu}\left[V_X dW_X + C_{XY}dW_Y\right] \qquad (5.21\text{a})$$

$$d\langle\hat{Y}_M\rangle = -\frac{\Gamma}{2}\langle\hat{Y}_M\rangle dt + 2\sqrt{\eta\mu}\left[V_Y dW_Y + C_{XY}dW_X\right], \qquad (5.21\text{b})$$

where, as usual, the variances $V_X \equiv \langle\hat{X}_M^2\rangle - \langle\hat{X}_M\rangle^2$ and $V_Y \equiv \langle\hat{Y}_M^2\rangle - \langle\hat{Y}_M\rangle^2$, and the covariance $C_{XY} \equiv \langle\hat{X}_M\hat{Y}_M + \hat{Y}_M\hat{X}_M\rangle/2 - \langle\hat{X}_M\rangle\langle\hat{Y}_M\rangle$.

Exercise 5.2 *Show these results.*

In the absence of measurement ($\mu = 0$) we see that, as expected, both means decay exponentially with time. This is essentially saying that, after a long evolution time, without any measurement, one's best guess is that the oscillator lies close to the origin in phase space. In the presence of measurement, Eqs. (5.21) each describe an Ornstein–Uhlenbeck process, which is the prototypical noisy relaxation process and essentially describes Brownian motion in the presence of friction. The presence of the measurement introduces noise, while information obtained through the measurement allows the oscillator to be localised in phase space, with its expected location determined by its variances and covariances and the measurement records dW_X and dW_Y.

5.2.3.2 Evolution of the variances and covariance

The precision with which the oscillator is localised is determined by its variances and covariances. Some care must be taken to calculate these correctly, taking into account the rules of Itō calculus. As mentioned earlier, where, in usual calculus, terms higher than first order in the increment (e.g., dt^2) are neglected, since $dW^2 = dt$ in Itō calculus, terms of order up to dW^2 must be kept.[4] One consequence of this is a modification of the usual chain rule of calculus to include second-order terms. Taylor expansion of the function $f(x, y)$ can be used to obtain the new two-variable chain rule

$$df(x, y) = \frac{\partial f}{\partial x}dx + \frac{\partial f}{\partial y}dy + \frac{1}{2}\frac{\partial^2 f}{\partial x^2} + \frac{1}{2}\frac{\partial^2 f}{\partial y^2}dy^2 + \frac{1}{2}\frac{\partial^2 f}{\partial x \partial y}dxdy. \tag{5.22}$$

Exercise 5.3 *Use the chain rule above to show that the increments of the variance V_X and covariance C_{XY} are*

$$dV_X = d\langle \hat{X}_M^2 \rangle - 2\langle \hat{X}_M \rangle d\langle \hat{X}_M \rangle - \left(d\langle \hat{X}_M \rangle \right)^2 \tag{5.23a}$$

$$dC_{XY} = \frac{1}{2}d\langle \hat{X}_M \hat{Y}_M + \hat{Y}_M \hat{X}_M \rangle - \langle \hat{Y}_M \rangle d\langle \hat{X}_M \rangle - \langle \hat{X}_M \rangle d\langle \hat{Y}_M \rangle$$
$$- d\langle \hat{X}_M \rangle d\langle \hat{Y}_M \rangle. \tag{5.23b}$$

After determining each of the relevant moments by applying Eq. (5.11) to Eq. (5.20), the above relations, along with an equivalent relation to Eq. (5.23a) for V_Y, allow the evolution of the covariance matrix of a harmonic oscillator under continuous measurement to be fully determined. The result, which you

---

[4]The rule is that terms of the form $dt^j dW^k$ can be neglected as long as $j + k/2 > 1$.

should convince yourself of, is

$$\dot{V}_X = -\Gamma V_X + \Gamma \left(\bar{n} + \frac{1}{2}\right) + \mu - 4\mu\eta \left(V_X^2 + C_{XY}^2\right) \quad (5.24a)$$

$$\dot{V}_Y = -\Gamma V_Y + \Gamma \left(\bar{n} + \frac{1}{2}\right) + \mu - 4\mu\eta \left(V_Y^2 + C_{XY}^2\right) \quad (5.24b)$$

$$\dot{C}_{XY} = -\left[\Gamma + 4\mu\eta \left(V_X + V_Y\right)\right] C_{XY} \quad (5.24c)$$

where we have neglected terms of order higher than dt and dW^2, and used the Itō rule as well as the following properties of Gaussian states [143]:

$$\langle \hat{X}_M^3 \rangle = 3\langle \hat{X}_M^2 \rangle V_X \quad (5.25a)$$

$$2\langle \hat{X}_M \rangle C_{XY} + \langle \hat{Y}_M \rangle \langle \hat{X}_M^2 \rangle = \langle \hat{X}_M \hat{Y}_M \hat{X}_M \rangle \quad (5.25b)$$

and equivalent expressions with $\hat{X}_M \leftrightarrow \hat{Y}_M$. Notice that these variances and covariance are independent of the measurement records W_X and W_Y. This is a result of the linearity of the system. We see from Eqs. (5.24) that, as might be expected, in the absence of measurement ($\mu = 0$) the variances and covariance all decay at rate Γ, while the variances both experience a balancing heating effect from the environment. The measurement introduces equal additional back-action heating to both quadrature variances, which arises due to radiation pressure noise, as we found in Section 3.2, as well as a nonlinear conditional damping effect, which also appears on the covariance, due to the knowledge gained from the measurement results.

It can be seen by inspection of Eqs. (5.24a) and (5.24b) that, without taking advantage of the measurement results (i.e., setting $\eta = 0$) the act of measurement raises the equilibrium phonon occupancy of the oscillator by μ/Γ. In Section 3.2 we identified this increase in phonon occupancy with the effective optomechanical cooperativity (Eq. (3.20)). This allows the measurement rate μ to be established in terms of measurable parameters:

$$\mu = \Gamma \left|C_{\text{eff}}\right|. \quad (5.26)$$

5.2.3.3 Steady-state solution

Taking the steady-state solutions of Eqs. (5.24) by setting the time derivatives to zero, it is easy to show that the covariance $C_{XY} = 0$, while the quadrature variances are

$$V_X = V_Y = \sqrt{\left(\frac{1}{8\eta \left|C_{\text{eff}}\right|}\right)^2 + \frac{2\bar{n} + 1}{8\eta \left|C_{\text{eff}}\right|} + \frac{1}{4\eta}} - \frac{1}{8\eta \left|C_{\text{eff}}\right|}. \quad (5.27)$$

Exercise 5.4 *Derive this expression.*

As discussed in [92] and illustrated in Fig. 5.1, this expression has three significant regimes, which we term the *back-action dominated regime*, the *classical measurement regime*, and the *weak measurement regime*.

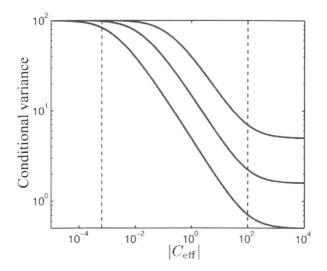

FIGURE 5.1 Conditional quadrature variances of a mechanical oscillator under continuous measurement as a function of the effective cooperativity. Here $\bar{n} = 100$, and $\eta = \{1, 0.1, 0.01\}$ for the lowermost to uppermost traces, respectively. The vertical dashed lines at $|C_{\text{eff}}| = 1/16\bar{n}$ and $|C_{\text{eff}}| = \bar{n}$ correspond to the thresholds between the weak measurement, classical measurement, and back-action dominated regimes, respectively, for $\eta = 1$.

In the back-action dominated regime $\{1/16\eta|C_{\text{eff}}|^2, (\bar{n} + 1/2)/|C_{\text{eff}}|\} \ll 1$, and the quadrature variances asymptote to

$$V_X = V_Y = \frac{1}{2\sqrt{\eta}}. \tag{5.28}$$

Here, the oscillator is localised by the measurement much faster than the delocalisation caused by the thermal bath, and the final quadrature variances are determined solely by the trade-off between measurement back-action and measurement conditioning. In the limit that the measurement has perfect efficiency, $V_X = V_Y = 1/2$, and the oscillator is localised to a pure Gaussian state with minimum uncertainty, i.e., it is conditioned into a zero entropy state, though at this stage it is not cooled since the mean quadrature values are nonzero and determined by the measurement records as dictated by Eqs. (5.21). The back-action dominated regime has been reached in the experiment of [312] where, by applying feedback similar to that discussed in Section 5.2.5, a feedback cooled mechanical occupancy of $\bar{n}_b = 5.3 \pm 0.6$ was achieved.

In the classical measurement regime, the thermal noise of the mechanical oscillator is resolved by the measurement but the localisation from the measurement does not occur much faster than delocalisation due to the thermal bath, with $1/16\eta(\bar{n}+1/2) \ll |C_{\text{eff}}| \ll \bar{n}+1/2$. The quadrature variances then become

$$V_X = V_Y = \sqrt{\frac{2\bar{n}+1}{8\eta\,|C_{\text{eff}}|}}. \tag{5.29}$$

Here, we see that the act of measurement reduces the variances below the thermal variance, with the localisation proportional to the inverse square root of the effective optomechanical cooperativity.

In the weak measurement regime, the measurement is insufficient even to resolve the thermal motion of the oscillator, with $1/16\eta|C_{\text{eff}}|^2 \gg \{1 + (\bar{n} + 1/2)/|C_{\text{eff}}|\}$. In this case, the quadrature variances are

$$V_X = V_Y = \bar{n} + |C_{\text{eff}}| + \frac{1}{2}, \tag{5.30}$$

with the measurement acting only, through the measurement back-action, to raise the temperature of the oscillator consistent with the result from Section 3.2.

It is interesting to observe that it is only in the far back-action dominated regime, where the unconditional temperature of the oscillator is increased well above the ground state, that a near-zero entropy conditional state can be achieved. So while, as we saw in Chapter 3, the standard quantum limit defines an optimal – noninfinite – interaction strength to achieve the best precision in measurements of the oscillator position or external forces using coherent light, the performance of conditioning and feedback cooling continue to improve indefinitely as the interaction strength increases.

Note that there is one further regime of interest that we have neglected in taking the rotating wave approximation. In the rotating wave approximation, the model is limited to not extend into the regime where the measurement is capable of resolving the zero-point motion of the oscillator in a time short compared with the mechanical oscillation period. This regime – which requires very strong optomechanical coupling – exhibits interesting behaviour, including dynamical mechanical squeezing. The reader is referred to [91] for an analysis valid in this regime.

5.2.4 Obtaining estimates for the quadratures

In the previous discussion we derived conditional variances for the quadratures of a mechanical oscillator undergoing continuous measurement. However, these variances can only be achieved in practise if the quadrature expectation values are optimally estimated from the measurement records. Using Eqs. (5.21) and (5.24), a process can be established to determine these optimal estimates. This is particularly simple in the long-time limit where the variances and covariance have reached steady state. In this case substituting for dW_X and dW_Y in Eqs. (5.21) in terms of the quadrature photocurrents i_X and i_Y from Eq. (5.19) and taking the Fourier transform, recognising that $C_{XY} = 0$ in the steady state, provides the algebraic relationship

$$\langle \hat{X}_M \rangle (\omega) = \left(\frac{2\sqrt{\eta\mu}}{\Gamma/2 + 2\eta\mu V_X - i\omega} \right) i_X(\omega) \qquad (5.31)$$

between $\langle \hat{X}_M \rangle$ and i_X here expressed in the frequency domain, and an identical relationship for $\langle \hat{Y}_M \rangle$ but with the subscript $X \to Y$. We see that, to obtain $\langle \hat{X}_M \rangle$, a simple Lorentzian filter should be applied to the X-quadrature photocurrent, with width given by $\Gamma + 4\eta\mu V_X$, where V_X is defined in the steady state in Eq. (5.27). Notice that the width of the filter increases as the measurement strength increases. This is reflective of the fact that, as the measurement strength increases, the thermal noise of the mechanical oscillator exceeds the measurement noise over a greater bandwidth, such that information about the mechanical quadratures is available over a broader range of frequencies.

5.2.5 Applying feedback to cool the oscillator

To feedback cool the oscillator, one takes the optimal estimates of the quadrature expectation values in Eq. (5.31) and applies them, via a feedback force, back onto the oscillator to displace it toward the origin. There are two sources of uncertainty in this process, the precision of the estimate itself and the precision of the feedback actuation. With careful photodetector engineering, an electronic noise floor more than an order of magnitude beneath the optical shot noise can be achieved in optomechanics experiments. The amplified output photocurrent is generally robust to other electronic noise sources. Consequently, noise in the feedback actuation process can usually be neglected. We do so here.

In an ideal feedback system, one would simultaneously displace the oscillator's mean position and momentum to the origin based on the quadrature estimates. However, while momentum displacements are naturally achieved by applying an impulse force to the oscillator, it is generally difficult to apply position displacements, in practise. The usual solution is to, rather than directly apply a position displacement, apply a momentum kick that is retarded in time to achieve the desired displacement after a quarter cycle of evolution of the oscillator. This approach is clearly prone to difficulties if the oscillator is highly damped. However, within the regime of validity of the rotating wave approximation, it does result in the desired position displacement [91, 92]. Expressed in terms of mechanical quadratures, the resulting interaction Hamiltonian is

$$H_I = \hbar f_X(t)\hat{X}_M + \hbar f_Y(t)\hat{Y}_M, \qquad (5.32)$$

where $f_X(t)$ and $f_Y(t)$ are feedback forces applied to each mechanical quadrature based on the measured quadrature photocurrents i_X and i_Y. One might naively think that the best strategy for feedback would be to apply a force proportional to $\langle \hat{X}_M \rangle$ and $\langle \hat{Y}_M \rangle$, respectively, to each quadrature. This turns out to be the optimal approach [91, 126].

Exercise 5.5 *Setting $f_X(t) = -\Gamma g \langle \hat{X}_M \rangle /2$ and $f_Y(t) = -\Gamma g \langle \hat{Y}_M \rangle /2$, where g is a dimensionless feedback gain parameter, apply the Hamiltonian of Eq. (5.32) in Eq. (5.20) and show that the feedback force modifies the evolution of the quadrature expectation values to*

$$d\langle \hat{X}_M \rangle = -\frac{\Gamma}{2}(1+g)\,\langle \hat{X}_M \rangle dt + 2\sqrt{\eta\mu}\,[V_X dW_X + C_{XY} dW_Y] \quad (5.33a)$$

$$d\langle \hat{Y}_M \rangle = -\frac{\Gamma}{2}(1+g)\,\langle \hat{Y}_M \rangle dt + 2\sqrt{\eta\mu}\,[V_Y dW_Y + C_{XY} dW_X], \quad (5.33b)$$

while the variances and covariance remain unchanged.

Comparing these expressions to Eqs. (5.21) for the evolution of the means without feedback, we see that the only effect of the feedback is to introduce additional damping to the means, with the decay rate increasing by a factor of $1 + g$. Inspection of Eq. (5.31) then shows that the filter function $h(\omega)$ which must be applied to the quadrature photocurrents to estimate the quadratures and implement the feedback is the Lorenzian

$$h(\omega) = \frac{2\sqrt{\eta\mu}}{\Gamma(1+g)/2 + 2\eta\mu V_X - i\omega}. \quad (5.34)$$

The bandwidth of the filter is $\Gamma(1+g) + 4\eta\mu V_X$ and increases with the gain g.

Once the variances and covariance have reached steady state, Eqs. (5.33) describe the diffusion of the mechanical quadrature means due to information extracted from the measurement. The final temperature that may be achieved in feedback cooling is limited both by the steady state quadrature variances and by this diffusion. The variance of the diffusion can be determined from Eqs. (5.33) using standard methods. In one such approach [110], one defines $y(t) = \langle \hat{X}_M \rangle e^{\Gamma(1+g)t/2}$. Remembering that $C_{XY} = 0$ in the steady state, the increment dy can then be shown to be $dy = 2\sqrt{\eta\mu} V_X e^{\Gamma(1+g)t/2} dW_X$. Formally integrating from $t' = 0$ to t and substituting back in terms of $\langle \hat{X}_M \rangle$ yields the temporal dynamics

$$\langle \hat{X}_M \rangle(t) = \langle \hat{X}_M \rangle(0)\,e^{-\Gamma(1+g)t/2} + 2\sqrt{\eta\mu}\,V_X \int_0^t e^{-\Gamma(1+g)(t-t')/2} dW_X(t').$$

In the long time limit the initial condition decays to zero, leaving only a Gaussian diffusion due to the measurement. Ensemble averaging over all possible realisations of the noise, the variance of this diffusion is[5]

$$
\begin{aligned}
\langle\!\langle\langle \hat{X}_M \rangle^2 \rangle\!\rangle - \langle\!\langle\langle \hat{X}_M \rangle\rangle\!\rangle^2 &= \langle\!\langle\langle \hat{X}_M \rangle^2 \rangle\!\rangle \\
&= \left\langle\!\!\left\langle \left(2\sqrt{\eta\mu} V_X \int_0^t e^{-\Gamma(1+g)(t-t')/2} dW_X(t') \right)^2 \right\rangle\!\!\right\rangle \\
&= \frac{4\eta\mu V_X^2}{\Gamma(1+g)},
\end{aligned}
\tag{5.36}
$$

with an identical expression, with $X \rightarrow Y$, for the Y-quadrature [92]. We see that the precision with which the feedback localises the mechanical oscillator to the origin depends on the steady-state variance of the oscillator, the measurement strength, and the feedback gain g, but crucially that the imprecision approaches zero as $g \rightarrow \infty$.

The total ensemble-averaged variance of each mechanical quadrature is just the sum of the conditional variance due to measurement and the variance of the mean due to imprecision in feedback. Within the rotating wave approximation, it is identical for both quadratures and given by

$$
\mathrm{var}[\hat{X}_M] = \mathrm{var}[\hat{Y}_M] = V_X + \langle\!\langle\langle \hat{X}_M \rangle^2 \rangle\!\rangle = \left(1 + \frac{4\eta\mu V_X}{\Gamma(1+g)} \right) V_X.
\tag{5.37}
$$

We see that, as long as $1 + g \gg 4\eta\mu V_X/\Gamma$, the conditional variance dominates, with the additional uncertainty due to the imprecision of the feedback becoming negligible.

The variances of Eq. (5.37) may be straightforwardly related to the mean occupancy of the oscillator since $\hat{n} = b^\dagger b = (\hat{X}_M^2 + \hat{Y}_M^2 - 1)/2$. We then find that

$$
\begin{aligned}
\langle\!\langle \bar{n}_b \rangle\!\rangle &= \frac{1}{2} \left(\mathrm{var}[\hat{X}_M] + \mathrm{var}[\hat{Y}_M] - 1 \right) \tag{5.38} \\
&= \left(1 + \frac{4\eta\mu V_X}{\Gamma(1+g)} \right) V_X - \frac{1}{2} \tag{5.39} \\
&= \left(1 + \frac{4\eta |C_{\mathrm{eff}}| V_X}{1+g} \right) V_X - \frac{1}{2}. \tag{5.40}
\end{aligned}
$$

5.3 BACK-ACTION EVADING MEASUREMENT

In Chapter 3 we examined the effect of quantum measurements on optomechanical systems, showing, among other things, that quantum back-action

---

[5] To obtain this result, we make use of the correlation property of Itô stochastic integrals [110], that, for arbitrary nonanticipating functions $G(t)$ and $H(t)$,

$$
\left\langle\!\!\left\langle \int_{t_0}^t G(t')dW(t') \int_{t_0}^t H(t')dW(t') \right\rangle\!\!\right\rangle = \int_{t_0}^t \langle\!\langle G(t')H(t') \rangle\!\rangle \, dt'.
\tag{5.35}
$$

Note that here a nonanticipating function is a function that is independent of the behaviour of the Wiener process at future times.

results in standard quantum limits to precision that certain measurements cannot exceed. For instance, in Section 3.1 we examined the case of position measurement on a free mass and showed that such a limit exists if there are no correlations between the two measurements or between either of the measurement and the initial momentum of the mechanical oscillator. However, a class of measurements exist where the back-action imparted from measurement of an observable is not dynamically linked to that observable. Such measurements were first proposed by Braginsky et al. [39, 40, 47] and Thorne et al. [284, 65] in the 1970s in the context of quantum optomechanics and are termed *back-action evading* or *quantum non-demolition (QND)* measurements. No standard quantum limit exists for measurements of this kind, which offers the prospect for greatly enhanced precision.

5.3.1 Momentum measurement on a free mass

The most basic example of a back-action evading measurement is measurement of the momentum of a free mass. While in the case of position measurement, the back-action from the measurement perturbs the momentum of the mass and therefore the position at later times, momentum measurement perturbs the position of the mass which – for a free mass – does not couple back into momentum, i.e., momentum is a constant of motion for a free mass and is termed a *good QND observable*, or simply a *QND observable*. This is immediately clear from the fact that the Hamiltonian of a free mass whose momentum is coupled linearly to some observable $\hat{M}$ of a measurement device with coupling strength λ

$$\hat{H} = \frac{\hat{p}^2}{2m} + \lambda \hat{p} \hat{M} \tag{5.41}$$

commutes with the momentum $\hat{p}$, with the momentum therefore being a constant of motion. Using the same approach as we used in Section 3.1 for differential position measurement, it is straightforward to show that differential momentum measurement on a free mass has no standard quantum limit.

Exercise 5.6 *Convince yourself of this.*

The situation is more complicated for measurements performed on a mechanical oscillator rather than a free mass, since then perturbations in position are dynamically linked to momentum. However, back-action evading measurements are still possible, with a range of such schemes proposed in the literature (for example, see [286, 70, 334, 321, 273]). In the next few subsections, we will introduce some of these schemes.

5.3.2 General back-action evading measurement

In Section 5.2 we examined the effect of continuous position measurement on the state of a mechanical oscillator within the rotating wave approximation.

We found that, as a result of back-action, at best, this continuous measurement conditions the state of the oscillator into a minimum uncertainty state with X- and Y-quadrature variances both equal to the zero-point variance of the oscillator. Alternatively, one could think about a noncontinuous measurement and whether this may allow the back-action to be evaded and one quadrature to be conditioned below the zero-point variance. A form of measurement that has obvious promise as a back-action evading measurement is a stroboscopic measurement, where the optical field is pulsed with a repetition rate equal to the mechanical frequency [284, 47, 44]. In this case, the measurement retrieves only information about one quadrature of the oscillator, while applying momentum kicks on the other. This has, indeed, been shown theoretically to allow the variance of one quadrature of the mechanical oscillator to be conditioned below the mechanical zero-point motion [41, 284]. A more general back-action evading measurement maybe modelled by introducing unequal measurement rates μ_X and μ_Y to our model of feedback cooling in Section 5.2.

Following the same procedure as in Section 5.2 but using measurement operators $\hat{c}_X \equiv \sqrt{\mu_X}\,\hat{X}_M$ and $\hat{c}_Y \equiv \sqrt{\mu_Y}\,\hat{Y}_M$ for the independent measurements of $\hat{X}_M$ and $\hat{Y}_M$, respectively, the quadrature photocurrents of Eq. (5.19) become

$$i_X(t)\,dt = \sqrt{\mu_X\eta}\langle\hat{X}_M\rangle\,dt + dW_X(t) \tag{5.42a}$$

$$i_Y(t)\,dt = \sqrt{\mu_Y\eta}\langle\hat{Y}_M\rangle\,dt + dW_Y(t), \tag{5.42b}$$

and the evolution of the mean and conditional variance of $\hat{X}_M$ are found to be

$$d\langle\hat{X}_M\rangle = -\frac{\Gamma}{2}\langle\hat{X}_M\rangle dt + 2\sqrt{\eta\mu_X}V_X dW_X + 2\sqrt{\eta\mu_Y}C_{XY}dW_Y \tag{5.43a}$$

$$\dot{V}_X = -\Gamma V_X + \Gamma\left(\bar{n} + \frac{1}{2}\right) + \mu_Y - 4\eta\left(\mu_X V_X^2 + \mu_Y C_{XY}^2\right), \tag{5.43b}$$

with identical results for $\hat{Y}_M$ but with $X \leftrightarrow Y$ throughout; similar to the case for feedback cooling, the covariance simply decays as $\dot{C}_{XY} = -[\Gamma + 4\eta(\mu_X V_X + \mu_X V_Y)]C_{XY}$, so that in the steady state $C_{XY} = 0$.

It is apparent from Eq. (5.43b) that, as should be expected, the back-action heating of the mechanical X-quadrature is proportional to the Y-quadrature measurement rate. The measurement-based conditioning in Eq. (5.43b) arises, in general, both from information extracted from the direct measurement of the X-quadrature, proportional to μ_X, and from information extracted from the Y-quadrature measurement due to correlations present between the two quadratures. However, since $C_{XY} = 0$ in the steady state, the Y-quadrature measurement makes no contribution in that limit. It is then straightforward to show that the steady-state X-quadrature conditional variance is

$$V_X = \sqrt{\left(\frac{\Gamma}{8\eta\mu_X}\right)^2 + (2\bar{n}+1)\frac{\Gamma}{8\eta\mu_X} + \frac{1}{4\eta}\left(\frac{\mu_Y}{\mu_X}\right)} - \frac{\Gamma}{8\eta\mu_X}, \tag{5.44}$$

with an identical result for V_Y but with the substitution $X \leftrightarrow Y$ throughout.

Similar to the case of feedback cooling, one can define a back-action dominated regime where $\mu_Y \gg \{\Gamma^2/16\eta\mu_X, (\bar{n}+1/2)\Gamma\}$. In this case the conditional variance becomes

$$V_X = \sqrt{\frac{1}{4\eta}\left(\frac{\mu_Y}{\mu_X}\right)}. \tag{5.45}$$

We see that V_X is no longer lower bounded by $1/2\eta^{1/2}$, as was the case for feedback cooling. Here, the precision of localisation of $\hat{X}_M$ improves as the quadrature measurement strengths become increasingly unbalanced. Indeed, so long as the back-action on $\hat{X}_M$ due to the Y-quadrature measurement remains dominant over other heating mechanisms so that Eq. (5.45) is a valid limit, as $\mu_Y/\mu_X \to 0$ we see that a perfect back-action evading measurement becomes possible with $V_X \to 0$. Since $V_X = \int_0^\infty d\omega\, \bar{S}_{XX}(\omega)/\pi$ (see e.g. Eq. (4.93)) this implies that the mechanical position quadrature power spectral density $\bar{S}_{XX}(\omega) \to 0$ at all frequencies, unconstrained by either of the standard quantum limit or zero-point limit introduced in Section 3.4.

The product of X- and Y-quadrature conditional variances is

$$V_X V_Y = \frac{1}{4\eta}, \tag{5.46}$$

which, for unit efficiency, saturates the Heisenberg uncertainty principle, resulting in a pure conditionally squeezed mechanical state.

5.3.3 Two-tone driving

A common approach to back-action evading measurement that is qualitatively similar to a stroboscopic measurement is to probe the mechanical oscillator with an optical field whose intensity varies sinusoidally at twice the mechanical resonance frequency, as shown in Fig. 5.2. In the frequency domain this field consists of two equal amplitude tones separated in frequency by 2Ω, hence the term *two-tone driving*. Two-tone driving is advantageous experimentally, since such fields are typically easier both to generate and to couple into the optical resonators currently necessary to achieve sufficient interaction strength to enter the quantum back-action dominated regime. A quantum treatment of cavity-enhanced back-action evasion via two-tone driving was first reported in [74]. Here, we proceed by using the results for a general back-action evading measurement in the previous subsection, while determining the mechanical quadrature measurement rates μ_X and μ_Y following an approach similar to that which was used to determine the radiation pressure heating due to a continuous measurement in Section 3.2.

5.3.3.1 Hamiltonian

To model two-tone driving, similar to our model of optomechanical entanglement in Section 4.4, we treat the optomechanical system in an interaction

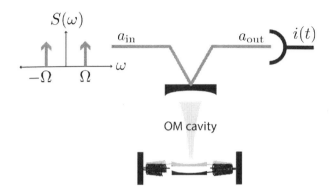

FIGURE 5.2   Schematic of a two-tone back-action evasion experiment. The incident optical field contains two bright coherent tones at $\pm\Omega$, as shown in the power spectral density $S(\omega)$. The phase quadrature of the output field is detected with homodyne detection, generating the photocurrent $i(t)$.

picture for both the light and the mechanical oscillator. The nonlinearised optomechanical Hamiltonian of Eq. (2.18) then becomes

$$\hat{H} = \hbar g_0 a^\dagger a \left(b^\dagger e^{-i\Omega t} + b e^{i\Omega t} \right), \qquad (5.47)$$

where here we have chosen the frame for the optical field to rotate at the optical cavity resonance frequency so that $\Delta = 0$. To introduce two-tone driving we choose a coherent amplitude that is sinusoidally varying at frequency Ω,

$$a \rightarrow 2\alpha \cos \Omega t + a, \qquad (5.48)$$

so that

$$a^\dagger a \rightarrow 4\alpha^2 \cos^2 \Omega t + \sqrt{2}\alpha \hat{X}_L \left(e^{i\Omega t} + e^{-i\Omega t} \right) + a^\dagger a, \qquad (5.49)$$

where we have introduced the subscript L to clearly distinguish the optical quadrature operators from the equivalent mechanical quadratures (labelled with the subscript M as usual). The first term in this expression makes clear that this choice of coherent amplitude does indeed produce two drive tones separated in frequency by 2Ω, since $2\cos^2 \Omega t = 1 + \cos 2\Omega t$.

Substituting in the result of Eq. (5.49), the Hamiltonian of Eq. (5.47) becomes

$$\hat{H} = \sqrt{2}\hbar g \hat{X}_L \left(e^{i\Omega t} + e^{-i\Omega t} \right) \left(b^\dagger e^{i\Omega t} + b e^{-i\Omega t} \right) \qquad (5.50)$$

where we have performed the usual linearisation, neglecting the term consisting of a product of three operators, and also neglected the coherent driving term. This term causes a constant displacement to the mechanical oscillator and coherently drives it at 2Ω. While this driving can have an effect on the mechanical dynamics, the 2Ω driving is suppressed for a high-quality mechanical oscillator since it is far-detuned from resonance, and the constant displacement only affects the overall detuning of the optical cavity, a static offset which can easily be corrected for experimentally (see Section 2.7). Furthermore, both effects are – at least in principle – perfectly deterministic and may be cancelled with judicious application of a classical force on the mechanical oscillator.

In the ideal good cavity limit $(\Omega \gg \kappa)$ the fast oscillating terms in Eq. (5.50) are suppressed, so that the Hamiltonian becomes

$$\hat{H} \approx \sqrt{2}\hbar g \hat{X}_L \left(b^\dagger + b \right) \tag{5.51}$$

$$= 2\hbar g \hat{X}_L \hat{X}_M. \tag{5.52}$$

While this may look similar to the linearised optomechanical Hamiltonian of Eq. (2.41), here the coupling to the light is via the X-quadrature of the mechanical oscillator, rather than its position, and furthermore, in this mechanical rotating frame there is no longer any oscillation at the mechanical resonance frequency. This is a significant difference, since it removes the coupling that transfers radiation pressure noise from the momentum of the oscillator onto the position and thereby makes the position a bad QND observable. Indeed, in analogy to momentum in the Hamiltonian of a free mass (Eq. (5.41)), in this Hamiltonian $\hat{X}_M$ is a constant of motion since $[\hat{X}_M, \hat{H}] = 0$ and is therefore a good QND observable.

Since the mechanical Y-quadrature is not coupled to the optical field, it is clear that, in this idealised good-cavity limit, the measurement rate $\mu_Y = 0$. It should therefore be expected that continuous measurement of the phase quadrature of the optical field will perform a back-action evading measurement of the mechanical X-quadrature.

5.3.3.2 Evolution of mechanical quadratures

The fast-rotating terms that we neglected in the Hamiltonian in the previous discussion are crucial to understanding the full effectiveness of a cavity optomechanical back-action evading measurement. Here, we return to the full linearised Hamiltonian of Eq. (5.50) and use the quantum Langevin approach to determine the evolution of the system including these terms.

Using Eqs. (1.135), the Hamiltonian can be expressed in terms of mechanical quadratures as

$$\hat{H} = 2\hbar g \hat{X}_L \left[\hat{X}_M \left(1 + \cos 2\Omega t \right) + \hat{Y}_M \sin 2\Omega t \right]. \tag{5.53}$$

Exercise 5.7 *Substitute this Hamiltonian into the quantum Langevin equation of Eq. (1.112) to show that, in the case of a high-quality oscillator where the rotating wave approximation is valid, the evolution of the mechanical* *quadratures is given by*

$$\dot{\hat{X}}_M = -\frac{\Gamma}{2}\hat{X}_M + \sqrt{\Gamma}\hat{X}_{M,\text{in}} + 2g\hat{X}_L \sin 2\Omega t \qquad (5.54\text{a})$$

$$\dot{\hat{Y}}_M = -\frac{\Gamma}{2}\hat{Y}_M + \sqrt{\Gamma}\hat{Y}_{M,\text{in}} - 2g\hat{X}_L\left(1 + \cos 2\Omega t\right). \qquad (5.54\text{b})$$

In the steady state, these equations can be straightforwardly solved in the usual manner by taking their Fourier transform and solving simultaneously. The result is

$$\hat{X}_M(\omega) = \chi(\omega)\left[\sqrt{\Gamma}\hat{X}_{M,\text{in}} + ig\left(\hat{X}_L(\omega - 2\Omega) - \hat{X}_L(\omega + 2\Omega)\right)\right] \qquad (5.55\text{a})$$

$$\hat{Y}_M(\omega) = \chi(\omega)\left[\sqrt{\Gamma}\hat{X}_{M,\text{in}} - g\left(2\hat{X}_L(\omega) + \hat{X}_L(\omega - 2\Omega) + \hat{X}_L(\omega + 2\Omega)\right)\right], (5.55\text{b})$$

where the mechanical susceptibility is $\chi(\omega) \equiv [\Gamma/2 - i\omega]^{-1}$, and we have used Eq. (3.11a) for the intracavity optical amplitude quadrature, as well as the Fourier transforms $\mathcal{F}\{\cos\Omega t\} = [\delta(\omega - \Omega) + \delta(\omega + \Omega)]/2$ and $\mathcal{F}\{\sin\Omega t\} = i[\delta(\omega - \Omega) - \delta(\omega + \Omega)]/2$. We can observe from these expressions that the mechanical X-quadrature is only driven by off-resonance fluctuations of the intracavity optical field which are suppressed by the filtering effect of the optical cavity. On the other hand, the mechanical Y-quadrature is driven by both on- and off-resonance optical fluctuations, with the on-resonance fluctuations being resonantly enhanced. One therefore expects that, since the back-action on the mechanical Y-quadrature is larger, the measurement will be biased towards the X-quadrature. As we saw in Section 5.3.2, measurement bias between the quadratures is a crucial feature required for back-action evading measurements.

5.3.3.3 *Mechanical quadrature measurement rates and conditional squeezing*

The power spectral densities of the mechanical quadratures can be calculated from Eqs. (5.55) using Eq. (1.42) and the rotating frame bath power spectral densities of Eqs. (1.118a). On the mechanical resonance ($\omega = 0$) they are

$$S_{X_M X_M}(0) = \frac{4}{\Gamma}\left[\bar{n} + \frac{1}{2} + \frac{C}{1 + 16(\Omega/\kappa)^2}\right] \qquad (5.56\text{a})$$

$$S_{Y_M Y_M}(0) = \frac{4}{\Gamma}\left[\bar{n} + \frac{1}{2} + C\left(2 + \frac{1}{1 + 16(\Omega/\kappa)^2}\right)\right], \qquad (5.56\text{b})$$

where $C = 4g^2/\kappa\Gamma$ is the optomechanical cooperativity defined in Eq. (3.14), and we have assumed that the optical driving field is in a coherent state ($\bar{n}_L = 0$). It is evident that two-tone driving indeed causes unbalanced back-action heating on the mechanical X- and Y-quadratures, which implies also that the measurement rates are unbalanced. By comparison with Eq. (5.43b), we can directly identify these measurement rates to be

$$\mu_X = \frac{C}{\Gamma}\left(2 + \frac{1}{1 + 16(\Omega/\kappa)^2}\right) = \frac{4g^2}{\kappa}\left(2 + \frac{1}{1 + 16(\Omega/\kappa)^2}\right) \quad (5.57a)$$

$$\mu_Y = \frac{C}{\Gamma}\left(\frac{1}{1 + 16(\Omega/\kappa)^2}\right) = \frac{4g^2}{\kappa}\left(\frac{1}{1 + 16(\Omega/\kappa)^2}\right). \quad (5.57b)$$

Clearly, in the good cavity limit ($\Omega \gg \kappa$), the mechanical Y-quadrature measurement rate approaches zero, while μ_X asymptotes to $8g^2/\kappa$, twice the bad-cavity measurement rate of a continuous measurement with the same cooperativity.[6] For a general cavity decay rate, Eqs. (5.57) can be used in Eq. (5.44) to determine the steady-state mechanical X-quadrature conditional variance achieved via two-tone driving. In the back-action dominated regime, Eq. (5.45) gives this variance to be

$$V_X = \sqrt{\frac{1}{4\eta}\left(\frac{1}{3 + 32(\Omega/\kappa)^2}\right)}. \quad (5.58)$$

Interestingly, quantum squeezing of the mechanical oscillator ($V_X < 1/2$) can be achieved in the back-action dominated regime for any resolved sideband factor (Ω/κ) as long as the measurement efficiency η exceeds $1/3$. Furthermore, with smaller efficiency, quantum squeezing can still be achieved in this regime as long as

$$\frac{\Omega}{\kappa} > \left(\frac{1 - 3\eta}{32\eta}\right)^{1/2}. \quad (5.59)$$

There have been several experimental demonstrations of back-action evasion using two-tone driving and stroboscopic measurement, including with superconducting microwave optomechanical systems [270] and with a magnetically polarised cloud of cold atoms [293]. In the latter case, the generation of a quantum squeezed state was verified, as shown in Fig. 5.3. Unconditional quantum squeezed states have also been observed using superconducting optomechanical systems via a modified scheme where the intensity of the two tones is unbalanced [320, 228, 171]. In such a scheme, a slightly higher red sideband intensity results in resolved sideband cooling (see Section 4.2.2) in parallel with an approximately back-action evading measurement [71, 165], allowing unconditional squeezed states to be produced without the requirement of feedback.

5.3.4 Detuned parametric amplification

We have just seen that, by applying measurements with unbalanced strengths to the mechanical X and Y quadratures, it is possible to perform a back-action evading measurement of one mechanical quadrature, localising it to

---

[6]This can be seen from Eq. (5.26), taking the limit $\Omega \ll \kappa$ so that $C_{\text{eff}} = C = 4g^2/\kappa\Gamma$.

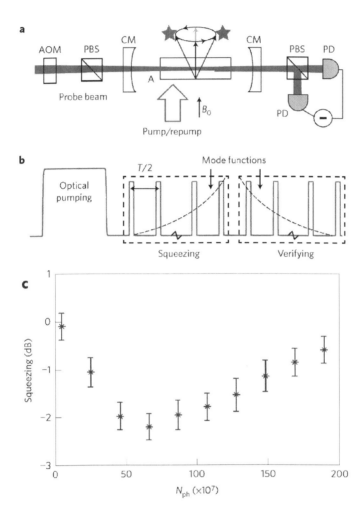

FIGURE 5.3 Stroboscopic back-action evading measurement achieved using a cloud of cold atoms as a magnetic oscillator. Adapted by permission from Macmillan Publishers Ltd: *Nature Physics* [293], copyright 2015. (a) Experimental apparatus. (b) Conditioning and verification protocol. (c) Resulting squeezing below the standard quantum limit. AOM: acousto-optic modulator, PBS: polarising beam splitter, CM: cavity mirror, PD: photodetector.

beneath its zero-point uncertainty. An alternative approach to back-action evading measurement is to introduce a nonlinearity to the Hamiltonian of the mechanical oscillator to phase sensitively amplify one quadrature of motion while deamplifying the other. This amplification process effectively unbalances the quadrature measurements even in the case of continuous measurement [274, 273], replacing the complexity of a stroboscopic measurement with the requirement of a mechanical nonlinearity.

5.3.4.1 Parametric amplification

We specifically consider parametric amplification of the motion of a mechanical oscillator, where an additional harmonic trapping potential is introduced for the mechanical oscillator the spring constant of which can be controlled in time. This can be achieved, for example, by introducing a spatially and temporally varying electric field which interacts with the mechanical oscillator through dipole forces [243]. The Hamiltonian of the mechanical oscillator then becomes

$$\hat{H} = \frac{\hat{p}^2}{2m} + \frac{\hat{q}^2}{2}\left(k + k_p(t)\right), \tag{5.60}$$

where k and $k_p(t)$ are the bare spring constant of the mechanical oscillator and the spring constant of the additional potential, respectively. The effect of the additional term can be understood simply – it dynamically alters the resonance frequency of the oscillator.

Exercise 5.8 *Show that, if $k_p(t)$ is stationary in time and much smaller than k, the Hamiltonian can be approximated by the frequency-shifted harmonic oscillator*

$$\hat{H}_{\text{stationary}} = \hbar\left(\Omega + \chi\right) b^\dagger b, \tag{5.61}$$

where the nonlinear coefficient $\chi \equiv k_p/2m\Omega$.

To parametrically drive the oscillator, the spring constant is sinusoidal modulated at a frequency close to twice the bare mechanical resonance frequency

$$k_p(t) \rightarrow k_p \cos 2\left[(\Omega + \Delta)t + \theta\right], \tag{5.62}$$

where 2Δ is the detuning of the parametric drive from 2Ω, not to be confused with the detuning of an optical cavity which we denote with the same symbol widely throughout the textbook, and θ is the phase of the drive. In a frame rotating at $\Omega + \Delta$, the Hamiltonian of Eq. (5.60) then becomes

$$\hat{H} = -\hbar\Delta b^\dagger b + \frac{\hbar\chi}{4}\left(\hat{b}^{\dagger 2}e^{2i\theta} + b^2 e^{-2i\theta}\right), \tag{5.63}$$

where χ is the nonlinear coefficient defined earlier, and we have neglected fast rotating terms using the standard rotating wave approximation, which is valid as long as $\Omega \gg \{\Gamma, \chi\}$. The nonlinear term in this equation should be reminiscent of a similar term for the optical field in our earlier ponderomotive squeezing Hamiltonian (see Eqs. (4.112) and (4.115)). Indeed, this term correspondingly causes mechanical squeezing (and amplification).

While in Section 4.5.3 we used the quantum Langevin approach to study ponderomotive squeezing, the stochastic master equation approach outlined in this chapter can equally well be used. Let us start by considering the unconditional dynamics including the interaction with the environment but no measurement ($\mu = 0$), and setting the drive phase to $\theta = -\pi/4$ for convenience. Using Eqs. (5.11) and (5.20) it is then easy to show that the mechanical quadrature means evolve as

$$\frac{d\langle \hat{X}_M \rangle}{dt} = -\left(\frac{\Gamma + \chi}{2}\right)\langle \hat{X}_M \rangle - \Delta \langle \hat{Y}_M \rangle \qquad (5.64a)$$

$$\frac{d\langle \hat{Y}_M \rangle}{dt} = -\left(\frac{\Gamma - \chi}{2}\right)\langle \hat{Y}_M \rangle + \Delta \langle \hat{X}_M \rangle. \qquad (5.64b)$$

Exercise 5.9 *Show these results.*

We see that, for this particular choice of drive phase, the parametric driving increases the decay rate of the mechanical X-quadrature, which results in squeezing, while decreasing the decay rate of the Y-quadrature by the same amount and thereby amplifying it. For $\chi > \Gamma$ the Y-quadrature experiences an exponential growth, rather than decay, leading to *parametric oscillation*. This phenomenon should be familiar to the reader, since it is the mechanism by which a swinger excites the oscillation of a swing. If $\Delta = 0$, the threshold for oscillation is exactly at $\chi = \Gamma$. However, since any detuning acts to couple energy from the Y-quadrature into the X-quadrature, the threshold is suppressed with increasing detuning, occurring in general at

$$\chi_{\text{threshold}} = \left[4\Delta^2 + \Gamma^2\right]^{1/2}. \qquad (5.65)$$

Quantum squeezing via detuned parametric amplification was treated in 1984 from the perspective of quantum optics [59]. Conditional squeezing by continuously monitoring the process was first considered more recently in [274, 275, 273], in the context of quantum optomechanics.

5.3.4.2 Back-action evasion

Since parametric driving acts to amplify one quadrature of the mechanical oscillator, signals on this quadrature may be amplified above the measurement noise, improving the performance of mechanical sensors [150]. However, since the parametric amplification directly modifies the dynamics of the amplified quadrature, this does not constitute a back-action evading measurement.[7] It is

---

[7]Put another way, the Hamiltonian of Eq. (5.63) does not commute, in general, with either $\hat{X}_M$ or $\hat{Y}_M$.

possible, however, to achieve a back-action evading measurement by choosing a special case for the detuning Δ [273]. Specifically, if we choose $\Delta = -\chi/2$, completing the square on Eq. (5.63) yields the simple Hamiltonian

$$\hat{H} = \frac{\hbar\chi}{2}\hat{X}_M^\theta{}^2. \tag{5.66}$$

This Hamiltonian is essentially identical to the Hamiltonian for a free mass (Eq. (5.41)). The quadrature $\hat{X}_M^\theta$ is a constant of motion like momentum, while the quadrature $\hat{Y}_M^\theta$ grows at a rate determined by $\hat{X}_M^\theta$ like position. Similar to the case of a free mass, were one to couple the oscillator to an ancilla system by including an interaction term of the form $\lambda\hat{X}_M^\theta\hat{M}$, it is clear that neither the parametric driving nor the coupling would affect the dynamics of the mechanical X^θ-quadrature since both terms commute with $\hat{X}_M^\theta$, while information about $\hat{X}_M^\theta$ would be encoded on the ancilla.

While the measurement described in the previous paragraph is the archetypal back-action evading measurement, it is not ideal from a practical standpoint since it requires *both* parametric driving and some form of stroboscopic measurement to achieve a single-quadrature measurement. The back-action evading measurement via two-tone driving discussed in Section 5.3.3, by contrast, requires only a stroboscopic measurement. Here, we would like to achieve a back-action evading measurement that relies only on parametric driving and continuous measurement. The essential idea can be understood by reference to Eqs. (5.64) for the unconditional dynamics of a detuned parametric oscillator. Equation (5.66) tells us that, for $\Delta = -\chi/2$, there exists a quadrature X^θ that is unaffected by the parametric drive, while Eqs. (5.64) tell us first that the presence of detuning introduces correlations between quadratures, and second that it amplifies one quadrature. Detuned parametric amplification thereby encodes the X^θ-quadrature on other quadratures of the oscillator. Amplification of the fluctuations on these quadratures then allows information to be extracted about the X^θ-quadrature with a weak continuous measurement. In the asymptotic limit of very large amplification, the measurement can be so weak that it introduces essentially no back-action upon the X^θ-quadrature.

Let us proceed to examine this effect by considering the evolution of the moments of an oscillator whose unitary dynamics are governed by Eq. (5.66) and in the presence of continuous measurement. Applying the approach outlined in Section 5.2, we obtain equations of motion for the quadrature variances and covariance identical to those in Eqs. (5.24) for feedback cooling, except for the additional terms [275]

$$\dot{V}_X = \dots + 2\chi\sin\theta\,[\sin\theta\,C_{XY} + \cos\theta\,V_X] \tag{5.67a}$$
$$\dot{V}_Y = \dots - 2\chi\cos\theta\,[\cos\theta\,C_{XY} + \sin\theta\,V_Y] \tag{5.67b}$$
$$\dot{C}_{XY} = \dots + \chi\sin^2\theta\,V_Y - \chi\cos^2\theta\,V_X. \tag{5.67c}$$

Exercise 5.10 *Show these results.*

Equations (5.67) are nonlinear equations of motion due to the variance-squared and covariance-squared measurement conditioning terms in Eqs. (5.24). This makes them difficult to solve in general. Substantial simplification is possible by choosing the parametric drive phase

$$\tan\theta = -\sqrt{\frac{V_X}{V_Y}}, \tag{5.68}$$

so that the terms introduced by the drive into the covariance evolution (Eq. (5.67c)) cancel. This can be done without loss of generality and has the effect only of orientating the quadratures of maximum squeezing and amplification to V_X and V_Y, respectively. With this choice of phase, the steady-state covariance $C_{XY} = 0$. Taking the limit of strong driving $V_Y \gg V_X$, so that $\cos\theta \approx 1$ and $\sin\theta \approx -\sqrt{V_X/V_Y}$, we find the coupled nonlinear equations

$$0 = -\Gamma V_X + \Gamma\left(\bar{n} + 1/2\right) + \mu - 4\mu\eta V_X^2 - 2\chi\sqrt{\frac{V_X^3}{V_Y}} \tag{5.69a}$$

$$0 = -\Gamma V_Y + \Gamma\left(\bar{n} + 1/2\right) + \mu - 4\mu\eta V_Y^2 + 2\chi\sqrt{V_X V_Y} \tag{5.69b}$$

for the steady-state variances. It should be apparent from these expressions that the parametric driving has enhanced the conditional damping of the mechanical X-quadrature and introduced antidamping to the Y-quadrature. The action of this is to conditionally localise the X-quadrature to better than is possible with continuous monitoring alone.

5.3.4.3 Conditional squeezing

Analytical solutions to Eqs. (5.69) have been obtained in [275] and we do not reproduce those here. Instead we consider the scenario where

$$V_X \ll \left\{\bar{n} + \frac{1}{2} + \frac{\mu}{\Gamma}, \sqrt{\frac{\Gamma\left(\bar{n} + 1/2\right)}{4\mu\eta}}\right\}. \tag{5.70}$$

This corresponds to an X-quadrature variance that is localised to better precision than both the back-action heated thermal variance and the variance achieved in the classical measurement regime of feedback cooling (see Section 5.2.3.3). In this case, Eq. (5.69a) yields the simple expression for the Y-quadrature variance

$$V_Y = \frac{4\chi^2 V_X^3}{\left[\Gamma\left(\bar{n} + 1/2\right) + \mu\right]^2}. \tag{5.71}$$

Using this result, and in the strong driving regime where $\chi^2 \gg (\Gamma(\bar{n}+1/2) + \mu)\mu\eta$, Eq. (5.69b) can then be solved for V_X with the result

$$V_X = \frac{1}{2\eta^{1/4}}\left[\frac{\left(\Gamma\left(\bar{n} + 1/2\right) + \mu\right)^3}{\chi^2\mu}\right]^{1/4}. \tag{5.72}$$

This expression has several notable characteristics. Firstly, we see that perfect localisation is possible ($V_X \to 0$) as the drive strength χ increases; secondly, this is true even in the limit of arbitrarily small measurement strength μ; thirdly, in the limit of perfect efficiency, quantum squeezing ($V_X < 1/2$) is achieved as long as the characteristic rate $\chi^{2/3}\mu^{1/3}$ is much greater than the combined thermal and back-action heating rates (($\Gamma(\bar{n} + 1/2) + \mu$)); and finally, there exists an optimal measurement strength beyond which V_X is dominated by measurement back-action noise and begins to degrade. These features contrast back-action evasion by two-tone optical driving, where the back-action dominated measurement regime is required to achieve appreciable squeezing and the squeezing approaches the asymptotic limit in Eq. (5.58) as the measurement strength increases. A further notable difference between the approaches is that parametric back-action evasion is less sensitive to measurement inefficiencies. This can be seen by the different scaling of the η pre-factors in Eqs. (5.58) and (5.72) and is a consequence of the parametric amplification which acts to amplify the mechanical signal above the vacuum noise introduced before measurement.

Conditional mechanical squeezing has been experimentally demonstrated using the protocol described here in the classical regime where the mechanical quadrature is localised beneath its equilibrium thermal noise but remains above the level of the zero-point motion [272] (see Fig. 5.4). In these experiments the mechanical oscillator was an atomic force microscope tip, read out through optical interferometry, and parametrically driven using a spatially and temporally varying electric field produced by a sharp electrode.

5.4 SURPASSING THE STANDARD QUANTUM LIMIT USING SQUEEZED LIGHT

In the previous section we saw how back-action evading measurements, where the observable being measured is a constant of motion of the system Hamiltonian, can be used to – as the name suggests – evade the back-action usually present for measurements on quantum systems and prepare states of a mechanical oscillator that are localised in one quadrature to better than the mechanical zero-point motion. This can be used to surpass (or more precisely avoid) the standard quantum limit to measurement precision. An alternative path to achieve measurement precision beyond the standard quantum limit is to utilise an ancilla system for the measurement that itself displays quantum correlations. We consider this class of measurements here.

5.4.1 Sub-standard quantum limit position measurement on a free mass

In our derivation of the standard quantum limit of position measurement on a free mass in Section 3.1, we explicitly assumed that there were no correlations between the measurement noise and the measurement back-action on the momentum of the mass. Here we briefly consider the more general case where

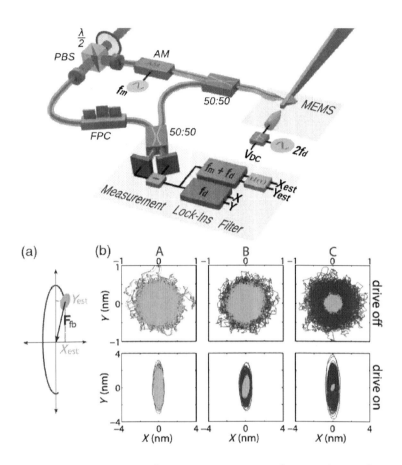

FIGURE 5.4 Back-action evading measurement of a mechanical oscillator in the classical regime, using detuned parametric amplification.
Reprinted figure with permission from [272]. Copyright 2013 by the
American Physical Society. *Top*: Experimental schematic. *Bottom:* (a)
Trajectory of state during measurement. (b) Error in unconditional (dark
grey) and conditional (light grey) estimates with and without parametric driving. PBS: polarising beam splitter, AM: amplitude modulator,
50:50: 50/50 beam splitter, f_m: modulation frequency, f_d: drive frequency, MEMS: atomic force microscope cantilever, $\lambda/2$: half-wave plate,
FPC: fibre polarisation controller, X_{est} (Y_{est}) estimated X-quadrature
(Y-quadrature).

quantum correlations can exist, to provide a simple demonstration of how a back-action evading measurement can be implemented by carefully engineering the interaction between a quantum system and the measurement system used to probe it.

Similar to Section 3.1, we consider two sequential position measurements on the free mass separated by a time τ. However, to allow for quantum correlations between the measurement noise and back-action noise, we consider that the sequential measurements are performed by coupling the position of the mass to some ancilla quantum system briefly at some initial time, decoupling it and waiting for a period of time, then briefly coupling it a second time, and finally making a measurement on the ancilla. Specifically, let us envision that the ancilla is a second free mass, with linear coupling introduced for short times around $t = 0$ and $t = \tau$ via a Hamiltonian proportional to $\hat{q}\,\hat{q}_a$, where $\hat{q}_a$ is the position operator of the ancilla mass. If the interactions are sufficiently short, it is straightforward to show, using the Heisenberg equation of motion (Eq. (1.20)), that the first interaction performs the symmetric transformations

$$
\begin{aligned}
\hat{p}(0_+) &= \hat{p}(0) + \lambda \hat{q}_a(0) & \text{(5.73a)} \\
\hat{p}_a(0_+) &= \hat{p}_a(0) + \lambda \hat{q}(0) & \text{(5.73b)}
\end{aligned}
$$

and leaves the positions of both masses unaffected, where $\hat{p}_a$ is the momentum of the ancilla, λ is some effective interaction strength, and the 0_+ is meant to indicate an infinitesimally short time after the interaction.

Exercise 5.11 *Show this result.*

We see from Eqs. (5.73) that the interaction both perturbs the momentum of the mass and encodes information about its position on the momentum of the ancilla. This may appear familiar to the reader since it has exactly the same form as the linearised optomechanical interaction, where the phase quadrature of the light is encoded with information about the mechanical position, and the mechanical momentum is encoded with information about the amplitude quadrature of the light. By inspection of Eq. (5.73b) we see that a direct noise-free measurement on the ancilla would allow the position of the mass to be determined with precision $\Delta q = \sigma[\hat{p}_a(0)]/|\lambda|$, while from Eq. (5.73a) we see that the back-action on the momentum of the mass is $\Delta p = |\lambda|\sigma[\hat{q}_a(0)]$. Hence, if the ancilla starts in a minimum uncertainty state, the product of measurement uncertainty and back-action $\Delta q \Delta p = \sigma[\hat{p}_a(0)]\sigma[\hat{q}_a(0)] = \hbar/2$, saturating the Heisenberg uncertainty principle.

In the same manner as in Section 3.1, the free Hamiltonian dynamics leave the momentum of both the mass and the ancilla unchanged, but after time τ couples the momentum of the mass into position as

$$
\hat{q}(\tau) = \hat{q}(0) + q_{\text{sig}} + \frac{\tau\lambda}{m}\hat{q}_a(0), \tag{5.74}
$$

where we have kept only the contribution from the measurement back-action, assuming as in Section 3.1 that, prior to the interaction, the mass is perfectly localised in momentum. We have also introduced an additional displacement q_{sig} occurring at some stage during the evolution which constitutes the signal to be measured – in gravitational wave detection, for instance, this would result from differential length contraction as the gravitational wave propagates through the detection device. Choosing the strength of the second interaction to be the same as the first but inverting the sign, the second interaction then transforms the momentum of the ancilla to

$$\hat{p}_a(\tau_+) \;=\; \hat{p}_a(0) + \lambda \hat{q}(0) - \lambda \hat{q}(\tau) \tag{5.75}$$

$$=\; \hat{p}_a(0) - \lambda q_{\text{sig}} - \frac{\tau \lambda^2}{m} \hat{q}_a(0). \tag{5.76}$$

A measurement of $\hat{p}_a(\tau_+)$ with outcome $\tilde{p}_a(\tau_+)$ then provides an estimate of the signal $\tilde{q}_{\text{sig}} = -\tilde{p}_a(\tau_+)/\lambda$ with uncertainty

$$\sigma[\tilde{q}_{\text{sig}}] \;=\; \sigma\left[\hat{p}_a(\tau_+)\right]/|\lambda| \tag{5.77}$$

$$=\; \frac{\tau|\lambda|}{m}\sigma\left[\hat{q}_a(0) - \frac{m}{\tau\lambda^2}\hat{p}_a(0)\right]. \tag{5.78}$$

If the ancilla is initially prepared in a minimum uncertainty state and its initial position and momentum are uncorrelated, Eq. (5.78) reduces to the position measurement uncertainty that would be achieved via two independent measurements of equal strength.[8] A standard quantum limit similar to Eq. (3.8) can then be directly derived. However, it is immediately apparent from Eq. (5.78) that correlations between the position and momentum of the ancilla can enable a reduction in uncertainty beneath the standard quantum limit. By inspection of the position evolution of a free mass (Eq. (3.3)), it can be identified that

$$\hat{q}_a(0) - \frac{m}{\tau\lambda^2}\hat{p}_a(0) = \hat{q}_a\left(-m\, m_a/\tau\lambda^2\right), \tag{5.79}$$

where m_a is the mass of the ancilla. Somewhat remarkably, the uncertainty in the measurement is dictated by the uncertainty in the position of the ancilla at the earlier time $t = -m\, m_a/\tau\lambda^2$. Localisation of the ancilla in position at this time – for instance, via a strong position measurement – will generate position-momentum correlations at time $t = 0$ that cancel the effects of back-action and measurement noise on the momentum of the ancilla. In principle, this allows arbitrarily accurate measurement of $\hat{q}_{\text{sig}}$, unconstrained by the standard quantum limit.

---

[8]The expression obtained is essentially the same as Eq. (3.6), with the only differences being 1) a factor of two that arises from the fact that here we consider two interactions of equal strength, whereas in Section 3.1 the second measurement was treated as a perfect noise-free measurement, corresponding to an infinite interaction strength, and 2) that here we have assumed that the free mass is initially perfectly localised in momentum – that assumption comes in later in Section 3.1.

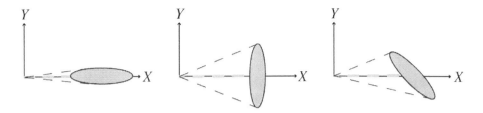

FIGURE 5.5 Ball and stick diagrams of phase (*left*), amplitude (*middle*), and $\pi/4$ (*right*) quadrature squeezed states.

Through this simple example, we have sought to demonstrate that careful preparation of an ancilla system that is used to perform a measurement on a quantum system can allow the standard quantum limit to be exceeded by an arbitrary degree. In what follows we consider some specific examples of approaches to achieve such standard quantum limit surpassing measurements that can be applied in cavity optomechnical systems. In these examples, the optical field plays the role of the ancilla, with correlations either prepared prior to injection of the field into the cavity optomechanical system or produced intrinsically by the optomechanical interaction.

5.4.2 Squeezed input field

In Chapter 3 we treated optical measurements of mechanical motion using coherent or thermal states of light. However, we found in Section 4.5.1 that it is possible to produce so-called squeezed states of light that exhibit noise in one quadrature beneath the noise on a coherent state (see Fig. 5.5). In the context of Section 4.5.1, these states were produced via the effective Kerr nonlinearity of the optomechanical interaction itself. Alternatively, squeezed states of light, produced externally by some other nonlinear process, can be injected into the optomechanical system to improve the precision of optical measurements of mechanical motion.

To approach this topic, we begin by revisiting the output optical field from a linearised optomechanical system with resonant optical drive ($\Delta = 0$) given in Eqs. (3.29). While having the advantage of simplicity, we note that Eqs. (3.29) are only valid in the unrealistic scenario of perfect cavity escape and detection efficiency. In situations involving squeezed states, inefficiencies have two effects. First, as usual, they decrease the magnitude of the mechanical signal imprinted on the measured optical field. Second, since squeezed states of light exhibit photon correlations, and inefficiencies degrade these correlations, inefficiencies degrade the quality of the squeezing. While this second effect does significantly alter the performance of squeezed-light-based measurements of mechanical motion, it does not change the essential qualitative behaviour presented here. To determine the quantitative effect of inefficiencies, the reader could perform a similar analysis to ours but including inefficiencies in the manner described in Appendix A.

Let us now consider a minimum-uncertainty squeezed input optical state with minimum variance V at some quadrature angle θ, and maximum variance $1/V$ at the orthogonal quadrature angle $\theta + \pi/2$. Such a state is not in thermal equilibrium and therefore has different correlation properties from those given for a thermal input state in Eqs. (1.117). Specifically,

$$\langle \hat{Y}_{\text{in}}^{\theta\,\dagger}(\omega)\hat{Y}_{\text{in}}^{\theta}(\omega') \rangle = V\delta(\omega - \omega') \tag{5.80a}$$

$$\langle \hat{X}_{\text{in}}^{\theta\,\dagger}(\omega)\hat{X}_{\text{in}}^{\theta}(\omega') \rangle = \frac{1}{V}\delta(\omega - \omega') \tag{5.80b}$$

$$\langle \hat{X}_{\text{in}}^{\theta\,\dagger}(\omega)\hat{Y}_{\text{in}}^{\theta}(\omega') \rangle = -\langle \hat{Y}_{\text{in}}^{\theta\,\dagger}(\omega)\hat{X}_{\text{in}}^{\theta}(\omega') \rangle = \frac{i}{2}\delta(\omega - \omega'). \tag{5.80c}$$

Using Eqs. (1.17) with the substitutions $\{\hat{Q}, \hat{P}\} \to \{\hat{X}, \hat{Y}\}$, the optical input quadrature operators may be re-expressed in terms of the rotated squeezed and antisqueezed quadrature operators as

$$\hat{X}_{\text{in}} = \cos\theta\,\hat{X}_{\text{in}}^{\theta} - \sin\theta\,\hat{Y}_{\text{in}}^{\theta} \tag{5.81a}$$

$$\hat{Y}_{\text{in}} = \cos\theta\,\hat{Y}_{\text{in}}^{\theta} + \sin\theta\,\hat{X}_{\text{in}}^{\theta}. \tag{5.81b}$$

The power spectral density of the output optical phase quadrature can be determined by substituting these expressions into Eq. (3.29b) and using Eqs. (1.43) and the correlation relations above, remembering that due to radiation pressure $\hat{Q}$ is a function of $\hat{X}_{\text{in}}$.

Exercise 5.12 *Show that the symmetrised power spectral density $\bar{S}_{Y_{\text{out}}Y_{\text{out}}}(\omega) = [S_{Y_{\text{out}}Y_{\text{out}}}(\omega) + S_{Y_{\text{out}}Y_{\text{out}}}(-\omega)]/2$ that would be obtained upon homodyne measurement of $\hat{Y}_{\text{out}}$ (given in Eq. (3.29b)) is*

$$\bar{S}_{Y_{\text{out}}Y_{\text{out}}}(\omega) = 4\Gamma\,|C_{\text{eff}}|\,\bar{S}_{Q^{(0)}Q^{(0)}}(\omega) \tag{5.82}$$

$$+ \left| \cos\theta - 4\Gamma\sin\theta\,|C_{\text{eff}}|\,\chi(\omega) \right|^2 V + \left| \sin\theta + 4\Gamma\cos\theta\,|C_{\text{eff}}|\,\chi(\omega) \right|^2 \frac{1}{4V},$$

where, as in our treatment of the standard quantum limit in Section 3.4 the unperturbed mechanical power spectral density $\bar{S}_{Q^{(0)}Q^{(0)}}(\omega) = |\chi(\omega)|^2(2\bar{n}(\omega) + 1)\omega/Q$. Show that taking the quantum optics approximation and the squeezed variance $V = 1/2$ retrieves Eq. (3.43), which we previously derived for the case of coherent optical driving.

Renormalising the expression from the above exercise in the dimensionless position units of the mechanical oscillator in the same way as in Section 3.4, we find

$$\bar{S}_{Q_{\text{det}}Q_{\text{det}}}(\omega) = \bar{S}_{Q^{(0)}Q^{(0)}}(\omega) + \bar{S}_{\text{det,sqz}}(\omega), \tag{5.83}$$

where the detection noise power spectral density is

$$\bar{S}_{\text{det,sqz}}(\omega) = \frac{1}{4\Gamma |C_{\text{eff}}|} \left[\left| \cos\theta - 4\Gamma \sin\theta |C_{\text{eff}}| \chi(\omega) \right|^2 V \right.$$
$$\left. + \left| \sin\theta + 4\Gamma \cos\theta |C_{\text{eff}}| \chi(\omega) \right|^2 \frac{1}{4V} \right]. \tag{5.84}$$

Equation (5.83) should be compared to the standard quantum limit of mechanical displacement measurement given in Eq. (3.65).

When $\theta = 0$ the input optical field is phase squeezed such that $S_{Y_{\text{in}}Y_{\text{in}}}(\omega) = V < 1/2$ (see Fig. 5.5 (left)). Equation (5.84) is then greatly simplified to

$$\bar{S}_{\text{det,sqz}}(\omega) = \frac{V}{4\Gamma |C_{\text{eff}}|} + \frac{\Gamma |C_{\text{eff}}|}{V} |\chi(\omega)|^2. \tag{5.85}$$

This expression bears a close resemblance to Eq. (3.51) for the detection noise power spectral density of a mechanical oscillator interrogated using coherent light. Apart from the missing η here, due to our taking the case of perfect efficiency, the only difference is that the squeezed variance V acts to modify the effective interaction strength between light and mechanics, with the effective cooperativity in Eq. (3.51) replaced with $C_{\text{eff}}/2V$. Since $V < 1/2$ for a squeezed state, we see that the act of phase squeezing is to increase the interaction strength. This makes sense, since phase squeezing both reduces the imprecision in a phase quadrature measurement and, through the complimentary increase in amplitude quadrature variance enforced by the Heisenberg uncertainty relation, increases the radiation pressure noise driving of the mechanical oscillator.

Exercise 5.13 *Show that amplitude squeezing ($\theta = \pi/2$) decreases the effective interaction strength.*

Since both phase and amplitude squeezing act only to modify the effective interaction strength, it is clear that they cannot be used to surpass the standard quantum limit. They do, however, alter the effective cooperativity (and therefore optical power) required to reach the standard quantum limit, as shown in Fig. 5.6a. This offers some advantages. If the total optical power that is available to be injected into the cavity optomechanical system is constrained to less than that required to reach the standard quantum limit with coherent light, phase squeezing offers a pathway to improved precision and to ultimately reach the standard quantum limit. On the other hand, if amplitude squeezing is used and the optical power is increased to reach the standard quantum limit, then the system is inherently more robust to optical detection inefficiencies. This is due to the fact that, with amplitude squeezing, the phase quadrature noise of the output field and the mechanical signal imprinted on the phase quadrature are both amplified above the vacuum fluctuations that are introduced due to inefficiencies.

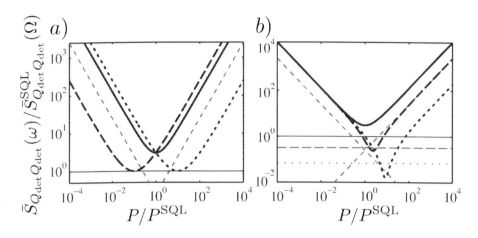

$\bar{S}_{Q_{\text{det}}Q_{\text{det}}}(\omega)/\bar{S}^{\text{SQL}}_{Q_{\text{det}}Q_{\text{det}}}(\Omega)$

P/P^{SQL}

FIGURE 5.6 Effect of squeezed light on the precision of mechanical displacement measurement. (a) On-resonance ($\omega = \Omega$) power spectral density of a measurement of $\hat{Q}$ using a squeezed input field as a function of drive power and normalised to the on-resonance standard quantum limit $\bar{S}^{\text{SQL}}_{Q_{\text{det}}Q_{\text{det}}}(\Omega) = 2/\Gamma$ from Eq. (3.65). Thick dashed, solid, and dotted curves were obtained for squeezing angles of $\theta = \{0, \pi/4, \pi/2\}$, respectively. Thin solid line: on-resonance standard quantum limit. (b) Power spectral density for a squeezing angle of $\theta = \pi/4$ and mechanical quality factor $Q = 100$, and at sideband frequencies $\omega = \{\Omega, \Omega + \Gamma, \Omega + 4\Gamma\}$ for the thick solid, dashed, and dotted curves, respectively. The thin solid, dashed, and dotted lines are the corresponding standard quantum limits. For both (a) and (b), we have chosen $\bar{n} = 0$, $\eta = 1$, and a squeezed quadrature variance of $V = 1/20$; the thin diagonal grey dashed lines are the on-resonance measurement imprecision and back-action noise for a coherent input state.

One might envision that, similar to our toy example of sub-standard quantum limited measurement on a free mass in Section 5.4.1, it would be possible to arrange for the measurement and back-action noise to destructively interfere by carefully balancing correlations between the phase and amplitude quadratures of the incident optical field, and thereby surpass the standard quantum limit. For any squeezing angle other than perfect amplitude or phase squeezing, the amplitude and phase quadrature fluctuations exhibit correlations (for example, see Fig. 5.5 (*right*)). Radiation pressure then correlates the motion of the mechanical oscillator to the phase quadrature of the optical field. The aim, then, would be to choose parameters so that, when the mechanical motion is imprinted onto the phase quadrature of the intracavity field, these correlations cancel, reducing the net detection noise. The potential for this to occur is evident in Eq. (5.84), where both terms in the square brackets have a coefficient that is the absolute-value-squared of a weighted sum of the measurement and back-action noise contributions. However, recognising that the on-resonance mechanical susceptibility, $\chi(\Omega) = i/\Gamma$, is imaginary, it is immediately clear from inspection of Eq. (5.84) that such cancellation is not possible at the mechanical resonance frequency. Consequently, the standard quantum limit cannot be surpassed at the mechanical resonance frequency for any choice of squeezing angle θ. Figure 5.6a shows the case of $\theta = \pi/4$ as an example. It can be seen that this squeezing angle leaves the optimum incident power unchanged but degrades the optimal sensitivity.

On the other hand, the mechanical susceptibility is approximately real when the condition $|\Omega^2 - \omega^2| \gg \omega\Gamma$ is met. If this condition is satisfied, it is possible to achieve the desired cancellation and thereby surpass the standard quantum limit. This occurs both at frequencies well above, and well below, the mechanical resonance frequency. The *free mass regime* where $\omega \gg \{\Omega, \Gamma\}$ is particularly important, having direct relevance to gravitational wave detection as discussed in Chapter 3. Fig. 5.6b shows the cancellation effect for the specific case of a squeezing phase angle of $\pi/4$, $V = 1/20$ and three different frequencies ω. The choice of squeezing angle $\tan\theta_{\mathrm{canc}} = -4\Gamma\Omega|C_{\mathrm{eff}}|/(\Omega^2 - \omega^2)$ approximately cancels the term from the antisqueezed quadrature[9] in the square brackets of Eq. (5.84) in the limit $|\Omega^2 - \omega^2| \gg \omega\Gamma$. One should notice that this squeezing angle is frequency dependent. As such, frequency-dependent squeezing is required to achieve broadband enhancement in precision. Such squeezing has been experimentally reported, for example, in [69].

Choosing the squeezing angle θ_{canc} from the previous paragraph and optimising the effective optomechanical cooperativity to minimise the total noise contribution from the squeezed quadrature ($|C_{\mathrm{eff}}^{\mathrm{opt}}| = |\Omega^2 - \omega^2|/4\Gamma\Omega$), the detection noise power spectral density of Eq. (5.84) becomes

$$\bar{S}_{\mathrm{det,sqz}}^{\theta_{\mathrm{canc}}}(\omega) \approx \frac{2\Omega V}{|\Omega^2 - \omega^2|}. \tag{5.86}$$

This approaches zero as $V \to 0$, in principle allowing a perfect measurement of the mechanical motion to be made both at $\omega = 0$ and at $\omega \to \infty$. At other frequencies, second-order terms proportional to $1/V$, that are not cancelled by our choice of squeezing angle, ultimately limit the measurement precision.

It is possible to simultaneously optimise the detection noise power spectral density of Eq. (5.84) over the squeezed variance V, squeezing angle θ and effective cooperativity C_{eff}. The result is that, the minimum $\bar{S}_{\mathrm{det,sqz}}(\omega)$ exactly saturates the zero-point limit $\bar{S}_{\mathrm{det}}^{\mathrm{ZPL}}(\omega)$ of Eq. (3.61). Since $\bar{S}_{\mathrm{det}}^{\mathrm{ZPL}}(\omega)$ is beneath the standard quantum limit at all off-resonance frequencies ($\omega \neq \Omega$), this shows that continuous measurements with non-classical light allow the off-resonant standard quantum limit to be surpassed. However, it also shows

---

[9]The term proportional to $1/V$.

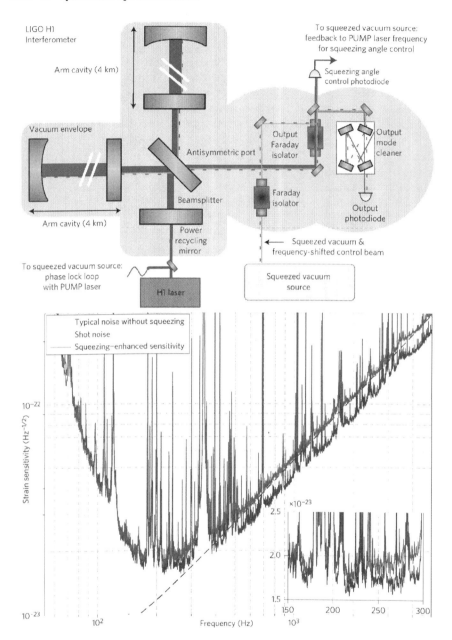

FIGURE 5.7 Improved measurement precision of the Hanford Laser Interferometer Gravitational Wave Observatory (LIGO) using squeezed light. Adapted by permission from Macmillan Publishers Ltd: *Nature Photonics* [1], copyright 2013. *Top*: Schematic of the interferometer. *Bottom*: Experimentally observed strain sensitivity with (light grey – bottom trace) and without (dark grey – top trace) squeezed light.

that – unlike the back-action evading measurements we treated earlier in this chapter – they do not allow arbitrarily good precision, being fundamentally constrained by the bath zero-point motion.

While we have shown here that, in both the free-mass and low frequency regimes, the introduction of correlations between the amplitude and phase quadratures of the incident optical field can greatly suppress the optical noise associated with measurements of mechanical motion, it is important to bear in mind that the response of an oscillator to external forces is greatly reduced in this regime. This is shown, for example, in the right-hand column of Fig. 3.7. As a result, in force sensing applications, generally, one would expect to achieve superior absolute sensitivity operating on resonance, and close to the standard quantum limit, than far off resonance, but beyond the standard quantum limit. In applications where the displacement is not introduced via a force on the mechanical oscillator, such as in gravitational wave detection, however, operating in the free mass regime offers the prospect for greatly enhanced sensitivity.

Many proof-of-principle experiments have been performed to demonstrate the enhanced measurement sensitivity available using squeezed light. The first experiments of this kind were interferometric measurements performed in the 1980s by the Kimble and Slusher laboratories[322, 128]. A significant constraint to the practical application of squeezed light in optical measurements has been the limited range of power constrained applications where sensitivity cannot be improved simply by injecting more light. Recently, squeezed light has found use in two classes of experiments that are significantly power constrained. The first is gravitational wave observatories, where, due to the exceptionally small signal size, current interferometers already operate close to their power limits [78, 1]. The second is nanoscale measurements of biological systems, where power is constrained by optical damage to, and photochemical intrusion into, the specimen [277, 278]. Figure 5.7 shows the effect of squeezing on the noise floor of the Laser Interferometer Gravitational Wave Observatory (LIGO) instrument at Hanford.

5.4.3 Ponderomotive squeezing

As we discovered in Section 4.5.3, it is possible to generate optical squeezing directly through the optomechanical interaction. It is then natural to ask, given the results in the previous section, whether this squeezing can intrinsically allow measurement precision exceeding the standard quantum limit. This turns out to be possible. The essential physics is that radiation pressure driving imprints optical amplitude quadrature noise on the mechanical oscillator, which is then mapped onto the phase quadrature of the optical field. The resulting noise correlations between the optical amplitude and phase quadratures allow the measurement noise to be reduced for an appropriate choice of homodyne detection quadrature (see Fig. 5.8). This technique, termed *variational measurement*, was first proposed in [303, 304].

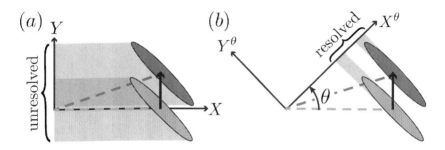

FIGURE 5.8 Ball and stick diagrams illustrating the use of ponderomotive squeezing to overcome the standard quantum limit. In the radiation pressure shot noise dominated regime, the precision of the measurement of mechanical motion based on the phase quadrature of light degrades (a). However, correlations are introduced such that superior signal to noise is possible when measuring an alternative optical quadrature (b). Here the bold arrow illustrates a displacement of the phase quadrature of the optical field due to motion of the mechanical oscillator.

To see this effect, we begin with Eq. (4.121) for the symmetrised power spectral density of an arbitrary quadrature θ of the output field from a cavity optomechanical system in the linearised regime and with on-resonance optical driving ($\Delta = 0$). Renormalising this spectrum in terms of the dimensionless mechanical position in the same way as in the previous section, we obtain for the mechanical position detection noise power spectrum

$$\bar{S}_{\det}^{\theta}(\omega) = \frac{1}{8\Gamma \, |C_{\mathrm{eff}}| \sin^2 \theta} + 2\Gamma |\chi(\omega)|^2 \, |C_{\mathrm{eff}}| + \frac{\chi(\omega) + \chi^*(\omega)}{2 \tan \theta}. \qquad (5.87)$$

As usual, it is apparent that there is a trade-off between measurement precision (the term proportional to C_{eff}^{-1}) and back-action heating (the term proportional to C_{eff}). In Section 3.4 we considered the case of a phase quadrature measurement, such that $\theta = \pi/2$. In this limit, we see that the measurement imprecision term in Eq. (5.87) is minimised, suggesting that phase quadrature measurement is optimal. However, this is not the case in general, since the final term in the expression, which is zero when $\theta = \pi/2$, can be negative at other phase angles. This is evident from the plot of the mechanical displacement noise power spectral densities $\bar{S}_{Q_{\det} Q_{\det}}^{\theta}(\omega) = \bar{S}_{QQ}^{(0)}(\omega) + \bar{S}_{\det}^{\theta}(\omega)$ in Fig. 5.9a, where the dashed curve for a homodyne phase angle $\theta = \pi/10$ drops below the dotted curve for a phase quadrature measurement at a narrow band of frequencies.

It is possible, in principle, to broaden the band of frequencies over which ponderomotive squeezing improves the measurement precision by introducing dispersion to the output field of the optomechanical system, such that different

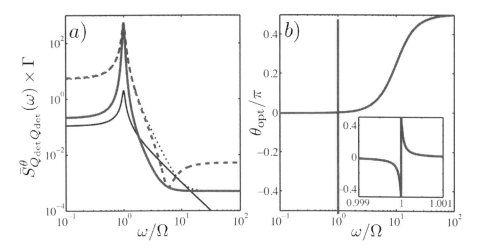

FIGURE 5.9 Surpassing the standard quantum limit using ponderomotive squeezing. (a) Mechanical displacement noise power spectral density. Dashed thick line: $\theta = \pi/10$; solid thick line: frequency-dependent optimal homodyne phase angle ($\theta = \theta_{\mathrm{opt}}$); dotted thin line: phase quadrature measurement ($\theta = 0$); solid line: standard quantum limit of displacement measurement (Eq. (3.65)). (b) Optimal homodyne phase angle θ_{opt} as a function of frequency. Inset: optimal homodyne phase angle close to the mechanical resonance frequency. For simplicity we have taken the effective cooperativity $C_{\mathrm{eff}}(\omega) = C$, which is valid as long as $\kappa \gg \omega$. Parameters: $P/P^{\mathrm{SQL}} = 1000$, $\bar{n} = 10$, $Q = 10$.

sideband frequencies experience different local oscillator phases [157] (see the bold solid curve in Fig. 5.9a).

Exercise 5.14 *Using Eq. (5.87), show that the power spectral density $\bar{S}^{\theta}_{\mathrm{det}}(\omega)$ is minimised for the local oscillator phase angle*

$$\tan \theta_{\mathrm{opt}} = \frac{-1}{2\Gamma \, |C_{\mathrm{eff}}| \, (\chi(\omega) + \chi^*(\omega))}, \tag{5.88}$$

and that for this phase angle

$$\bar{S}^{\theta_{\mathrm{opt}}}_{\mathrm{det}}(\omega) = \frac{1}{8\Gamma \, |C_{\mathrm{eff}}|} + 2\Gamma |\chi(\omega)|^4 \, |C_{\mathrm{eff}}| \left(\frac{\omega}{Q}\right)^2. \tag{5.89}$$

The optimal local oscillator phase exhibits a sharp dispersive feature around the mechanical resonance of width approximately determined by $\Gamma/|C_{\mathrm{eff}}|$, as well as a second broader feature, as shown in Fig. 5.9b. The sharp feature leads to challenging requirements for the dispersive element inserted into the output optical field. One attractive proposed method to achieve these requirements is to utilise the dispersive feature from optomechanically induced transparency discussed in Section 4.3 [185, 233].

Choosing the optimal interaction strength in Eq. (5.89)

$$|C_{\text{eff}}|_{\text{opt}} = \frac{Q}{4\Gamma |\chi(\omega)|^2 \omega} \tag{5.90}$$

we find – just as we did for measurement enhancement via incident squeezing in the previous subsection – that the minimum achievable mechanical position detection noise power spectral density exactly saturates the zero-point limit of Eq. (3.61). Consequently, ponderomotive squeezing allows the standard quantum limit to be reached at the mechanical resonance frequency and allows precision that, while fundamentally bounded at the level of zero-point motion, exceeds the standard quantum limited precision at all frequencies away from resonance.

CHAPTER 6

Single-photon optomechanics

CONTENTS

6.1	Optomechanical photon blockade	....	194
6.2	Single-photon states	....	196
	6.2.1	Quantum Langevin equation method	197
	6.2.2	Cascaded master equation method	199
	6.2.3	Fock state master equation method	200
6.3	Single-photon pulse incident on a single-sided optomechanical cavity	....	202
6.4	Double-cavity optomechanical system	....	206
	6.4.1	Single-photon input	....	206
	6.4.2	Two-photon input	....	213
6.5	Macroscopic superposition states using single-photon optomechanics	....	219
	6.5.1	Conditional entangled state of two mechanical resonators	....	225
	6.5.2	Photon blockade with two single-photon pulses	....	227
6.6	Single-sideband-photon optomechanics	....	230

In most experiments to date the vacuum optomechanical coupling rate g_0 has been small compared to the dissipative rates of the system. In Chapter 2 we showed how to compensate for this by linearising the nonlinear dynamics about the steady-state amplitude α_{ss} in the cavity to give an effective coupling constant $g \equiv g_0 \alpha_{ss}$. This approach fails in regimes where g_0 is sufficiently large for the presence of a single-photon to significantly alter the system dynamics. It also fails when the cavity is not driven by a coherent field, for example, a sequence of single-photon states, as in that case the average amplitude of the field is zero. In this chapter we consider what happens when the optomechanical coupling rate is sufficiently large that a single-photon can be used to control the mechanical motion. This is now possible in optomechanics experiments using ensembles of cold atoms [51]. It has not yet been achieved in more macroscopic solid-state optomechanical systems. However, rapid experimental progress is being made in engineering large single-photon optomechanical coupling in technologies such as optomechanical crystals at optical frequencies[99] and bulk-element superconducting resonators at microwave frequencies[227]. Indeed, evidence of the effect of single microwave photons on the dynamics of the mechanical resonator has been reported in [227].

The *single-photon strong coupling regime* where $g_0 > \{\kappa, \Gamma\}$ and the regime of *optomechanical photon blockade* are two particularly notable regimes where the linearised optomechanical Hamiltonian fails. This chapter introduces some examples of optomechanical protocols that can be implemented in these regimes. We will need methods for dealing with Fock state inputs to optical cavities. This can be done using quantum Langevin equations provided we are able to evaluate the correlation functions for the input field when that field is prepared in such a nonclassical state as single-photon pulse. If we wish to use the Schrödinger picture, we need to find an analog of the master equation (see Chapter 5) that can describe single-photon input states. This can be done using the Fock state master equation method devised by Combes and collaborators [127]. We will consider a simple example to show how the radiation pressure interaction leads to entanglement between single photons emitted from the cavity and the mechanical degree of freedom. We go on to show how this interaction can be used in a double cavity optomechanical system to stochastically drive the mechanical system to an energy eigenstate. In the final example we show how the mechanical degree of freedom in the double cavity system can be used to implement a controlled beam splitter interaction and show how the semiclassical limit is obtained for mechanical systems driven into large amplitude coherent motion. This provides an explicit example for how quantum control can be reduced to classical control in a suitable limit.

6.1 OPTOMECHANICAL PHOTON BLOCKADE

In optomechanical photon blockade, a single-photon absorbed by the cavity can give a sufficiently large impulse to the mechanical element that its subsequent motion can detune the cavity, making it less likely that a second photon will be absorbed.

The polaron transformation introduced in Section 4.5.2 provides an elegant approach to understanding the physics of optomechanical photon blockade [234]. We saw there that the radiation pressure interaction can be removed and replaced with an effective Kerr nonlinearity for the cavity field. A single intracavity optical photon then acts to shift the resonance frequency of the cavity by $\chi_0 = g_0^2/\Omega$, where as usual Ω is the mechanical resonance frequency.

We found in Section 4.5.2 that, in the absence of coherent driving and bath interactions, the polaron transform diagonalises the optomechanical Hamiltonian. The energy eigenstates are then product states of photon and phonon number $|E_{nm}\rangle = |n\rangle_a \otimes |m\rangle_b$ where $E_{nm} = \hbar(\Delta n - \chi_0 n^2) + \hbar\Omega m$. With a fixed

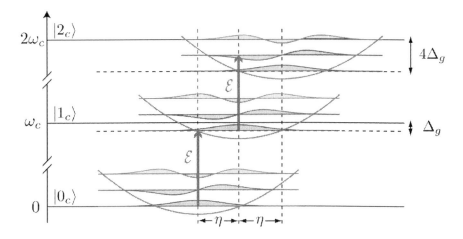

FIGURE 6.1 Energy level diagram of an optomechanical system, showing the mechanical displacements and optical energy level shifts that result from optical excitation and therefore illustrating the central concepts behind optomechanical photon blockade. Here $|n_c\rangle$ labels the energy of the nth eigenstate of the optical cavity in the absence of any optomechanical interaction $(g_0 = 0)$, $\eta = g_0/\Omega$, $\Delta_g = \chi_0$, and ω_c is the optical cavity resonance frequency $(\Omega_c$ in our notation). Reprinted with permission from [234] Copyright 2011 by the American Physical Society.

value for m, the energy levels are not equally spaced and the gap between successive energy levels diminishes as n increases (see Fig. 6.1). When the driving laser is switched on and tuned to resonance with the cavity $(\Delta = 0)$ so that the $n = 0 \to 1$ transition is resonant, the effect of the anharmonicity is to make the $n = 1 \to 2$ transition off-resonant. This transition will therefore have a reduced transition probability provided that the width of the cavity resonance is much less that the anharmonicity, $\kappa \ll \chi_0$. This is the idea of a one-photon blockade: the cavity can easily absorb one photon but having done so the probability per unit time to absorb a second is suppressed. In order to absorb a second photon, the first photon must be emitted from the cavity. This leads to correlations in the photon emission rate from the cavity and will be seen in the second-order correlation function $g^{(2)}(0)$ for photon counting of emission events from the cavity. Unlike the linearised model of optomechanical systems which we have examined up until now, the states exiting the optomechanical system here are not constrained to be Gaussian. Indeed, in the limit of strong photon blockade of the optomechanical system, a weak coherent incident pulse of light may be transformed into an output single-photon. Such states are highly non-Gaussian and display strong Wigner negativity.

6.2   SINGLE-PHOTON STATES

We saw in the previous section that, within the regime of photon blockade, highly nonclassical states of light may be produced via the optomechanical interaction. In what follows we will consider the reverse scenario where photon number states of light are injected into the optomechanical system. Let us consider the case in which the field driving the cavity is made up of a sequence of pulses with exactly N photons per pulse. Mostly we will be interested in the case $N = 1$, a single-photon state. As we saw in Chapter 2, the state of the field driving the cavity is specified by giving the state of the input field operators $a_{\text{in}}(t)$. Until now this field was assumed to be Gaussian with nonzero mean field and typically stationary. Single-photon states are not like this. We first need to understand how to treat the input quantum stochastic process in the case that the cavity is driven by a single-photon state.

We consider the single-photon state as a superposition of a single excitation over many frequencies with a spectral modeshape defined by the amplitude $\xi(\omega)$ in an interaction picture rotating at the carrier frequency Ω_L. The state can then be defined as

$$|1_\xi\rangle \equiv \int_{-\infty}^{\infty} d\omega \; \xi(\omega) a_{\text{in}}^\dagger(\omega)|0\rangle, \qquad (6.1)$$

where $\xi(\omega)$ is normalised such that $\int_{-\infty}^{\infty} d\omega \, |\xi(\omega)|^2 = 2\pi$. Here, $|0\rangle$ is the global vacuum state of the electromagnetic field, i.e., it is the state annihilated by the positive frequency components of the field. The action $\hat{a}_{\text{in}}^\dagger(\omega)|0\rangle$ is to introduce a photon in the frequency component ω, while leaving all other frequency components in their ground states. The state $|1_\xi\rangle$ is thus a coherent superposition of a single excitation in all frequency modes, weighted by probability amplitudes given by $\xi(\omega)$.

Exercise 6.1 *Using Eq. (6.1) and the properties of ladder operators in Section 1.1.1, show that the average field amplitude of a single-photon state is zero*

$$\langle 1_\xi|a_{\text{in}}(t)|1_\xi\rangle = 0. \qquad (6.2)$$

Then show that

$$a_{\text{in}}(t)|1_\xi\rangle = \xi(t)|0\rangle, \qquad (6.3)$$

where

$$\xi(t) = \int_{-\infty}^{\infty} d\omega \; e^{-i\omega t}\xi(\omega) \qquad (6.4)$$

is the Fourier transform of $\xi(\omega)$.

Exercise 6.2 *The probability per unit time to detect a single-photon on an ideal detector is proportional to $n(t) = \langle a_{\text{in}}^\dagger(t)a_{\text{in}}(t)\rangle$. Show that, for a single-photon state,*

$$n(t) = |\xi(t)|^2. \qquad (6.5)$$

6.2.1 Quantum Langevin equation method

Despite the vanishing of the average field amplitude of a single-photon state, we can use much of our classical intuition to understand how linear optical systems respond to single-photon driving. Suppose we consider a single-sided optical cavity with cavity damping rate $\sqrt{\kappa}$. In Chapter 1 we saw that the quantum stochastic differential equation for the intracavity field operator is given by

$$\dot{a} = -(\kappa/2 + i\Delta)a + \sqrt{\kappa}a_{\text{in}}(t), \tag{6.6}$$

where, as usual, $\Delta \equiv \Omega_c - \Omega_L$ is the detuning between the cavity resonance and the carrier frequency of the single-photon pulse. The solution is

$$a(t) = a(0)e^{-(\kappa/2+i\Delta)t} + \sqrt{\kappa}\int_0^t dt' e^{-(\kappa/2+i\Delta)(t-t')}a_{\text{in}}(t'). \tag{6.7}$$

While the single-photon input does not contribute to the average field,

$$\langle a(t) \rangle = \langle a(0) \rangle e^{-(\kappa/2+i\Delta)t}, \tag{6.8}$$

it does change the average photon number in the cavity,

$$\langle a^\dagger(t)a(t) \rangle = \langle a^\dagger(0)a(0) \rangle e^{-\kappa t} + \kappa e^{-\kappa t}\left|\int_0^t dt'\, e^{(\kappa/2+i\Delta)t'}\xi(t')\right|^2. \tag{6.9}$$

As a specific example, a cavity single-photon source with an emission rate γ has a probability amplitude function of the form

$$\xi(t) = \sqrt{\gamma}e^{-\gamma t/2}H(t), \tag{6.10}$$

where $H(t)$ is the Heaviside step function. This form could describe the photon emitted from a single-sided optical cavity with linewidth γ excited to a single-photon state at $t = 0$. Let us also assume the cavity is empty at time $t = 0$. Then

$$\langle a^\dagger(t)a(t) \rangle = \frac{4\kappa\gamma}{(\kappa-\gamma)^2 + 4\Delta^2}\left|e^{-(\gamma/2-i\Delta)t} - e^{-\kappa t/2}\right|^2. \tag{6.11}$$

This is plotted in Fig. 6.2. We see that the mean intracavity photon number rises after some delay, and then falls as the photons leak away. As the detuning increases, the peak intracavity photon number decreases as more photons are reflected from the increasingly off-resonance cavity.

We can also compute the rate of photon counting for the output field from the cavity. Using the input-output relations from Chapter 1, $a_{\text{out}}(t) = a_{\text{in}}(t) - \sqrt{\kappa}a(t)$, we see that

$$\langle a_{\text{out}}^\dagger(t)a_{\text{out}}(t) \rangle = \langle a_{\text{in}}^\dagger(t)a_{\text{in}}(t) \rangle + \kappa\langle a^\dagger(t)a(t) \rangle - \sqrt{\kappa}\langle a_{\text{in}}^\dagger(t)a(t) + a^\dagger(t)a_{\text{in}}(t) \rangle. \tag{6.12}$$

The last term is an interference term. If we detect a photon outside the cavity,

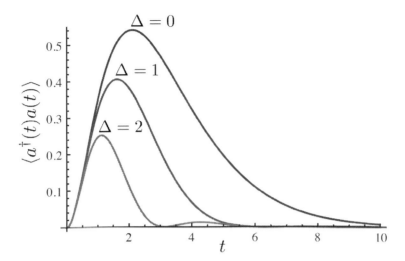

FIGURE 6.2 The intracavity mean photon number for a single-sided cavity versus time for a initially empty cavity driven by a single-photon pulse with carrier frequency detuned from the cavity by $\Delta = \Omega_c - \Omega_L$. We chose units such that the cavity amplitude decay rate is $\kappa = 1$. Various values of detuning are shown.

we cannot know if it was detected after being reflected from the cavity mirror, or if it first entered the cavity, and then was emitted. These two pathways interfere, and the last term represents this. Using the previous example, we see that

$$
\begin{aligned}
\langle a_{\text{out}}^{\dagger}(t)a_{\text{out}}(t)\rangle \;=\; & \gamma e^{-\gamma t}H(t) + \frac{4\kappa^2\gamma}{(\kappa-\gamma)^2+4\Delta^2}\left|e^{-(\gamma/2-i\Delta)t} - e^{-\kappa t/2}\right|^2 \\
& -\frac{4\kappa\gamma(\kappa-\gamma)}{(\kappa-\gamma)^2+4\Delta^2}\left[e^{-\gamma t} - e^{-(\kappa+\gamma)t/2}\cos\Delta t\right] \\
& +\frac{8\Delta\kappa\gamma}{(\kappa-\gamma)^2+4\Delta^2}e^{-(\kappa+\gamma)t/2}\sin\Delta t.
\end{aligned}
\tag{6.13}
$$

The output counting rate is shown in Fig. 6.3, where we can clearly see the effect of the interference term as a zero in the count rate. Photons counted after this time are very likely to have been emitted from inside the cavity.

As we have seen in earlier chapters, it is often easier to consider the output field in terms of its Fourier components. Inspection of Eq. (6.7) indicates that, if we neglect transients, the response of the cavity is a convolution of the input field operator and an exponential. Thus

$$
a(\omega) = \frac{\sqrt{\kappa}\,a_{\text{in}}(\omega)}{\kappa/2 - i(\omega - \Delta)}.
\tag{6.14}
$$

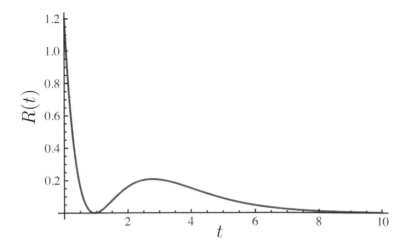

FIGURE 6.3 The photon number detection rate, $R(t) = \langle a^{\dagger}_{\text{out}}(t) a_{\text{out}}(t) \rangle$, outside of a single sided cavity versus time for a initially empty cavity driven by a single-photon pulse with carrier frequency on resonance with the cavity ($\Delta = 0$). We choose units such that the source and cavity amplitude decay rates are $\gamma = 1$ and $\kappa = 1.2$, respectively. The zero in the count rate is due to the interference between the two indistinguishable histories for a photon counted outside that cavity: reflection from the cavity mirror, and emission from within the cavity.

Using the Fourier space version of the input-output relations, we then find that

$$a_{\text{out}}(\omega) = \left[\frac{\kappa/2 + i(\omega - \Delta)}{\kappa/2 - i(\omega - \Delta)} \right] a_{\text{in}}(\omega). \tag{6.15}$$

In Fourier space, this is simply a frequency-dependent phase shift. In the case of a single-photon input, it is not difficult to see that the output field (in the long time limit) is also a single-photon with frequency-dependent amplitude given by

$$\xi_{\text{out}}(\omega) = \left[\frac{\kappa/2 + i(\omega - \Delta)}{\kappa/2 - i(\omega - \Delta)} \right] \xi_{\text{in}}(\omega). \tag{6.16}$$

6.2.2 Cascaded master equation method

There is an alternative way to model the single-photon input in the Schrödinger picture by explicitly including the single-photon source. As an example, we will assume that the source is a cavity source of the kind described in [211]. This can be modelled as a single-sided cavity that, at time $t = 0$, is excited to a single-photon state. The field that leaves this source cavity is then irreversibly coupled[1] into the optomechanical cavity. We model this kind of single-photon excitation process using the theory of cascaded quantum systems [58, 112] to give a master equation for the dynamics of the optomechanical system plus the source cavity

$$\dot{\rho} = \frac{\sqrt{\kappa\gamma}}{2}[ca^{\dagger} - c^{\dagger}a, \rho] + \mathcal{D}[J]\rho + \gamma\mathcal{D}[c]\rho + \kappa\mathcal{D}[a]\rho, \qquad (6.17)$$

where ρ is the joint state of the source and optomechanical cavity, $c, c^{\dagger}$ are the annihilation and creation operators for photon number states in the source cavity, the emission superoperator $\mathcal{D}$ is defined in Section 5.1.3, and the detection (or "jump") operator is $J \equiv \sqrt{\gamma}c + \sqrt{\kappa}a$, which consists of the sum of two terms: the first term shows that the photon can be detected directly from the source before it is transmitted to cavity a, and the second describes a photodetection event for a photon emitted for the cavity mode a. We have also assumed that the source cavity is on resonance with the receiver cavity, a.

The rate at which photons are detected after the optomechanical cavity is given by

$$R(t) = \langle J^{\dagger}J\rangle = \gamma\langle c^{\dagger}c\rangle + \kappa\langle a^{\dagger}a\rangle + \sqrt{\kappa\gamma}\langle a^{\dagger}c + c^{\dagger}a\rangle . \qquad (6.18)$$

The last term in the second expression is the same interference term discussed previously (see Eq. (6.12)). This is easily verified by solving the equations of motion for the three sets of quadratic moments, $\langle a^{\dagger}a\rangle, \langle c^{\dagger}c\rangle, \langle a^{\dagger}c + c^{\dagger}a\rangle$.

Exercise 6.3 *Derive the set of coupled equations of motion for each of these quadratic moments and verify that, for $t \geq 0$,*

$$\langle c^{\dagger}c\rangle = e^{-\gamma t} \qquad (6.19)$$

$$\langle a^{\dagger}a\rangle = \frac{4\kappa\gamma}{(\kappa - \gamma)^2}\left(e^{-\gamma t/2} - e^{-\kappa t/2}\right)^2 \qquad (6.20)$$

$$\langle a^{\dagger}c + c^{\dagger}a\rangle = -\frac{4\sqrt{\kappa\gamma}}{\gamma - \kappa}e^{-\gamma t/2}\left(e^{-\kappa t/2} - e^{\kappa t/2}\right) . \qquad (6.21)$$

6.2.3 Fock state master equation method

An alternative approach to the quantum Langevin method has been developed by Combes and co-workers [127]. This is based on a generalised master equation approach. Master equations have been widely used in quantum optics to describe coherent input fields together with Gaussian noise. A single-photon input is quite different; it is time dependent and thus nonstationary and does not correspond to Gaussian noise.

---

[1] This requires a circulator to prevent the field reflected from the cavity from propagating back into the source.

Consider, again, the case of a single-photon pulse input to an empty single-sided cavity. If the initial state of the cavity system is $|\eta\rangle$ and the initial state of the external input field is the one-photon state $|1_\xi\rangle$, then the joint initial state is $|\eta 1_\epsilon\rangle = |\eta\rangle \otimes |1_\epsilon\rangle$. The total state of the external field and cavity field at time $t > 0$ is $|\Psi(t)\rangle = U(t)|\eta\rangle \otimes |1_\epsilon\rangle$. As the interaction proceeds, the external field and the cavity field become entangled states, with one photon in the external field and no photons in the cavity superposed with no photons in the external field and one photon in the cavity. Initially, the cavity field is in the vacuum state. It then becomes excited before eventually decaying back to the vacuum as the photon is emitted into the external field.

Let $\hat{A}$ be a *system* operator (for example, $a^\dagger a$) and define the Heisenberg picture operator

$$j_t(\hat{A}) = U^\dagger(t)\left(\hat{A} \otimes \mathbb{I}_f\right) U(t) , \tag{6.22}$$

where $\mathbb{I}_f$ is the identity operator acting on the field. Note that, because $U(t)$ describes an interaction between the system and the field external to the cavity, $j_t(\hat{A})$ is, generally, a joint system-field operator. How does this operator evolve in time?

The objective is to find the moments of such time-evolved operators. For initial state $|\eta 1_\epsilon\rangle = |\eta\rangle \otimes |1_\epsilon\rangle$, we can define matrix elements of $j_t(\hat{A})$ in the Heisenberg picture as

$$\varpi_t^{mn}(\hat{A}) = \langle \eta, m_\xi | j_t(\hat{A}) | \eta, n_\xi \rangle . \tag{6.23}$$

For example, if $\hat{A} = a^\dagger a$, then $\varpi_t^{11}(a^\dagger a) \equiv \langle \hat{n}(t) \rangle$ is simply the mean photon number inside the cavity at time t. Equivalently, we can work in the Schrödinger picture in which states evolve with time. To this end, we define the time-dependent system operators $\rho_{m,n}(t)$ to give the same moments,

$$\text{Tr}[\rho_{mn}(t)\hat{A}] = \varpi_t^{mn}(\hat{A}) . \tag{6.24}$$

Taking the time derivative of this expression and using the cyclic property of trace, we find the *Fock state master equation* for an optical cavity driven by number state input fields with temporal modeshape $\xi(t)$,

$$
\begin{aligned}
\dot{\rho}_{mn} &= \frac{1}{i\hbar}[\hat{H}, \rho_{mn}] + \kappa \mathcal{D}[a]\rho_{mn} \\
&\quad + \sqrt{\kappa m}\xi(t)[\rho_{m-1n}, a^\dagger] + \sqrt{\kappa n}\xi^*(t)[a, \rho_{mn-1}] \\
&\equiv \mathcal{L}[\rho_{mn}] + \sqrt{\kappa m}\xi(t)[\rho_{m-1n}, a^\dagger] + \sqrt{\kappa n}\xi^*(t)[a, \rho_{mn-1}] , \quad (6.25)
\end{aligned}
$$

where $\mathcal{L}$ is the Liouvillian superoperator (see also Section 5.1.3), whose action on an operator $\hat{A}$ is defined, within the rotating wave approximation and assuming that the optical bath has zero occupancy, as

$$\mathcal{L}[\hat{A}] \equiv \frac{1}{i\hbar}[\hat{H}, \hat{A}] + \kappa \mathcal{D}[a]\hat{A}. \tag{6.26}$$

Just as in Eq. (6.23), the indices m and n take the values of the possible single-photon excitations in the external field, in this case simply $1, 0$.

6.3 SINGLE-PHOTON PULSE INCIDENT ON A SINGLE-SIDED OPTOMECHANICAL CAVITY

As a first example of single-photon optomechanics, we consider a single-sided cavity with single-photon pulse incident. From Chapter 1, the quantum Langevin equations for the cavity amplitude operator a and the mirror displacement amplitude b are

$$\dot{a} = -i(\Delta + g_0(b + b^\dagger))a - \frac{\kappa}{2}a + \sqrt{\kappa}a_{\text{in}} \qquad (6.27)$$

$$\dot{b} = -i\Omega b - ig_0 a^\dagger a . \qquad (6.28)$$

For simplicity, we have neglected the mechanical damping. This is, of course, only valid if the mechanical reservoir is at a very low temperature and even then assumes that the mechanical damping is slow compared to all other time scales.

Let us begin by assuming that the time duration of the both the single-photon pulse (γ^{-1}), and the cavity decay (κ^{-1}), is much shorter than the mechanical period Ω^{-1} (i.e., the bad cavity limit). As a first approximation, we thus regard the mirror as stationery over the time during which the cavity fills and empties. We also assume that $\kappa \gg g_0$. We can then take the mechanical position $\hat{Q} \equiv (b + b^\dagger)/\sqrt{2}$ to be a constant, and integrate the field Langevin equation directly. In the Fourier domain, neglecting transients, the output optical field operator is then given by Eq. (6.15) with $\Delta \to \Delta + g_0(b + b^\dagger)$. In the corresponding Schrödinger picture, if the state of the system is pure this implies that the output field is entangled with the mechanical degree of freedom. In other words, the output field is correlated with the position of the mirror at the time of its interaction with the input single-photon pulse.

How can one extract the mechanical displacement information from the output field? Standard homodyne detection will not work here, as the average field of a single-photon is zero (see Exercise 6.1).[2] Another way is to consider a Hong–Ou–Mandle (HOM) interferometer, in which the output field from the cavity is mixed on a 50/50 beam splitter with an identical input photon that has been passed through an identical cavity but without any mechanical element. In a HOM interferometer, if the two photons incident on the beam splitter are identical, the coincidence rate for photodetections at each of the output ports goes to zero [241]. More generally, the coincidence rate is given by

$$C = \frac{1}{2} - \frac{1}{8\pi^2} \iint d\omega \, d\omega' \, \xi_{\text{out},1}(\omega)\xi^*_{\text{out},1}(\omega')\xi^*_{\text{out},2}(\omega)\xi_{\text{out},2}(\omega') , \qquad (6.29)$$

where $\xi_{\text{out},1}(\omega)$ and $\xi_{\text{out},2}(\omega)$ are the output spectral modeshape functions of the two photons. The spectral modeshape function of the photon that has not interacted with the mechanical element, $\xi_{\text{out},2}(\omega)$, is given directly by

---

[2]Although state tomography based on homodyne detection could be used.

Eq. (6.16), while the spectral modeshape function of the photon that has interacted with the mechanical element, $\xi_{\text{out},1}(\omega)$, is given by Eq. (6.16) with the replacement $\Delta \to \Delta + \sqrt{2}g_0\hat{Q}$.

Exercise 6.4 (a) Taking the case of zero detuning ($\Delta = 0$), show that

$$C = \frac{1}{2} - \frac{1}{8\pi^2} \left| \int d\omega \, |\xi_{\text{in}}(\omega)|^2 \left[\frac{\kappa/2 + i(\omega - \sqrt{2}g_0\hat{Q})}{\kappa/2 - i(\omega - \sqrt{2}g_0\hat{Q})} \right] \left(\frac{\kappa/2 - i\omega}{\kappa/2 + i\omega} \right) \right|^2 .$$

(b) Expand this expression to second order in $g_0\hat{Q}$ and trace over the mechanical degrees of freedom, to show that the expectation value of the coincidence rate is

$$\langle C \rangle = Rg_0^2 \langle \hat{Q}^2 \rangle, \tag{6.30}$$

where

$$R = \kappa^2 \left[\frac{1}{2\pi} \int d\omega \frac{|\xi_{\text{in}}(\omega)|^2}{(\kappa^2/4 + \omega^2)^2} - \left(\frac{1}{2\pi} \int d\omega \frac{|\xi_{\text{in}}(\omega)|^2}{\kappa^2/4 + \omega^2} \right)^2 \right] \tag{6.31}$$

quantifies sensitivity of the overlap between the two single photon pulses to mechanical displacements, which depends both on the input pulse modeshape and the decay rate, κ, of the optomechanical cavity.

To describe the system in the resolved sideband limit of $\kappa \ll \Omega$, we will use the Fock state master equation method. It is easier to solve this problem in the interaction picture, for which the Hamiltonian is

$$\hat{H}_I(t) = \hbar g_0 a^\dagger a(be^{-i\Omega t} + b^\dagger e^{+i\Omega t}) . \tag{6.32}$$

From Eq. (6.25), we see that $\dot{\rho}_{00} = \mathcal{L}\rho_{00}$, so that the initial state is a fixed point for ρ_{00}. Thus

$$\rho_{00}(t) = \rho_{00}(0) = |0\rangle_a\langle 0| \otimes |0\rangle_b\langle 0| . \tag{6.33}$$

Substituting this into the equation for $\rho_{10}(t)$, we find that $\rho_{10}(t)$ satisfies the same equation as $\rho_{00}(t)$ but with an inhomogeneous source term of the form $-\xi(t)|1\rangle_a\langle 0| \otimes |0\rangle_b\langle 0|$:

$$\dot{\rho}_{10} = \mathcal{L}[\rho_{10}] - \sqrt{\kappa}\xi(t)\hat{S}_{10} , \tag{6.34}$$

where $\hat{S}_{10} \equiv |1\rangle_a\langle 0|\otimes|0\rangle_b\langle 0|$ and, using Eq. (6.26), the action of the Liouvillian superoperator is

$$\mathcal{L}[\hat{A}] = -ig_0 \left[a^\dagger a(be^{-i\Omega t} + b^\dagger e^{i\Omega t}), \hat{A} \right] + \kappa \left(a\hat{A}a^\dagger - \frac{1}{2}a^\dagger a\hat{A} - \frac{1}{2}\hat{A}a^\dagger a \right) . \tag{6.35}$$

The solution to this equation is

$$\rho_{10}(t) = -\sqrt{\kappa} \int_0^t dt' \, \xi(t')e^{-\kappa(t-t')/2}\hat{R}^\dagger(t)\hat{R}(t')\hat{S}_{10}, \tag{6.36}$$

where $\hat{R}(t)$ is given by

$$
\begin{aligned}
\hat{R}(t) \;=\;& 1 + ig_0 \int_0^t dt_1 (b(t_1) + b^\dagger(t_1)) \\
& + (ig_0)^2 \int_0^t dt_2 \int_0^{t_2} dt_1 (b(t_1) + b^\dagger(t_1))(b(t_2) + b^\dagger(t_2)) + (ig)^3 \cdots \\
\equiv\;& \mathcal{A} : \exp\left[ig_0 \int_0^t dt' (b(t') + b^\dagger(t')) \right],
\end{aligned}
$$

which defines the anti-time-ordering operator, with $b(t) = be^{-i\Omega t}$. Clearly, $\hat{R}^{-1}(t) = \hat{R}^\dagger(t)$.

The top level equation is

$$
\dot{\rho}_{11} = \mathcal{L}[\rho_{11}] + \sqrt{\kappa}\xi^*(t)[a, \rho_{10}] + \sqrt{\kappa}\xi(t)[\rho_{01}, a^\dagger] . \tag{6.37}
$$

Substituting into this equation of motion the solution for ρ_{10} in Eq. (6.36), we obtain

$$
\begin{aligned}
\dot{\rho}_{11} \;=\;& \mathcal{L}[\rho_{11}] + \kappa \left[|1\rangle_a\langle 1| - |0\rangle_a\langle 0| \right] \tag{6.38} \\
& \times \left[\int_0^t dt' \, \xi^*(t)\xi(t')e^{-\kappa(t-t')/2}\hat{R}^\dagger(t)\hat{R}(t')|0\rangle_b\langle 0| + h.c. \right] .
\end{aligned}
$$

Multiplying each side by $a^\dagger a$ and taking the trace (see Eq. (6.23)), we see that the mean photon number in the cavity is given by the solution to the equation

$$
\frac{d\langle a^\dagger a\rangle}{dt} = -\kappa\langle a^\dagger a\rangle + \kappa \left(\int_0^t dt' \, \xi^*(t)\xi(t')\langle\psi_b(t)|\psi_b(t')\rangle e^{-\kappa(t-t')/2} + c.c. \right), \tag{6.39}
$$

where

$$
|\psi_b(t)\rangle = \hat{R}(t)|0\rangle_b . \tag{6.40}
$$

The equation of motion for the cavity amplitude, obtained in the same fashion, does not have any contribution from the single-photon driving term, and is given by

$$
\frac{d\langle a\rangle}{dt} = -ig_0\langle a(be^{-i\Omega t} + b^\dagger e^{i\Omega t})\rangle . \tag{6.41}
$$

This indicates an effective phase modulation of the single-photon state inside the cavity due to the oscillatory motion of the mirror. Indeed, this is what is indicated in Eq. (6.39). Inspection of the integrand in the second term in Eq. (6.39) shows that the input field two-time correlation function $\langle a_{\text{in}}^\dagger(t)a_{\text{in}}(t')\rangle$ is effectively modified by the mechanical response,

$$
\langle a_{\text{in}}^\dagger(t)a_{\text{in}}(t')\rangle = \xi^*(t)\xi(t) \to \xi^*(t)\xi(t')\langle\psi_b(t)|\psi_b(t')\rangle . \tag{6.42}
$$

In the case of an optomechanical cavity, the effect of the mechanical phase modulation on the single-photon is captured by the function $\langle\psi_b(t)|\psi_b(t')\rangle$.

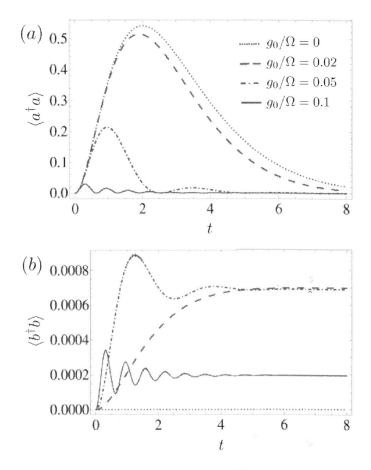

FIGURE 6.4 The intracavity photon number (a) and the mean vibrational quantum number of the mechanical resonator (b) as a function of time for an optomechanical cavity driven by a single-photon and with increasing values of the coupling constant in units of the mechanical frequency, $g_0/\Omega = 0.0, 0.02, 0.05, 0.1$. Here, $\kappa/\Omega = \gamma/\Omega = 0.001$ with γ being the decay rate of the single-photon source cavity, as previously defined.

The state, to lowest order, is close to an oscillator coherent state with the amplitude $\beta(t)$. This is the semiclassical amplitude of an oscillator, with an initial zero position and momentum, subject to a forcing term of the form in Eq. (6.32) with the cavity in a one-photon eigenstate. The transformation of the input single-photon amplitudes then appears as a phase modulation, averaged over the initial vacuum state of the mechanical resonator. In Fig. 6.4(a) we show the intracavity photon number as a function of time for increasing values of g_0/Ω. We see that, when this ratio is large, the excitation number of the cavity field is low and even goes to zero for certain times. At these times the photon is reflected from the cavity. These oscillations are evidence of the strong self phase modulation of the field in the cavity as the mechanical resonator begins to move. This is evident in Fig. 6.4(b), where we plot the average excitation energy in the mechanical resonator. The oscillations are at the mechanical frequency and in phase with the mean photon number in the cavity. For low values of the ratio g_0/Ω, the mechanical system is simply driven to a nonzero amplitude coherent state.

6.4 DOUBLE-CAVITY OPTOMECHANICAL SYSTEM

We now turn to consideration of a coupled-cavity system as shown in Fig. 6.5, in which the rate at which photons are coherently transferred between cavities depends upon the displacement of a mechanical degree of freedom. An example of this kind of system has been developed by the Painter group [67]. Another example is based on a single bulk flexural mode driven by the opposing radiation pressure forces of two optical cavity modes. If the cavity modes are coupled, transformation to normal modes leads to a model in which the normal mode coupling is modulated by the mechanical displacement [184].

The description of this system is given in terms of the Hamiltonian

$$\hat{H} = \hbar\Omega_1 a_1^\dagger a_1 + \hbar\Omega_2 a_2^\dagger a_2 + \hbar\Omega b^\dagger b + \hbar g_0 (b + b^\dagger)(a_1^\dagger a_2 + a_1 a_2^\dagger), \qquad (6.43)$$

where a_1 and a_2 are the annihilation operators for the fields in cavities 1 and 2, Ω_1 and Ω_2 are their respective resonance frequencies, Ω is, as usual, the mechanical resonance frequency, and we assume optomechanical coupling rates of equal magnitude g_0 between the mechanics and each of optical cavities. We will further assume that the system is so designed that $\Omega_2 = \Omega_1 + \Omega$. The optomechanical interaction picture Hamiltonian, including only resonant terms, is

$$\hat{H}_I = \hbar g_0 (b^\dagger a_1^\dagger a_2 + b a_1 a_2^\dagger). \qquad (6.44)$$

This represents a kind of coherent Raman process whereby one photon from cavity 2 is transferred into cavity 1, simultaneously exciting one phonon in the mechanical degree of freedom.

6.4.1 Single-photon input

We will first assume that one cavity is driven by a single photon while the other has vacuum input. In that case, at most, there is one photon in the system at any time. It is convenient to use a dual-rail qubit encoding for the single-photon states of each cavity,

$$|0\rangle \equiv |1\rangle_1 |0\rangle_2 \qquad (6.45)$$
$$|1\rangle \equiv |0\rangle_1 |1\rangle_2, \qquad (6.46)$$

where $|n\rangle_i$ are photon number eigenstates for cavity a_i. In this restricted subspace, we can define

$$a_1^\dagger a_2 = |0\rangle\langle 1| \equiv \sigma_-, \qquad (6.47)$$

$$a_1 a_2^\dagger = |1\rangle\langle 0| \equiv \sigma_+, \qquad (6.48)$$

where $\sigma_\pm$ are the usual raising and lowering operators for a pseudo-spin system. The optomechanical interaction Hamiltonian can then be written as a Jaynes–Cummings Hamiltonian

$$\hat{H}_I = \hbar g_0 (b\sigma_+ + b^\dagger \sigma_-). \qquad (6.49)$$

Each cavity mode is treated as a single-sided cavity with the photon decay rates given by κ_j with $j = 1, 2$. The single photon source, modelled as a source cavity prepared with a single photon, is irreversibly coupled into the input of the cavity 2. As before, the emission rate from the source cavity is γ. The input single-photon states thus constitute a pulse with an exponential

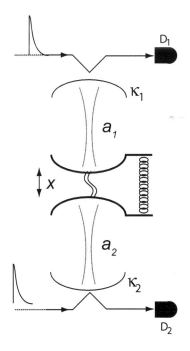

FIGURE 6.5 A representation of two coherently coupled cavities with the coupling rate proportional to a mechanical displacement. Each cavity is coupled on one side to a waveguide mode with, at most, one photon. D_1 and D_2 represent single-photon counters.

temporal profile with lifetime γ^{-1}. In this example we will proceed using the cascaded master equation method (see Section 6.2.2).

The Jaynes–Cummings representation in Eq. (6.49) allows an interesting parallel to be drawn with the approach to cavity quantum electrodynamics of the Haroche group [133], allowing the system to be configured to yield information on the state of the mechanical system. We will first review the simple case in which interactions are deterministic. In the Haroche approach the qubit is realised as a single two-level atom, while the bosonic degree of freedom is a microwave cavity field. In our case, the qubit is a two-mode optical field while the bosonic degree of freedom represents a mechanical mode. The mathematical description is the same.

Let the qubit begin in the state $|1\rangle$, while the mechanical state is arbitrary. The state of the system in the interaction picture then evolves, as always, as

$$|\psi(t)\rangle = \hat{U}(t)|\psi(0)\rangle, \tag{6.50}$$

where the time evolution unitary $\hat{U}(t) = \exp(-i\hat{H}_I t/\hbar)$ with the Hamiltonian $\hat{H}_I$ given in Eq. (6.49). After an interaction time τ, we make a projective measurement of the qubit state, resulting in a single binary number x and collapsing the qubit to the state $|x\rangle_\sigma$, where here we use the subscript σ to denote the qubit. The resulting conditional (unnormalised) state for the bosonic degree of freedom is

$$
\begin{aligned}
|\tilde{\psi}^{(x)}\rangle &= {}_\sigma\langle x|\psi(t)\rangle && (6.51)\\
&= {}_\sigma\langle x|\hat{U}(\tau)|1\rangle_\sigma|\psi(0)\rangle_b && (6.52)\\
&\equiv \hat{E}(x)|\psi(0)\rangle_b, && (6.53)
\end{aligned}
$$

where we have define the Krauss measurement operators [210]

$$
\begin{aligned}
\hat{E}(x) &\equiv {}_\sigma\langle x|\hat{U}(\tau)|1\rangle_\sigma && (6.54)\\
&= {}_\sigma\langle x|e^{-i\theta(b\sigma_+ + b^\dagger \sigma_-)}|1\rangle_\sigma && (6.55)
\end{aligned}
$$

with $\theta \equiv g_0\tau$. It is a relatively straightforward to show that

$$
\begin{aligned}
\hat{E}(1) &= \cos(\theta\sqrt{bb^\dagger}) && (6.56)\\
\hat{E}(0) &= -ib^\dagger(bb^\dagger)^{-1/2}\sin(\theta\sqrt{bb^\dagger}). && (6.57)
\end{aligned}
$$

Exercise 6.5 *Obtain these expressions for yourself.*

The probability of obtaining the result x is simply the required normalisation of the unnormalised conditional state

$$p(x) = \langle\tilde{\psi}^{(x)}|\tilde{\psi}^{(x)}\rangle = {}_b\langle\psi(0)|\hat{E}^\dagger(x)\hat{E}(x)|\psi(0)\rangle_b. \tag{6.58}$$

As $\sum_x \hat{E}^\dagger(x)\hat{E}(x) = \mathbb{1}$, the identity operator, this probability distribution is normalised.

Clearly there is at most one bit of information per trial. Since the positive operator valued measure, $\hat{E}^\dagger(x)\hat{E}(x)$, commutes with the number operator, the measurement constitutes a very coarse-grained measurement of phonon number. It is not a quantum nondemolition measurement (see Section 5.3), as the interaction Hamiltonian Eq. (6.17) does not commute with the phonon number operator.

If the initial state of the mechanics is a coherent state

$$|\psi(0)\rangle = |\beta\rangle_b \tag{6.59}$$

$$= e^{-|\beta|^2/2} \sum_{n=0}^{\infty} \frac{\beta^n}{\sqrt{n!}} |n\rangle_b \tag{6.60}$$

with initial Poissonian number distribution

$$P_0(n) = e^{-|\beta|^2} \frac{|\beta|^{2n}}{n!}, \tag{6.61}$$

and the measurement result is $x = 1$, the conditional number distribution after the measurement is

$$P_1(n) = [p(1)]^{-1} \cos^2(\theta\sqrt{n+1}) P_0(n). \tag{6.62}$$

Iterating the measurement, using the conditional state from the first measurement as the initial state for the bosonic degree of freedom for the next measurement, and varying the value of θ, we can trace the progress in the conditional number distribution, as shown in Fig 6.6. In a sequence of measurements, for appropriate values of θ, the conditional state can approach a number state.

Now we return to the model of interest in which the interaction of the qubit with the mechanics is governed by the stochastic process of photon absorption and emission from the optical cavities. For simplicity, we neglect mechanical damping and thermalisation, the effects of which are discussed in [26]. We first compute the conditional state conditioned on no detections, either at D_1 or D_2, up to time t. The quantum theory of continuous photon counting [265] shows that this (unnormalised) conditional state is determined by

$$\tilde{\rho}^{(0,0)}(t) = \mathcal{S}(t)\rho(0), \tag{6.63}$$

where the superscript is defined by (n_1, n_2) where n_i is the count number recorded at detector D_i, and $\mathcal{S}(t)\rho(0) = \tilde{\rho}^{(0,0)}(t)$ is the conditional state given no counts up to time t and is given by solving

$$\frac{d\tilde{\rho}}{dt} = -i(\hat{K}\tilde{\rho} - \tilde{\rho}\hat{K}^\dagger). \tag{6.64}$$

The non-Hermitian operator $\hat{K}$ is given by

$$\hat{K} = g_0(b^\dagger a_1^\dagger a_2 + b a_1 a_2^\dagger) - i\sqrt{\gamma\kappa_2}(c a_2^\dagger - c^\dagger a_2)/2 - iJ^\dagger J/2 - i\kappa_1 a_1^\dagger a_1/2, \tag{6.65}$$

where, as before, c and $c^\dagger$ are the annihilation and creation operators for photons in the source cavity, and $J = \sqrt{\gamma}c + \sqrt{\kappa_2}a_2$ is the detection operator. The normalisation of this state is simply the probability for no counts up to time t,

$$p(n_1 = 0, n_2 = 0, t) = \text{tr}[\tilde{\rho}^{(0,0)}(t)]. \tag{6.66}$$

Note that, if the initial state is pure, we need only to solve the effective Schrödinger equation

$$\frac{d|\tilde{\psi}^{(0)}\rangle}{dt} = -i\hat{K}|\tilde{\psi}^{(0)}\rangle \tag{6.67}$$

to give

$$\tilde{\rho}^{(0,0)}(t) = |\tilde{\psi}^{(0}(t)\rangle\langle\tilde{\psi}^{(0)}(t)|. \tag{6.68}$$

We now ask for the conditional state of the system given that one photon is counted at D_2 in time t to $t + dt$. Such an event means that no photon has decayed through the output of cavity a_1. This conditional state is

$$\tilde{\rho}^{(0,1)}(t) = J\mathcal{S}(t)\rho(0)J^\dagger. \tag{6.69}$$

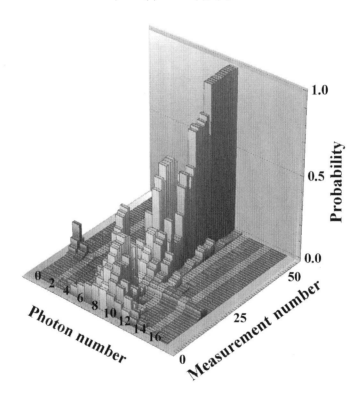

FIGURE 6.6 Photon number probabilities versus photon number and number of readout, all with $x = 1$, $\beta = 3$ and for various values of θ. From Basiri-Esfahani et al. [26].

If the initial state is a pure state, this conditional state is also a pure state,

$$|\tilde{\psi}^{(0,1)}\rangle = J|\tilde{\psi}^{(0)}(t)\rangle = \sqrt{\gamma}c|\tilde{\psi}^{(0)}(t)\rangle + \sqrt{\kappa_2}a_2|\tilde{\psi}^{(0)}(t)\rangle, \tag{6.70}$$

where, as usual, γ and κ are the linewidths of the source and receiver cavity, respectively. This is a superposition of the two ways in which a photon can be counted: direct reflection from the cavity or emission from inside the cavity. This leads to an interference term in the detection rate, as discussed already in Section 6.2.1.

If the photon decays through cavity a_1, it can never be detected at D_2. However, if we do not monitor this output, we have no way of knowing when this happens. Note also that, once a photon is lost, the operation $\mathcal{S}$ acts trivially as the identity. Clearly, to keep the error low we need to ensure $\kappa_1 \ll \kappa_2$. If we do monitor this output, the loss is heralded and therefore can be conditionally evaded.

At each time step we can compute the detection rate to generate a random detection time t_1 in a time interval which starts just after the minimum and ends close to where the detection rate is nearly zero. This choice assures us that, with high probability, we are detecting a photon from cavity a_2 after it has interacted with the mechanical system, not one which is reflected off the mirror directly from the source. At a detection event we then act with the jump operator J on $|\psi^{(0)}(t_1)\rangle$, to get the conditional state of the system $|\tilde{\psi}^{(1)}(t_1)\rangle$ given that one photon is counted in time t_1 to $t_1 + dt_1$. The phonon number distribution for the conditional state now becomes

$$P_n^1(t_1) = P_n^0(t_0)P(n, t_1), \tag{6.71}$$

where

$$P(n, t_1) = \langle \tilde{\psi}^{(1)}(t_1)|\tilde{\psi}^{(1)}(t_1)\rangle \tag{6.72}$$

and $P_n^0(t_0)$ is the prior number distribution before the measurement. We will repeat the measurement process, preparing another single-photon in the source and using as a new initial state the conditional state left from the previous measurement. Given the state of the system after each measurement, we can recalculate the detection rate, from which we again sample a random detection time. The phonon number distribution function after r measurements is given by the interactive map

$$P_n^r(t_r) = P_n^{r-1}(t_{r-1})P(n, t_r). \tag{6.73}$$

The details can be found in [26].

Each measurement provides partial information about phonon number (typically less than one bit per trial). The procedure can be explained by looking at Eq. (6.73) and is quite similar to the simple model with deterministic interaction times discussed earlier. After the rth detection event, the phonon number distribution is multiplied by a filter function $P(n, t_r)$, which, for appropriate values of t_r, suppresses the probability of certain values of n.

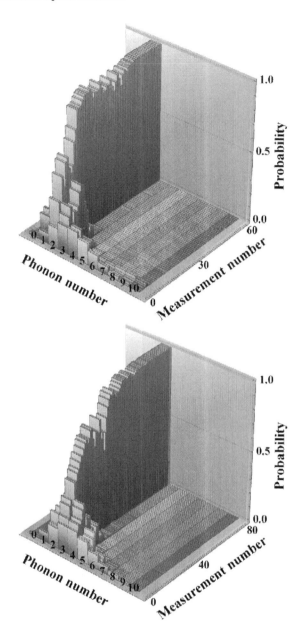

FIGURE 6.7 Phonon number distribution function histograms after successive measurements for $\beta = 2$ $(\bar{n}_b = 4)$, $\kappa_2 = g_0$, $\gamma = 0.9g_0$, $\kappa_1 = 0.2g_0$ and random detection times. As the number of measurements is increased, the phonon number distribution evolves from a Poissonian distribution into number state distributions with (*top*) n=2 and (*bottom*) n=3. From Basiri-Esfahani et al. [26].

As the measurement process continues more information is obtained, leading to a gradual collapse of the distribution onto a single number state. Figure 6.7 shows the phonon number distributions for simulations from [26] of two experiments with 60 and 80 successive measurements, respectively. As discussed earlier, to obtain appreciable information in a single trial and therefore reach (or get close to) a Fock state within a reasonable number of trials, the optomechanical system must operate in the strong coupling regime for which g_0 is of the order of κ_2. Hence, for these simulations the value $\kappa_2 = g_0$ is chosen.

6.4.2 Two-photon input

We now consider the case with two-photon input, a single-photon in each of the two input waveguides, including a delay time τ between the two single-photon generation events. This input encodes not a qubit but a qutrit. To see this, we introduce the bosonic representation of angular momentum. Define:

$$\hat{J}_+ = a_1 a_2^\dagger = \hat{J}_-^\dagger, \tag{6.74}$$

$$\hat{J}_z = \frac{1}{2}(a_2^\dagger a_2 - a_1^\dagger a_1), \tag{6.75}$$

with the corresponding Casimir invariant given by $\hat{J}^2 = \frac{\hat{N}}{2}(\frac{\hat{N}}{2} + 1)$ where $\hat{N} = a_2^\dagger a_2 + a_1^\dagger a_1$. The interaction Hamiltonian can then be written as

$$\hat{H}_I = \hbar g_0(\hat{J}_+ b + \hat{J}_- b^\dagger). \tag{6.76}$$

As there are now at most two photons in the optical cavities, we are restricted to the subspaces $0 \le N \le 2$. Cavity decay causes incoherent transitions between these subspaces. The case $N = 1$ corresponds to the Jaynes–Cummings model for the interaction of a single qubit with a bosonic degree of freedom, discussed in the previous section. The case $N = 2$ corresponds to a qutrit interacting with a bosonic degree of freedom.

We could now proceed to consider the conditional state of the mechanics conditioned on counting output photon numbers. This goes through much the same as the single-photon case except that now each trial gives more than a single bit of information, as we are using a qutrit rather than a qubit. However, we will not pursue this here. Instead we will ask the converse question: how does the mechanics control the state of the light?

The measurement scenario corresponds to a *quantum* controller in which the light and mechanics become entangled. Making measurements on the light gives information on the mechanics. There is another important limit in which the mechanics changes the state of the light but there is very little entanglement between them. This is the classical control limit. What do we need to change to move from a quantum to a classical controller?

If the mechanical degree of freedom is prepared in a coherent state $|\beta\rangle$, we can use a canonical transformation $b \to \bar{b} + \beta$ to include this amplitude in the Hamiltonian (see also Section 2.7), and the initial mechanical state becomes the ground state. We can define a semiclassical limit as $g_0 \to 0, \beta \to \infty$ such that $g_0\beta \equiv \bar{g}$ is a constant. The interaction Hamiltonian becomes

$$\hat{H}_I = 2\hbar\bar{g}(a_1^\dagger a_2 + a_1 a_2^\dagger) + \hbar\lambda(\hat{J}_+\bar{b} + \hat{J}_-\bar{b}^\dagger), \tag{6.77}$$

where $\lambda = \bar{g}/\beta \ll 1$ is a perturbation parameter.

To zeroth order in λ, the unitary evolution is simply the beam spitter unitary

$$\hat{U}_0(t) = e^{-i\bar{g}t(a_1^\dagger a_2 + a_1 a_2^\dagger)}, \tag{6.78}$$

where $\theta \equiv 2\bar{g}t$. In terms of the $su(2)$ representation this is simply a rotation about the x direction. The beam splitter interaction does not entangle the optical fields with the mechanics. Corrections due to first and higher order in λ represent residual entanglement between the optical and mechanical degrees of freedom. In this section we will consider those corrections and discuss an experimental protocol to probe them.

We saw in Section 6.3 that a single-photon pulse incident on a single optomechanical cavity could be used to measure the mechanical displacement using HOM interference. As a measurement necessarily requires that the photon becomes entangled with the mechanics, this is, in effect, a measurement of that entanglement. We will use a similar approach based on HOM interference to obtain an experimental signature of the entanglement between the mechanics and the two optical fields. The idea is to set up a mechanically controlled beam splitter interaction between the input and output photon pairs to produce HOM interference as a function of the coherent excitation of the mechanics.

There are two ways in which the mechanics can be prepared in a coherent state. In both cases we assume a laser cooling scheme has first prepared the mechanics in the ground state (see Section 4.2). In one approach, it is subjected to a classical resonant force which drives it to a steady state which is a coherent state. In the second approach, a strong continuous coherent optical field can be injected in one of the input optical waveguides to implement a beam splitter interaction between the other optical mode and the mechanics.[3] A strong coherent pulse on the second optical input can then be transferred to a coherent excitation of the mechanics. In this protocol the mechanical degree of freedom is acting as a quantum memory [212]. In both cases it is possible to prepare coherent states with different values of β with which the emergence of the semiclassical limit, and corrections due to residual entanglement between the optics and the mechanics, can be probed.

To consider the semiclassical regime, we use the quantum Langevin equa-

---

[3]This is closely related to the concept of optomechanically induced transparency discussed in Section 4.3

tions for the optical fields

$$\frac{da_1(t)}{dt} = -i\bar{g}a_2(t) - \frac{\kappa_1}{2}a_1(t) + \sqrt{\kappa_1}a_{1,\text{in}}(t) \tag{6.79a}$$

$$\frac{da_2(t)}{dt} = -i\bar{g}a_1(t) - \frac{\kappa_2}{2}a_2(t) + \sqrt{\kappa_2}a_{2,\text{in}}(t), \tag{6.79b}$$

with single-photon input states for which the amplitude functions are given by $\xi(t), \eta(t)$ for $a_{1,\text{in}}(t)$ and $a_{2,\text{in}}(t)$, respectively. This is a linear system which is easily solved. We can then use the input-output relations to compute the joint photon counting probabilities at the output from each cavity mode.

Exercise 6.6 *Show that the solutions to the quantum Langevin equations, with $\kappa_1 = \kappa_2 = \kappa$, are given by*

$$
\begin{aligned}
a_1(t) &= \sqrt{\kappa}\Big[A(t)\int_0^t dt'\Big(C(t')a_{1,\text{in}}(t') + D(t')a_{2,\text{in}}(t')\Big) \\
&\quad + B(t)\int_0^t dt'\Big(D(t')a_{1,\text{in}}(t') + C(t')a_{2,\text{in}}(t')\Big)\Big] \\
a_2(t) &= \sqrt{\kappa}\Big[B(t)\int_0^t dt'\Big(C(t')a_{1,\text{in}}(t') + D(t')a_{2,\text{in}}(t')\Big) \\
&\quad + A(t)\int_0^t dt'\Big(D(t')a_{1,\text{in}}(t') + C(t')a_{2,\text{in}}(t')\Big)\Big], \tag{6.80}
\end{aligned}
$$

where $A(t) = e^{-\kappa t/2}\cos(\bar{g}t)$, $B(t) = -ie^{-\kappa t/2}\sin(\bar{g}t)$, $C(t) = e^{\kappa t/2}\cos(\bar{g}t)$, and $D(t) = ie^{\kappa t/2}\sin(\bar{g}t)$.

As an example, we can compute the effective transmisivity and reflectivity coefficients when only one photon is incident on the system prepared in the state $\xi(t) = \sqrt{\bar{\gamma}}e^{-\gamma t/2}H(t)$. These are defined by

$$R = \int_0^\infty \langle a_{1,\text{out}}^\dagger a_{1,\text{out}t}\rangle_t \, dt,$$

$$T = \int_0^\infty \langle a_{2,\text{out}}^\dagger a_{2,\text{out}}\rangle_t \, dt. \tag{6.81}$$

In the symmetric case for which $\kappa_1 = \kappa_2 = \kappa$ these are given by

$$
\begin{aligned}
T &= \frac{8\kappa\bar{g}^2(\gamma + 2\kappa)}{(4\bar{g}^2 + \kappa^2)(4\bar{g}^2 + (\gamma + \kappa)^2)}, \\
R &= 1 - T. \tag{6.82}
\end{aligned}
$$

In Fig. 6.8 we show the reflectivity, R, of the effective beam spitter as a function of κ and $\bar{g}$ in units of γ. By adjusting κ and γ, the reflectivity can be tuned from very small values to $R = 1$. This is a kind of mode matching between the input pulse and the response of the coupled cavity system. There

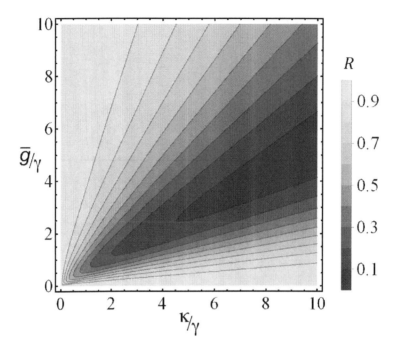

FIGURE 6.8 The reflection coefficient R versus normalised optical cavity damping rate κ and cavity coupling rate $\bar{g}$ normalised to the input photon bandwidth γ.

are two branches for each R value. The lower branch would be more appropriate for experimental prospects, since it allows the beam splitter to be tuned using smaller values of $\bar{g}$.

We now consider how this device functions as a HOM interferometer. Identical single-photon pulses are injected into each of the cavity mode inputs but with a time delay τ between them.[4] The pulse shapes are then taken as $\xi(t) = \sqrt{\gamma_1} e^{-\frac{1}{2}\gamma_1 t}$ for a_1 and $\eta(t) = \sqrt{\gamma_2} e^{-\frac{1}{2}\gamma_2(t-\tau)}$ for a_2. The joint statistics of photons counted by detectors D_1 and D_2 is given by the second-order correlation function

$$P_{1,1}(\tau) = \frac{1}{N_1 N_2} \int_0^\infty \int_0^\infty \langle a_{1,\text{out}}^\dagger(t) a_{2,\text{out}}^\dagger(t') a_{2,\text{out}}(t') a_{1,\text{out}}(t) \rangle dt dt'.$$

(6.83)

---

[4]Note that, here, τ is the delay between the two single-photon input pulses, not between detection events.

with

$$N_k = \int_0^\infty dt \, \langle a_{k,out}^\dagger(t') a_{k,out}(t') \rangle. \tag{6.84}$$

This second-order correlation function gives the probability of detecting a photon at detector D_1 at any time t between 0 to infinity *and* the other photon at detector D_2 at some time t' from 0 to infinity. This expression can be analytically calculated for the initial state $|\psi(0)\rangle = |1_{a_1,\xi} 1_{a_2,\eta}\rangle$ by applying the input-output relations (see Section 1.4.3) and the solutions to Eqs. (6.79) to give

$$P_{1,1}(\tau) = \frac{e^{-3\tau(\kappa+\gamma)/2}}{A} \Big(B e^{3\tau(\kappa+\gamma)/2} \tag{6.85}$$
$$+ C e^{-\tau(3\kappa+\gamma)/2} + D e^{\tau(\kappa+3\gamma)/2} + E e^{\tau(\kappa+\gamma)} \Big),$$

where

$$A = (4\bar{g}^2 + \kappa^2)^2 \Big(16\bar{g}^4 + (\gamma^2 - \kappa^2)^2 + 8\bar{g}^2(\gamma^2 + \kappa^2) \Big)^2,$$

$$
\begin{aligned}
B = \; & (4\bar{g}^2 + (\gamma - \kappa)^2)^2 \Big(256\bar{g}^8 + \kappa^4(\gamma + \kappa)^4 \\
& + 8\bar{g}^2(\gamma^2 - 2\kappa^2)(16\bar{g}^4 + \kappa^2(\gamma + \kappa)^2) \\
& + 16\bar{g}^4(\gamma^4 + 2\gamma^2\kappa^2 + 20\gamma\kappa^3 + 22\kappa^4) \Big),
\end{aligned}
\tag{6.86}
$$

$$C = -32\bar{g}^2\kappa^2(4\bar{g}^2 + \gamma^2 - \kappa^2)^2(4\bar{g}^2 + \kappa^2)^2,$$
$$D = -32\bar{g}^2\gamma^2\kappa^2 F^2,$$
$$E = -64\bar{g}^2\gamma\kappa^2(4\bar{g}^2 + \gamma^2 - \kappa^2)(4\bar{g}^2 + \kappa^2)F,$$

and

$$F = \kappa(-12\bar{g}^2 - \gamma^2 + \kappa^2)\cos(\bar{g}\tau) + 2\bar{g}(4\bar{g}^2 + \gamma^2 - 3\kappa^2)\sin(\bar{g}\tau).$$

Figure 6.9 shows the HOM dip in $P_{1,1}(\tau)$ (at $\tau = 0$) versus κ and $\bar{g}$ in units of γ.

Now that we have seen how the semiclassical (i.e., with $b \to \beta$) HOM interference will work, we can return to include the mechanical degree of freedom explicitly and evaluate the residual entanglement as reflected by a decrease in the HOM visibility. We will use the Fock state master equation method developed by Combes and co-workers [23] and introduced in Section 6.2.3. For two optical cavities, this takes the form

$$
\begin{aligned}
\frac{d}{dt}\rho_{m,n;p,q}(t) = \; & -i[\hat{H}, \rho_{m,n;p,q}] + (\mathcal{D}[L_1] + \mathcal{D}[L_2])\rho_{m,n;p,q} \\
& + \sqrt{m}\xi(t)[\rho_{m-1,n;p,q}, L_1^\dagger] + \sqrt{p}\eta(t)[\rho_{m,n;p-1,q}, L_2^\dagger] \\
& + \sqrt{n}\xi^*(t)[L_1, \rho_{m,n-1;p,q}] \\
& + \sqrt{q}\eta^*(t)[L_2, \rho_{m,n;p,q-1}],
\end{aligned}
\tag{6.87}
$$

where $\hat{H}$ is the Hamiltonian given in Eq. (6.44), and $L_i = \sqrt{\kappa_i}a_i$.

The generalised system density operator $\rho_{m,n;p,q}$ is a joint density matrix on the system and the input field. The first two subscripts m, n refer to the photon number basis for cavity-1, the top cavity in Fig. 6.5, and the second two subscripts p, q refer to the photon number basis for cavity-2, the bottom cavity in Fig. 6.5 . Hence, each of these indices takes the values 0 and 1. This gives us a hierarchy of differential equations. Define

$$\rho_{field}(0) = \sum_{m,n,p,q=0}^{\infty} c_{m,n;p,q}|n_\xi;q_\eta\rangle\langle m_\xi;p_\eta|, \tag{6.88}$$

where $c_{1,1;0,0} = c_{0,0;1,1} = \frac{1}{2}$, $c_{0,1;1,0} = \frac{1}{2}e^{-i\theta}$, $c_{1,0;0,1} = \frac{1}{2}e^{i\theta}$, and the other coefficients $c_{m,n;p,q} = 0$. As the input field is given by Eq. (6.88), the total state of the system now becomes [23]

$$\rho_{system}(t) = \sum_{m,n,p,q=0}^{\infty} c^*_{m,n;p,q}\rho_{m,n;p,q}(t). \tag{6.89}$$

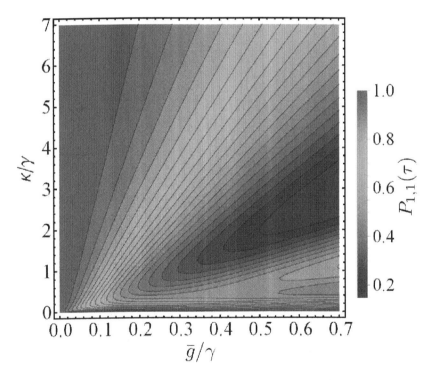

FIGURE 6.9 Contour plot of the joint probability $P_{1,1}(\tau)$ at zero delay, $\tau = 0$, versus κ/γ and $\bar{g}/\gamma$.

This hierarchy of differential equations can be solved for the two modes with $\xi(t) = \eta(t) = \sqrt{\gamma}e^{-\frac{1}{2}\kappa(t-\tau)}$. The solutions for $\rho_{1,1;0,0}(t), \rho_{0,0;1,1}(t), \rho_{0,1;1,0}(t)$, and $\rho_{1,0;0,1}(t)$ then provide the detection probabilities at detectors 1 and 2.

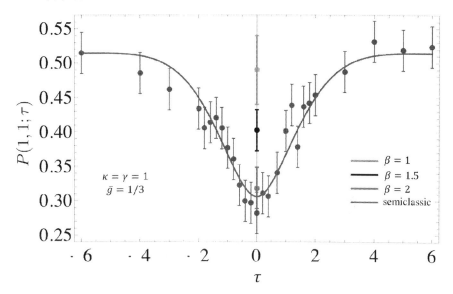

FIGURE 6.10 HOM interference via a mechanically controlled beam splitter. We plot $P_{1,1}$ obtained by a stochastic simulation of the Fock state master equation versus the time delay between the single-photon input pulses for various values of the coherent excitation of the mechanical resonator. Reproduced with permission from [27].

In Fig. 6.10 we show the probability $P_{1,1}(\tau)$ of detecting one photon at each detector as a function of τ for various values of β, and $\gamma = \kappa = 1$, and $\bar{g} = 1/3$. We see that, as the coherent state amplitude increases, the visibility pattern approaches the semiclassical limit [27]. For $\beta = 12$, which is not shown in this figure, the maximum visibility equals 0.99.

6.5 MACROSCOPIC SUPERPOSITION STATES USING SINGLE-PHOTON OPTOMECHANICS

In Chapter 10 we discuss possible optomechanical tests of gravitational decoherence. The key requirement is the ability to prepare a mechanical system, an optomechanical mirror for example, in a superposition of two distinct positions. The resulting gravitational field from such a mass distribution is uncertain and there are good arguments (see Chapter 10) for suspecting that this will lead to a source of noise and decoherence. An early proposal for using optomechanics to prepare such a superposition was given by Bouwmeester

[193, 221]. In this section we give a more detailed analysis of a related scheme that explicitly includes the stochastic nature of photon absorption and emission from optomechanical cavities. Our analysis is based on the discussion in Akram et al. [6].

Consider the optomechanical system in Fig. 6.11. A single-photon source is used to drive an optomechanical cavity and the output mode from the cavity is then directed onto a photon counter. As the single-photon can either be reflected directly from the cavity to the detector or absorbed into the cavity and then transmitted to the detector, the mechanical element can be left in a superposition state of being kicked or not kicked by the interaction with the photon, conditional upon photon detection that does not distinguish the two possible single-photon histories.

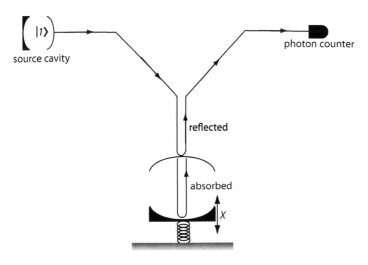

FIGURE 6.11   An optomechanical scheme for creating a superposition state of mechanical motion.

We will use the cascaded system method (see Section 6.2.2). The master equation for the cascaded system of the source cavity and the optomechanical system is

$$\frac{d\rho}{dt} = -\frac{i}{\hbar}[\hat{H}, \rho] + \mathcal{D}[J]\rho, \qquad (6.90)$$

where

$$\hat{H} = \hat{H}_{om} + \hat{H}_{cas} \qquad (6.91)$$

with

$$\hat{H}_{om} = \hbar\Omega b^{\dagger}b + \hbar g_0 a^{\dagger}a(b + b^{\dagger}). \qquad (6.92)$$

Here, as usual, $a, a^{\dagger}$ are the annihilation and creation operators for the optical resonator and $b, b^{\dagger}$ are the annihilation and creation operators for the mechanical resonator with frequency Ω.

$$\hat{H}_{cas} = -i\sqrt{\kappa\gamma}(ca^\dagger - c^\dagger a)/2, \tag{6.93}$$

and the jump operator describing photodetection is given, again, by

$$J = \sqrt{\gamma}c + \sqrt{\kappa}a \tag{6.94}$$

with $c, c^\dagger$ being the annihilation and creation operators for the source cavity, with decay constant γ. The decay rate of the optical resonator for the optomechanical system is κ.

As our scheme has only a single-photon, and there is no other channel for the photon to be lost, if the initial state of the system is pure, the conditional state of the system conditioned on a photon detection event at any time will also be pure. The unnormalised conditional state of the system, $|\tilde{\Psi}^{(0)}\rangle$, given no count up to time t, is given by solving

$$\frac{d|\tilde{\Psi}^{(0)}(t)\rangle}{dt} = \left(-i\hat{H}/\hbar - \frac{1}{2}J^\dagger J\right)|\tilde{\Psi}^{(0)}(t)\rangle. \tag{6.95}$$

The initial state of the system is

$$|\tilde{\Psi}(0)\rangle = |1\rangle_c|0\rangle_a|0\rangle_b , \tag{6.96}$$

i.e., one photon in the source, and the cavity and mechanical oscillator both in their ground states. We expand the conditional state of the optomechanical system as

$$|\tilde{\Psi}^{(0)}(t)\rangle = |1\rangle|\phi_1(t)\rangle_b + |2\rangle|\phi_2(t)\rangle_b \tag{6.97}$$

with

$$|1\rangle = |0\rangle_c|1\rangle_a \tag{6.98}$$
$$|2\rangle = |1\rangle_c|0\rangle_a . \tag{6.99}$$

Before the photon is counted, it is in a superposition of inside and outside the cavity.

We will proceed using the canonical polaron transformation discussed in Sections 6.1 and 4.5.2 to transform the Hamiltonian in Eq. (6.91) to the polaron picture

$$\hat{\bar{H}} \equiv \hat{S}\hat{H}\hat{S}^\dagger$$
$$= \hbar\Omega b^\dagger b - \hbar\chi_0(a^\dagger a)^2 - i\hbar\sqrt{\kappa\gamma}(ca^\dagger D(\beta) - c^\dagger a D^\dagger(\beta))/2, \tag{6.100}$$

where $\hat{S}$ is defined in Eq. (4.110), the single-photon optical frequency shift $\chi_0 = g_0^2/\Omega$ as defined in Eq. (4.113), $D(\beta)$ is a displacement operator with $\beta = -g_0/\Omega$, and, as usual, operators in the polaron frame are identified with bar accents. In the polaron picture, the state is written

$$|\tilde{\Psi}^{(0)}(t)\rangle_P = \hat{S}|\tilde{\Psi}^{(0)}(t)\rangle, \tag{6.101}$$

so that, in this picture, the evolution of unnormalised conditional state of the optomechanical system described by Eq. (6.95) becomes

$$\frac{|\tilde{\Psi}^{(0)}(t)\rangle_P}{dt} = \left(-i\hat{\bar{H}}/\hbar - \frac{1}{2}\bar{J}^\dagger \bar{J}\right)|\tilde{\Psi}^{(0)}(t)\rangle \qquad (6.102)$$

where $\bar{J} \equiv \hat{S}J\hat{S}^\dagger$; while the conditional state of Eq. (6.97), itself, becomes

$$
\begin{aligned}
|\tilde{\Psi}^{(0)}(t)\rangle_P &= |1\rangle|\bar{\phi}_1(t)\rangle_b + |2\rangle|\bar{\phi}_2(t)\rangle_b \qquad &(6.103)\\
&= |1\rangle D(\beta)|\phi_1(t)\rangle_b + |2\rangle|\phi_2(t)\rangle_b \ . \qquad &(6.104)
\end{aligned}
$$

From Eq. (6.104) we can see that the state of the optomechanical system in the original picture can be determined by quantifying

$$|\bar{\phi}_1(t)\rangle_b = D(\beta)|\phi_1(t)\rangle_b \ , \qquad (6.105)$$

and then transforming back to the original picture by inverting the canonical transformation $\hat{S}$. The details of this process can be found in [6]. The result is that unnormalised conditional state, given that no photons are counted up to time t, is

$$|\tilde{\Psi}^{(0)}(t)\rangle = |1\rangle|\phi_1(t)\rangle_b + |2\rangle|0\rangle_b e^{-\gamma t/2}, \qquad (6.106)$$

where

$$
\begin{aligned}
|\phi_1(t)\rangle_b &= D^\dagger(\beta)\hat{R}(t)D(\beta)|0\rangle_b \qquad &(6.107)\\
|\phi_2(t)\rangle_b &= e^{-\gamma t/2}|0\rangle_b \qquad &(6.108)
\end{aligned}
$$

with

$$\hat{R}(t) = \sum_{n=0}^{\infty} \frac{1 - e^{-(i\Omega n - i\chi_0 + (\kappa-\gamma)/2)t}}{i\Omega n - i\chi_0 + (\kappa - \gamma)/2}|n\rangle\langle n|. \qquad (6.109)$$

It should be noted here that the term $\Omega n - i\chi_0$ in the denominator of this equation will lead to a photon blockade (see Section 6.1).

If the photon is counted between t and $t + dt$, the resulting conditional state is found by applying the jump operator J so that the (unnormalised) conditional state of the mechanics given that no photons are counted up to time t *and* one photon is counted between t and $t + dt$ is

$$|\tilde{\Phi}^{(1)}(t)\rangle = \sqrt{\kappa}|\phi_1(t)\rangle + \sqrt{\gamma}e^{-\gamma t/2}|0\rangle, \qquad (6.110)$$

where, for succinctness, we have dropped the subscript b. The first term is given in Eq. (6.107) and the tilde signifies that the state is unnormalised. The normalisation of the conditional state is simply the rate for photon counts and is given by

$$
\begin{aligned}
R_1(t) &= \langle\tilde{\Phi}^{(1)}(t)|\tilde{\Phi}^{(1)}(t)\rangle \qquad &(6.111)\\
&= \kappa\langle\phi_1(t)|\phi_1(t)\rangle + \gamma e^{-\gamma t} + \sqrt{\kappa\gamma}e^{-\gamma t/2}(\langle 0|\phi_1(t)\rangle + c.c.).
\end{aligned}
$$

The first term is the rate to count photons *given* that they come from the optomechanical cavity (the "receiver"). The second term is the rate to count photons given that they come straight from the source reflected from the receiver. The last term arises due to interference between photons reflected and those transmitted from inside the optomechanical cavity. Thus the mean number of photons inside the optomechanical cavity, prior to the detection, is just

$$\langle a^\dagger a \rangle(t) = \langle \phi_1(t)|\phi_1(t)\rangle . \tag{6.112}$$

If we assume that the mechanics starts in the ground state, we can use Eq. (6.107) to show that

$$R_1(t) = \kappa \langle \beta | \hat{R}^\dagger(t)\hat{R}(t)|\beta\rangle + \gamma e^{-\gamma t} + \sqrt{\kappa\gamma} e^{-\gamma t/2}(\langle \beta|\hat{R}(t)|\beta\rangle + c.c), \tag{6.113}$$

where $|\beta\rangle$ is a coherent state of the mechanical oscillator. The mean photon number in the optomechanical cavity prior to detection is then

$$\langle a^\dagger a \rangle = \langle \beta|\hat{R}^\dagger(t)\hat{R}(t)|\beta\rangle = \sum_{n=0}^{\infty} e^{-|\beta|^2} \frac{|\beta|^{2n}}{n!}|r_n(t)|^2, \tag{6.114}$$

where

$$r_n(t) = \frac{\left(1 - e^{-(i\Omega n - i\chi_0 + (\kappa-\gamma)/2)t}\right)}{i\Omega n - i\chi_0 + (\kappa - \gamma)/2}. \tag{6.115}$$

The interference term in the single-photon count rate (the last term in Eq. (6.111)) is determined by

$$\langle \beta|\hat{R}(t)|\beta\rangle = \sum_{n=0}^{\infty} e^{-|\beta|^2} \frac{|\beta|^{2n}}{n!} r_n(t) . \tag{6.116}$$

If a photocount is recorded at short times, it will most likely correspond to photons reflected off the optomechanical cavity without interacting with the mechanical system. Hence postselection on rare late detection events ensures both that the photon entered the optomechanical system, and that it interacted for a prolonged period. In these circumstances, even if the bare optomechanical coupling is small, postselection would lead to an effectively enhanced optomechanical interaction, resulting in a significant momentum kick to the mirror, as we now show.

Let us assume that the mechanical resonator starts in the ground state. The conditional mean amplitude given a photon count at time t is

$$\langle \Phi^{(1)}(t)|b|\Phi^{(1)}(t)\rangle = [R_1(t)]^{-1} \kappa \langle \phi_1(t)|b|\phi_1(t)\rangle, \tag{6.117}$$

where the normalisation is given by the single-photon count rate in Eq. (6.111).

Exercise 6.7 *Show this result from Eq. (6.110) for the unnormalised conditional state of the mechanics.*

Using Eq. (6.107), we find

$$\langle \Phi^{(1)}(t)|b|\Phi^{(1)}(t)\rangle = [R_1(t)]^{-1}\left(\frac{g_0}{\Omega}\right)\sum_{n=0}^{\infty} e^{-|\beta|^2}\frac{|\beta|^{2n}}{n!}r_n^*(t)[r_n(t) - r_{n+1}(t)],$$

(6.118)

where $r_n(t)$ is given in Eq. (6.115). If the detection time t is long, $r_n(t)$ is maximised due to the exponential component in Eq. (6.115) reducing to a very small value. Hence the amplitude of the moment, $|\langle b \rangle|$, and thus the amplitude of the momentum, $|i\langle b^\dagger - b\rangle|$, increases as the detection time increases, corresponding to displaced conditional mechanical state, and an effective enhancement of the optomechanical cooperativity.

In Fig. 6.12 we show the conditional momentum as a function of time for different values of the mechanical frequency Ω. Initially, as the photon enters the optomechanical cavity, it circulates with the optomechanical system, inducing an optomechanical interaction which imparts a momentum kick on the mechanical mode. As we have shown above, postselecting on late detection times of the interacting photon will impart a large momentum kick to the mechanical oscillator. In fact, from Eq. (6.92), in the semiclassical limit the momentum can be approximated as $-i\langle b - b^\dagger\rangle = \frac{-2g_0}{\Omega}\sin(\Omega t)$. In the limit that $\Omega t \ll 1$, where the time is short compared to the mechanical period but long compared to the cavity decay rate, $-i\langle b - b^\dagger\rangle \approx -2g_0 t$.

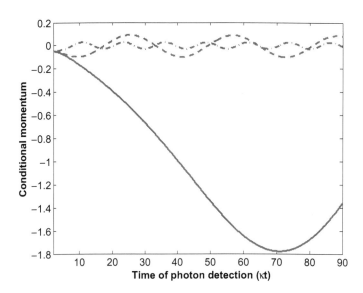

FIGURE 6.12 The conditional momentum of the mechanical oscillator versus detection time, for $\gamma/\kappa = 2$, $g_0/\kappa = 0.01$ and varying Ω/κ: 0.5 (dashed-dotted curve), 0.2 (dashed curve), and 0.02 (solid curve). From Akram et al. [6], with permission.

6.5.1 Conditional entangled state of two mechanical resonators

We are now in a position to consider a version of the Bouwmeester proposal [193, 221], shown in Fig. 6.13. The input and output beam splitters mean that we need to modify the jump operators that describe detection events:

$$J_{D_1} = \sqrt{\gamma}c + \frac{\sqrt{\kappa_1}a_1 + \sqrt{\kappa_2}a_2}{2} \tag{6.119}$$

$$J_{D_2} = \frac{\sqrt{\kappa_1}a_1 - \sqrt{\kappa_2}a_2}{2}, \tag{6.120}$$

where κ_i is the decay rate of each optical resonator, and the phase differences across the branches of the beam splitter are set up such that there is a null detection rate at the port D_2.

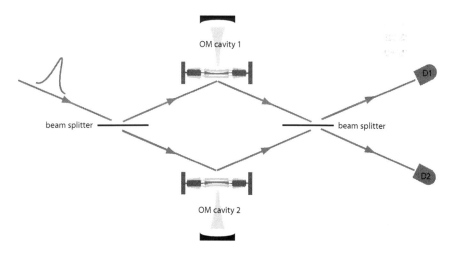

OM cavity 1

beam splitter

beam splitter

OM cavity 2

D1

D2

FIGURE 6.13 Two optomechanical cavities set up as a Mach-Zhender interferometer with detection ports D_1 and D_2. The beam splitters are considered to have 50:50 transmission and reflection.

The no-jump dynamics of the system is governed by a conditional Schrödinger's equation similar to Eq. (6.95), where in this case

$$|\tilde{\Psi}^{(0)}(t)\rangle = \exp\left[-\frac{i}{\hbar}Ht - \frac{1}{2}J_{D_1}^\dagger J_{D_1}t - \frac{1}{2}J_{D_2}^\dagger J_{D_2}t\right]|\tilde{\Psi}(0)\rangle, \tag{6.121}$$

and in the Schrödinger picture evolves as

$$\frac{d|\tilde{\Psi}^{(0)}(t)\rangle}{dt} = -i\left(\frac{\hat{H}}{\hbar} - \frac{i}{2}J_{D_1}^\dagger J_{D_1} - \frac{i}{2}J_{D_2}^\dagger J_{D_2}\right)|\tilde{\Psi}^{(0)}(t)\rangle. \tag{6.122}$$

Here, the basis of the system is defined by the three different cavities in which the photon could exist before being counted, i.e., the source cavity, optomechanical cavity 1, or optomechanical cavity 2, such that, at $t = 0$, the initial state of the composite system is

$$|\tilde{\Psi}(0)\rangle = (|1\rangle_c|0\rangle_{a_1}|0\rangle_{a_2})\,|0\rangle_{b_1}|0\rangle_{b_2}, \qquad (6.123)$$

with the mechanical oscillators taken to be in their ground states at $t = 0$.

We proceed as for the single-cavity case, so that a photon count at D_1 gives the unnormalised conditional state of the composite mechanical system as

$$
\begin{aligned}
|\tilde{\Phi}^{D_1}(t)\rangle &= \sqrt{\gamma}e^{-\gamma t}\bigg(-\frac{\kappa_1}{2}D_1^\dagger(\beta_1)\hat{R}_1 D_1(\beta_1) \\
&\quad -\frac{\kappa_2}{2}D_2^\dagger(\beta_2)\hat{R}_2 D_2(\beta_2) + 1 \bigg)|0\rangle_{b_1}|0\rangle_{b_2},
\end{aligned}
\qquad (6.124)
$$

where the $\hat{R}_i$ are each given as in Eq. (6.109) and $\beta_i = g_{0_i}/\Omega_i$, with g_{0_i} and Ω_i, respectively, being the vacuum optomechanical coupling rate and resonance frequency of mechanical oscillator i. It is clear that the conditional state has the general form of an entangled cat state:

$$|\psi\rangle = c_1|\alpha\rangle_{b_1}|0\rangle_{b_2} + c_2|0\rangle_{b_1}|\alpha\rangle_{b_2} + c_3|0\rangle_{b_1}|0\rangle_{b_2}, \qquad (6.125)$$

where, as usual, $|\alpha\rangle$ denotes a coherent state with complex amplitude α, and c_1, c_2, and c_3 are complex coefficients.

In order to validate this conditional state, we calculate the corresponding Wigner function. The total Wigner function is a function on the four-dimensional phase space. We will project it into a two-dimensional phase space defined by $p_1 = p_2 = 0$. It is instructive to calculate the mean amplitude in phase space experienced by each mechanical mode $\langle b_k\rangle$ as a result of its interaction with the single-photon. In Fig. 6.14(a), we plot the real and imaginary parts of the mean amplitude in phase space, $\langle b_k\rangle$, for a mechanical mode which has been driven by a single-photon for one complete mechanical period, T_m. After the mechanical mode has interacted with the photon for a quarter cycle $t_1 = T_m/4$, it is displaced such that its imaginary component is zero, and the real component is maximum. It goes through a phase change after $t_1 = T_m/4$, such that, at half cycle of the mechanical period, $t_1 = T_m/2$, the imaginary part of $\langle b_k\rangle$ is now maximum, whereas the real part is zero. At this time, the mechanical system has a maximum allowed amplitude in phase space for the parameters chosen. Hence, if detected at this specific time, one would expect it to exhibit its optimal nonclassical behaviour as an entangled cat state. For detection times $t > T_m/2$, the complete cycle is closed. Therefore, in order to probe the nonclassicality of the conditional state, we focus on its Wigner function at half period. Care must be taken to find the right projection into a two-dimensional subspace. In Fig. 6.14(b), we plot a projection

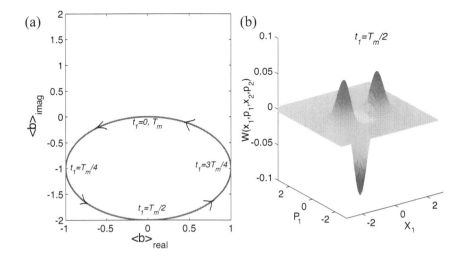

FIGURE 6.14 (a) A parametric plot of the real and imaginary components of the mean amplitude in phase space for a mechanical oscillator. (b) A projection of the Wigner function of the combined conditional state of two identical mechanical modes versus P_1 and X_1. The state is conditioned on a detection at half cycle of the mechanical period. Parameters are chosen in units of optical decay rate; $\kappa_1 = \kappa_2 = \kappa, g_0/\kappa = 0.02, \Omega/\kappa = 0.02, \gamma/\kappa = 2/\kappa$. From Akram et al. [6], with permission.

of the conditional Wigner in which two peaks are evident, with an interference between them resulting in negativity in the Wigner function. This indicates a conditional cat state of the cavity field.

6.5.2 Photon blockade with two single-photon pulses

As we have just seen, conditioned on a single-photon detection, the effective interaction between the optical cavity field and the mechanical system can lead to a large effective interaction time resulting in large changes in the momentum of the mechanical system. In Section 6.1 we saw how the optomechanical interaction can lead to a single-photon blockade. We return to this here for the case of driving with two consecutive single-photon pulses.

Consider the two-photon excitation protocol shown in Fig. 6.15. We will calculate the rate of detection of the second photon as a function of the delay time, T_d, and duration of its interaction with the mechanical oscillator, τ. The conditional (normalised) state of the mechanical system, given that the first photon was counted at time t_1 in Eq. (6.110), can be written in the form

$$|\Phi^{(1)}(t_1)\rangle = \sqrt{\kappa}|T\rangle + \sqrt{\gamma}|R\rangle, \qquad (6.126)$$

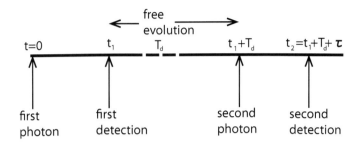

FIGURE 6.15 A protocol for exciting an optomechanical cavity with two consecutive single-photon pulses. The first photon prepared in the source cavity at time $t = 0$ interacts with the mechanics and is detected at time t_1. The mechanical resonator then evolves freely for a time delay, T_d, between detection of the first photon at t_1 and preparation of the second photon. The second photon arrives at $t_1 + T_d$ and interacts for a time τ before being detected at time $t_2 = t_1 + T_d + \tau$. From Akram et al. [6], with permission.

which is a superposition of two histories: detection after transmission through the cavity $|T\rangle$ and detection after reflection from the cavity $|R\rangle$.

This state evolves for period T_d, at the end of which the source is prepared with another single-photon. The second photon interacts for a time τ with the optomechanical cavity and is then detected at a time $t_2 = t_1 + T_d + \tau$. The final conditional state of the system is a result of applying the jump operator twice to the initial state at $t_1 = 0$. There are four indistinguishable temporal histories for the two detection events. The conditional (unnormalised) state of the mechanical resonator is given by

$$|\Phi^{(2)}(t_2 : T_d : t_1 : 0)\rangle = \kappa|TT\rangle + \gamma|RR\rangle + \sqrt{\kappa\gamma}(|RT\rangle + |TR\rangle). \quad (6.127)$$

The *conditional* rate of detection of the second photon R_2 can be evaluated as

$$
\begin{aligned}
R_2(t_2, T_d, t_1) &= \langle\phi^{(2)}(t_2 : T_d : t_1 : 0)|\phi^{(2)}(t_2 : T_d : t_1 : 0)\rangle \quad (6.128) \\
&= \kappa^2\langle TT|TT\rangle + \gamma^2\langle RR|RR\rangle \\
&\quad + \kappa\gamma\left[\langle RT|RT\rangle + \langle TR|TR\rangle\right] \\
&\quad + \kappa\gamma\left[\langle TT|RR\rangle + \langle RT|TR\rangle + c.c.)\right] \\
&\quad + \kappa\sqrt{\kappa\gamma}\left[\langle TT|RT\rangle + \langle TT|TR\rangle + c.c.\right] \\
&\quad + \gamma\sqrt{\kappa\gamma}\left[\langle RR|RT\rangle + \langle RR|TR\rangle + c.c.\right].
\end{aligned}
$$

Explicit expressions for each doubly conditioned part $|XY\rangle$ of the state $|\Phi^{(2)}\rangle$ as well as for each term of Eq. (6.128) are given by Akram et al. [6]. The total

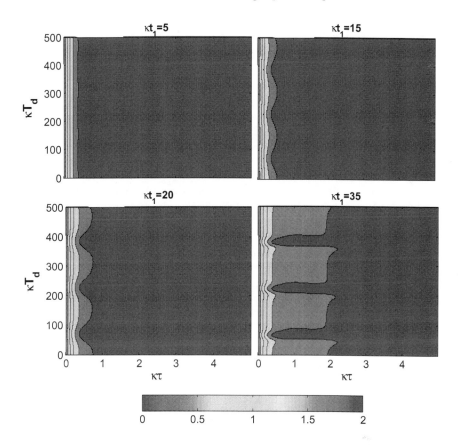

FIGURE 6.16   The total conditional rate of detection of the second photon as a function of its optomechanical interaction time $\kappa\tau$ and free evolution κT_d for $\gamma/\kappa = 2$, $g_0/\kappa = 0.05$, $\Omega/\kappa = 0.02$ and different detection times of the first photon κt_1. From Akram et al. [6], with permission.

conditional rate of detection for the second photon is shown in Fig. 6.16. If the first photon is detected early, Fig. 6.16 shows that the profile for the total conditional rate of the second photon remains independent of the delay time T_d. This is because the first is unlikely to have excited the mechanical motion. On the other hand, if the first photon is detected late, we observe a periodic modulation of the rate of detection for the second photon with respect to the delay time T_d. This indicates that the first photon significantly excited the mechanical motion of the mirror, which then moves to detune the cavity from resonance with the second photon: the first photon detection has primed a photon blockade for the absorption by the cavity of the second photon. This effect is maximised at a wait time of a quarter cycle after arrival of the first photon.

6.6 SINGLE-SIDEBAND-PHOTON OPTOMECHANICS

For the majority of this chapter, we have examined quantum optomechanical systems in the single-photon strong coupling regime ($g_0 > \{\kappa, \Gamma\}$), where the presence of a single photon can significantly change the system dynamics. This is, generally, a challenging regime to achieve. In Chapters 3–5, we considered approaches to quantum optomechanics that use a bright coherent field to boost the effective interaction strength between the light and mechanical oscillator. In this case, when outside of the single-photon strong coupling regime, the optomechanical dynamics are well described by a linearised Hamiltonian (See Section 2.7). As such, it is not possible to directly generate states of either the light, or the mechanical oscillator, that exhibit Wigner-negativity. However, it is possible to generate states with Wigner-negativity outside of the single-photon strong coupling regime by combining bright coherent driving at a laser frequency Ω_L with single-photon counting of the output light at sideband frequencies $\Omega_L \pm \Omega$ [291, 108, 7].

The essential physics is that the interaction between the light and the mechanical oscillator induces scattering events that either (a) generate a phonon in the mechanical oscillator and Raman scatter a photon from Ω_L to $\Omega_L - \Omega$, or, (b) subtract a phonon from the mechanical oscillator and anti-Raman scatter a photon from Ω_L to $\Omega_L + \Omega$. A sideband photon counting event can then condition the state of the mechanical oscillator into a nonclassical state [291]. For instance, if the mechanical oscillator is initially cooled to its ground state (see Section 4.2), the mechanical oscillator can be conditioned into a single-phonon Fock state via detection of a photon with frequency $\Omega_L - \Omega$ [291, 108]. This approach has been used to generate nonclassical states of ensembles of cold atoms [168, 289] and, more recently, 40 THz optical phonons in bulk diamond crystals [173]. In the context of cavity optomechanics, phonon counting experiments have been performed indirectly by counting sideband photons [76, 172], though nonclassical state generation via this method has yet to be achieved.

CHAPTER 7

Nonlinear optomechanics

CONTENTS

7.1   Duffing nonlinearity 231
7.2   Quantum Duffing oscillator 233
     7.2.1   Quantum tunnelling in the Duffing oscillator. 237
     7.2.2   Squeezing in the driven Duffing oscillator 239
7.3   Nonlinear damping 240
7.4   Self-pulsing and limit cycles 241
     7.4.1   Hopf bifurcation from a quadratic optomechanical
             interaction 242
7.5   Nonlinear measurement of a mechanical resonator 249

The optomechanical interaction is already nonlinear: it is quadratic in the field amplitudes and linear in the mechanical oscillator amplitudes. We saw this, for instance, in Sections 4.5.2 and 6.1, where the canonical (polaron) transformation was used to decouple the mechanical and optical degrees of freedom, thereby introducing a quartic nonlinearity in the field. In this chapter we will discuss alternative forms of nonlinearity in both the potential energy stored in the mechanical degree of freedom and in the coupling of the mechanical element to the cavity field.

7.1   DUFFING NONLINEARITY

One can expand the elastic potential energy of a mechanical resonator to fourth order in the mechanical displacement. The resulting model is known as the Duffing oscillator and exhibits a fixed-point bifurcation as the strength of the mechanical driving is increased [209, 62]. In the quantum regime, nanomechanical resonators have provided a clear setting for the Duffing mechanical nonlinearity [11]. In the optical regime, a study of the role of the Duffing nonlinearity was presented by Zaitsev et al. [330]. Electrostatic tuning of the Duffing nonlinearity for optomechanical systems has been proposed by Rips et al. [239]. If a system with a Duffing nonlinearity is harmonically driven, parametric resonance can result, as discussed in Chapter 5.

The Hamiltonian for a mechanical system with a Duffing nonlinearity is [62, 20]

$$\hat{H} = \frac{\hat{p}^2}{2m} + \frac{m\Omega^2}{2}(\hat{q}^2 + \frac{\lambda}{2}\hat{q}^4) - f_0 \cos(\omega_p t)\hat{q}, \qquad (7.1)$$

where m is the effective mass of the mechanical resonator, Ω is the linear resonator frequency taking into account the applied strain, and $f_0 \cos(\omega_p t)$ is a harmonic driving force at the pump frequency of ω_p and amplitude f_0. The nonlinearity parameter is λ with units of inverse length squared. Depending on the sign of λ, the Duffing nonlinearity can be either *hardening*, in which case the resonant frequency is increased due to the nonlinearity, or *softening*, with the nonlinearity decreasing the resonant frequency. In practice, which type of nonlinearity is exhibited depends on the strain conditions of the mechanical resonator [62]. In what follows, we specifically consider the case of a hardening nonlinearity, where $\lambda > 0$.

The classical equation of motion for this system, including dissipation, is

$$\ddot{q} + \Gamma\dot{q} + \Omega^2(q + \lambda q^3) = E \cos(\omega_p t) , \qquad (7.2)$$

where $E = f_0/m$ and, as usual, Γ is the rate of energy dissipation. We look for approximate long time solutions (i.e., after transients have died out) with the harmonic ansatz

$$q(t) = \frac{1}{2}A(t)e^{i\omega_p t} + \text{c.c.} . \qquad (7.3)$$

We will assume that the nonlinearity is weak so that

$$\lambda a_c^2 \ll 1, \qquad (7.4)$$

where a_c is a suitable length scale defined below. As we will see, this requires a large quality factor $Q \equiv \Omega/\Gamma$, in other words, weak damping. Substituting Eq. (7.3) into the equation of motion, we find that, in the long time limit

$$A\left((\Omega^2 - \omega_p^2) + 3\lambda\Omega^2|A|^2/4 + i\Omega\omega_p/Q\right) = E \qquad (7.5)$$

and thus the amplitude of the motion in the long time limit is given by the solutions to

$$|A|^2 \left[\left((\Omega^2 - \omega_p^2) + 3\lambda\Omega^2|A|^2/4\right)^2 + \Omega^2\omega_p^2/Q^2\right] = E^2. \qquad (7.6)$$

We now take $\omega_p = \Omega + \delta$ and assume that $\omega_p \approx \Omega \gg \delta$. The resonance curve is then well approximated by

$$\left(\delta - \frac{3\lambda\Omega|A|^2}{8}\right)^2 + \frac{\Omega^2}{4Q^2} = \frac{E^2}{|A|^2\Omega^2} . \qquad (7.7)$$

In Fig. 7.1 we plot $|A|$ versus δ for different values of the force E (see also

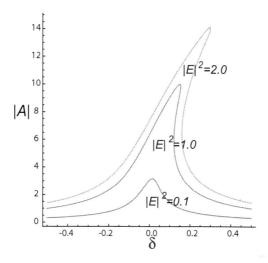

FIGURE 7.1 The stationary amplitude of the displacement, $|A|$, versus detuning, $\delta = \omega_p - \Omega$, for different values of the force, E. The units of time have been chosen such that $\Omega = 1$.

the optomechanical equivalent in Fig 2.3). There is a critical value for the forcing term E_c above which the long time limit is multivalued. As we will see in more detail below, not all of these solutions are stable. For now, we note the key feature: there is a pulling of the resonance to higher frequencies, known as hardening. As mentioned earlier, if $\lambda < 0$, the nonlinearity would be softening. The maximum value of $|A|$ at which the resonance amplitude has an infinite slope defines a critical amplitude a_c, which we can use as a convenient length scale. The value of a_c is found to be [163]

$$a_c^2 = \frac{2\sqrt{3}}{9\lambda Q} .\tag{7.8}$$

Comparing this to Eq. (7.4), we see that an assumption of small nonlinearity then requires that $Q \gg 1$ in order for the harmonic approximation (Eq. (7.3)) to be valid.

7.2 QUANTUM DUFFING OSCILLATOR

We turn now to a quantum description of the Duffing oscillator. The position and momentum operators of the resonator may be written in terms of the raising and lowering operators $b^\dagger$ and b via Eqs. (1.9). Substituting these expressions into the Hamiltonian of Eq. (7.1), and redefining the drive amplitude $f_0 = -2\hbar\epsilon/x_{xp}$ to match convention, where, as always, x_{xp} is the zero-point motion of the oscillator, we find

$$\hat{H} = \hbar\Omega b^\dagger b + \hbar\frac{\chi}{12}(b + b^\dagger)^4 + 2\hbar\epsilon\cos(\omega_p t)(b + b^\dagger), \tag{7.9}$$

where χ gives the nonlinear dispersion in terms of the nonlinearity parameter λ as

$$\chi = \frac{3\hbar\lambda}{8m}. \tag{7.10}$$

In the example of a doubly clamped platinum beam in [163], $\chi \sim 3.4 \times 10^{-4}$ s^{-1}.

Moving to an interaction picture at the pump frequency, ω_p, and assuming that $\{\Omega, \omega_p\} \gg \chi$ to neglect rapidly oscillating terms in the quartic term, the Hamiltonian may be approximated by

$$\hat{H}_I = \hbar\Delta b^\dagger b + \hbar\frac{\chi}{2}(b^\dagger)^2 b^2 + \hbar\epsilon(b + b^\dagger), \tag{7.11}$$

where $\Delta = \Omega - \omega_p$ is the detuning of the resonator from the pump. In this form we see that the nonlinearity is similar to the effective nonlinearity due to the radiation pressure coupling after the polaron transformation (see Sections 4.5.2 and 6.1).

We include dissipation in the usual way by a Markov master equation (see Section 5.1). In the interaction picture this is

$$\dot{\rho} = -\frac{i}{\hbar}[\hat{H}_I, \rho] + \frac{\Gamma}{2}(\bar{n}+1)(2b\rho b^\dagger - b^\dagger b\rho - \rho b^\dagger b) + \frac{\Gamma\bar{n}}{2}(2b^\dagger\rho b - bb^\dagger\rho - \rho bb^\dagger), \tag{7.12}$$

where, as usual, Γ is the rate of energy loss from the resonator and $\bar{n}$ is the mean phonon number in a bath oscillator at frequency Ω. We will usually assume low temperature operation so that $\bar{n} \approx 0$. This model was introduced long ago in quantum optics to describe optical bistability due to a Kerr nonlinear medium (see section 4.5.2) [96]. The system has a long time solution, a steady state in the rotating frame, that can change stability as the pump field is varied.

In the high-Q limit, using the quantum Langevin equation of Eq. (1.112) and the above interaction Hamiltonian, the quantum stochastic differential equation for this system can be shown to be

$$\dot{b} = -i\epsilon - \left(\Gamma/2 + i(\Delta + \chi b^\dagger b)\right) b + \sqrt{\Gamma} b_{\text{in}}, \tag{7.13}$$

where the noise correlation functions are given in Eqs. (1.115).

Exercise 7.1 *Derive this result.*

The corresponding semiclassical equation

$$\dot{\alpha} = -i\epsilon - \left[\Gamma/2 + i(\Delta + \chi|\alpha|^2)\right]\alpha \tag{7.14}$$

can be found by taking average values of both sides and factorising the moment, $\langle b^\dagger b^2 \rangle = |\alpha|^2 \alpha$ with $\alpha \equiv \langle b \rangle$.[1] For a more rigorous justification of this semiclassical equation, we refer the reader to Section 7.2.1, which uses the Fokker-Planck equation for the positive P-function as in Drummond and Walls [96]. The semiclassical steady state of Eq. (7.14) is defined by $\dot{\alpha} = 0$, which determines the fixed points (also called critical points) α_0. These critical points are given by the solution to the state equation

$$I_p = n_0 \left[\frac{\Gamma^2}{4} + (\Delta + \chi n_0)^2 \right], \tag{7.15}$$

where the pump strength $I_p \equiv \epsilon^2$, while $n_0 \equiv |\alpha_0|^2$ determines the average energy in the nanomechanical resonator. This equation corresponds to Eq. (7.7) in our dimensionless units. I_p is a cubic in n_0 with turning points at the values of n_0 that satisfy

$$\frac{dI_p}{dn_0} = \frac{\Gamma^2}{4} + (\Delta + 3\chi n_0)(\Delta + \chi n_0) = 0. \tag{7.16}$$

As a function of the pump strength, the average vibrational excitation number n_0 is multivalued. Eq. (7.16) defines values at which the slope diverges, indicative of a change in stability.

In Fig 7.2 we plot n_0 versus the pump strength $|\epsilon|^2$ for various values of the detuning, Δ. For negative detuning, $\Delta < 0$, n_0 becomes a multivalued function of ϵ. This is a manifestation of the nonlinear hardening of the spring constant that we saw for the resonance equation (Eq. (7.7)). Not all the steady-state solutions are stable. To determine stability, we linearise the equations of motion around the fixed points by writing $\alpha(t) = \alpha_0 + \delta\alpha(t)$. The equations of motion for the fluctuation field $\delta\alpha(t)$ are then given by

$$\frac{d}{dt} \begin{pmatrix} \delta\alpha \\ \delta\alpha^* \end{pmatrix} = \mathbf{M} \begin{pmatrix} \delta\alpha \\ \delta\alpha^* \end{pmatrix}, \tag{7.17}$$

where

$$\mathbf{M} = \begin{pmatrix} -\Gamma/2 - i(\Delta + 2\chi n_0) & -iG \\ iG^* & -\Gamma/2 + i(\Delta + 2\chi n_0) \end{pmatrix} \tag{7.18}$$

with $G \equiv \chi\alpha_0^2$, with α_0 being the solution to the cubic

$$\alpha_0 [\Gamma/2 + i(\Delta + \chi n_0)] = -i\epsilon \tag{7.19}$$

obtained by taking the steady state limit of Eq. (7.14) (i.e., setting $\dot{\alpha} = 0$). We can then reexpress α_0 as $\alpha_0 = \sqrt{n_0}e^{i\phi_0}$, where

$$\tan\phi_0 = \frac{\Gamma}{2\Delta + 2\chi n_0}. \tag{7.20}$$

---

[1]Note that, while throughout the majority of this text α is used to denote the *optical* coherent amplitude, here we use it to denote the *mechanical* coherent amplitude.

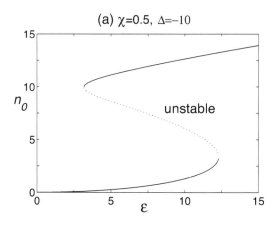

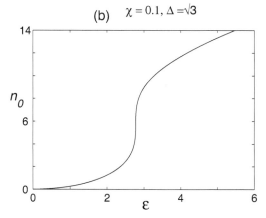

FIGURE 7.2 Plots of the mean vibrational excitation number of the nanomechanical resonator, n_0, versus pump field amplitude, ϵ ,for $\Gamma = 2.0$. The unstable branch shown in (a) is absent from (b) due to different values of pump field detuning and Duffing nonlinearity.

Since we have taken ϵ as real, ϕ_0 is the phase shift of the oscillator field from the pump field.

The stability is determined by the eigenvalues $\lambda^{\pm}$ of the matrix $\mathbf{M}$.

Exercise 7.2 *Show that these eigenvalues are*

$$\lambda^{\pm} = -\frac{\Gamma}{2} \pm i\sqrt{(\Delta + 3\chi n_0)(\Delta + \chi n_0)} \ . \tag{7.21}$$

A stable steady state requires that the real parts of the eigenvalues must be negative.

Exercise 7.3 *Convince yourself that the fixed points are stable outside, and unstable between, the turning points of the state equation (Eq. (7.15)).*

In Figure 7.2 we show the unstable fixed points as a dashed line in plot (a). Note that one of the eigenvalues vanishes at the turning (or bifurcation) points and the linearised analysis breaks down at these points. The turning points are also often referred to as *switching points* since, when the driving power increases beyond them, the steady-state value makes a rapid transition to the new stable steady-state on either the upper or lower branch. The bistable steady-state discussed here is analogous to the optomechanical bistability in Section 2.6.2.

When classical noise, such as thermal noise, is included the result discussed in the previous paragraph must be qualified somewhat, since noise can cause the system to switch before the average driving power reaches the turning point; this is called thermal activation. Quantum noise at zero temperature, however, can also cause the system to switch from one stable branch to another through dissipative quantum tunnelling. We now discuss this highly nonclassical possibility in more detail.

7.2.1 Quantum tunnelling in the Duffing oscillator.

Drummond and Walls [96] found an exact steady-state solution to the master equation Eq. (7.12) in the case of zero temperature $\bar{n} = 0$. This was done using a particular representation of the density operator known as the positive P-representation defined by

$$\rho(t) = \int \mu(\alpha, \beta) P(\alpha, \beta, t) \frac{|\alpha\rangle\langle\beta^*|}{\langle\beta^*|\alpha\rangle}, \qquad (7.22)$$

where $|\alpha\rangle, |\beta\rangle$ are oscillator coherent states and $\mu(\alpha, \beta)$ is an appropriate integration measure chosen to ensure convergence of the normalisation integral corresponding to the condition $\mathrm{tr}\rho = 1$. It is possible to see that, in this representation, direct integration gives normally ordered moments in the form

$$\langle (b^\dagger)^m b^n \rangle = \mathrm{tr}[(b^\dagger)^m b^n \rho] = \int \mu(\alpha, \beta)(\beta)^m \alpha^n P(\alpha, \beta, t) . \qquad (7.23)$$

Substitution in the master equation, Eq. (7.12), leads to an equivalent equation of motion for $P(\alpha, \beta, t)$,

$$
\begin{aligned}
\frac{\partial P(\alpha, \beta, t)}{\partial t} &= \partial_\alpha \left[(\Gamma/2 + i\Delta)\alpha + i\epsilon + i\chi\beta^2\alpha \right] P(\alpha, \beta, t) \qquad (7.24) \\
&\quad + \partial_\beta \left[(\Gamma/2 - i\Delta)\beta - i\epsilon - i\chi\alpha^2\beta \right] P(\alpha, \beta, t) \\
&\quad - \left[i\frac{\chi}{2}\partial_{\alpha\alpha}^2\alpha^2 - i\frac{\chi}{2}\partial_{\beta\beta}^2\beta^2 \right] P(\alpha, \beta, t).
\end{aligned}
$$

This is in the form of a Fokker-Planck equation for a nonlinear stochastic system. If we neglect the second-order derivatives, the *noise terms*, the first-order derivatives give the systematic part of the equations of motion on the manifold $\alpha^* = \beta$; i.e., they give the semiclassical equations of motion. Comparing the argument of the partial derivative with respect to α to the semiclassical equation of motion (Eq. (7.14)) we derived earlier by factorisation, we find that they are identical.

It is possible to find the general steady-state solution for Eq. (7.24), since this equation satisfies the potential conditions [111]. The result is

$$P_{ss}(\alpha, \beta) = \mathcal{N}e^{-V(\alpha,\beta)}, \tag{7.25}$$

where

$$V(\alpha, \beta) = -2\alpha\beta - \lambda\ln(\chi\alpha^2) - \lambda\ln(\chi\beta^2) - \frac{2\epsilon}{\chi\alpha} - \frac{2\epsilon}{\chi\beta} \tag{7.26}$$

with

$$\lambda = \frac{(\Gamma/2 + i\Delta)}{2\chi} - 1 \tag{7.27}$$

and $\mathcal{N}$ is the normalisation calculated by choosing an appropriate contour in the domain of α, β. The steady-state moments can be calculated using Eq. (7.23). The explicit expressions are given in Drummond and Walls [96]. For our purposes, here we reproduce a plot of $|\langle b \rangle|$ versus the pump amplitude ϵ, superimposed on the semiclassical study state curves; see Fig. 7.3. Note that the quantum steady-state amplitude does *not* exhibit bistability. The reason for this is that the steady-state quantum amplitude corresponds to an ensemble average over a steady-state phase-space distribution that has support on *both* fixed points in the bistable region of the curve. On the other hand, the semiclassical steady state given in Eq. (7.19) refers to deterministic solutions to the equations of motion.

We can further elucidate this explanation by considering a large number of experiments on a single system wherein we slowly sweep the drive strength ϵ. When the drive strength enters the bistable region, there is a finite probability per unit time that the system will *jump* from one fixed point to the other and back again, driven by the quantum noise implicit in the second-order differential terms in Eq. (7.24). As such transitions are stochastic, over many trials they will occur for different values of the driving strength. As a result, the steady-state ensemble average shows a smooth monotonic variation of the average amplitude $|\langle b \rangle|$. The stochastic transitions discussed here occur for zero temperature and are thus uniquely quantum mechanical in origin and should not be confused with thermal activation over a barrier. In analogy to a similar phenomenon in conservative double well systems, we refer to this kind of stochastic switching between fixed points as dissipative quantum tunnelling.

P-function (and Q-function) methods for the Duffing model posed in the form of Eq. (7.9) were first given by Vogel and Risken [302]. Dykman [98] has given a careful treatment of dissipative quantum tunnelling in a periodically driven Duffing oscillator which he calls *quantum activation* to distinguish this zero-temperature phenomenon from standard thermal activation in bistable

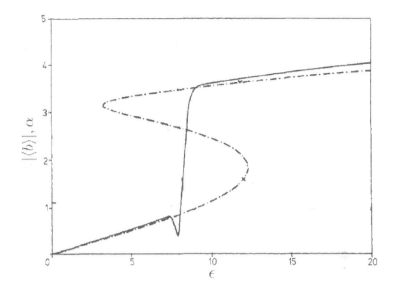

FIGURE 7.3 Plots of the quantum steady state mean amplitude of the nanomechanical resonator amplitude, $|\langle b \rangle|$ (solid curve), and the steady state semiclassical amplitude, α (dot-dashed curve), versus drive amplitude ϵ for the same parameters as in Fig 7.2(a). From Drummond and Walls [96]. IOP Publishing. Reproduced with permission. All rights reserved.

systems. Dykman's analysis is carried out using the Wigner quasi-probability representation of the density operator rather than the positive P representation we have discussed here. Also using Wigner function moments and the numerical solution of the master equation, Katz et al. [151] studied the quantum limit of switching between fixed points in a Duffing resonator at finite temperature. To our knowledge, dissipative quantum tunnelling (quantum activation) has not yet been observed in mechanical Duffing oscillators although it has been observed in superconducting quantum circuits containing a Josephson junction [298].

7.2.2 Squeezing in the driven Duffing oscillator

While the exact steady-state solution obtained in the previous section reveals dissipative quantum tunnelling, it does not enable us to calculate dynamical features such as the noise power spectrum and two-time correlation functions for the displacement amplitude. To calculate these we resort to an approximation scheme based on linearisation around the stable steady states. We will follow the treatment of Babourina-Brooks et al. [20].

As discussed in Section 2.7, the idea of linearisation is to make a canonical transformation that displaces the canonical operators by the semiclassical steady-state values of the mean amplitudes. Thus

$$b \rightarrow b + \alpha_0, \tag{7.28}$$

where α_0 is the solution to Eq. (7.19), the steady-state mean amplitude of the resonator. In the P-representation this is equivalent to the linearisation used to test for stability of the steady state in Eq. (7.17). The corresponding linearised interaction picture Hamiltonian is then quadratic in the annihilation and creation operators

$$\hat{H} = -\hbar(\Delta + 2\chi n_0)b^\dagger b + \hbar G(b^2 + b^{\dagger\,2}), \tag{7.29}$$

where, as before, $G = \chi \alpha_0^2$ and $n_0 = |\alpha_0|^2$, and in the linearisation procedure we have neglected terms higher than second order in the operators. This result is similar to the treatment of squeezing via the optomechanical interaction in Section 4.5.2.2.

Exercise 7.4 *Derive this Hamiltonian, starting from Eq. (7.11).*

The corresponding quantum stochastic differential equations are

$$\frac{d}{dt}\begin{pmatrix} b \\ b^\dagger \end{pmatrix} = \mathbf{M}\begin{pmatrix} b \\ b^\dagger \end{pmatrix} + \sqrt{\Gamma}\begin{pmatrix} b_{\mathrm{in}}(t) \\ b_{\mathrm{in}}^\dagger(t) \end{pmatrix}, \tag{7.30}$$

where $\mathbf{M}$ is given by Eq. (7.18) and the noise correlation functions are unchanged by the displacement transformation. To make a connection between the nonlinear oscillator, itself, and what is actually measured in the laboratory, we need a transducer for the displacement. In Chapter 3 we saw that the transducer is the optical field exiting the cavity.

The occurrence of an effective parametric amplification in the driven Duffing oscillator can be used to amplify a weak probe signal [20]. Antoni et al. [15] demonstrated this experimentally using an optomechanical system fabricated as a suspended photonic-crystal (PhC) nanomembrane. This device had mechanical resonance frequencies around 1 MHz and quality factors (Q) of about 5000. They measured a Duffing parameter of $\lambda \approx 10^{12}$ m^{-2}.

7.3 NONLINEAR DAMPING

Many Duffing nonlinear nanomechanical and micromechanical oscillators exhibit nonlinear damping for which the damping rate is not a constant (as we have assumed so far) but rather depends on the degree of excitation of the resonator [331]. A common assumption in such a model is that, in addition to the standard linear coupling of the resonator displacement to a thermal bath, there is a term proportional to the square of the resonator displacement in the system-bath coupling. This leads to energy-dependent dissipation.

A detailed theoretical model of this situation can be quite complex since, in general, the resulting dynamics can be non-Markovian. A particularly simple model results in the weak coupling high-frequency limit described by the Markov master equation in the interaction picture

$$\frac{d\rho}{dt} = -\frac{i}{\hbar}[\hat{H}_I, \rho] + \Gamma \mathcal{D}[b]\rho + \Gamma_2 \mathcal{D}[b^2], \tag{7.31}$$

where $\hat{H}_I$ is the system Hamiltonian in the interaction picture, Γ is the usual linear damping rate, and Γ_2 is the nonlinear damping rate. Similar models have been introduced in quantum optics to describe two-photon absorption [79]. We can get a good idea of how the model behaves by simply computing the equation of motion for $\langle b \rangle$:

$$\frac{d\langle b \rangle}{dt} = -i(\ldots) - \frac{\Gamma}{2}\langle b \rangle - \Gamma_2 \langle b^\dagger b^2 \rangle. \tag{7.32}$$

The semiclassical equation of motion can be found by factoring all moments and writing $\alpha = \langle b \rangle$,

$$\dot{\alpha} = -i(\ldots) - \frac{\Gamma}{2}\alpha - \Gamma_2 |\alpha|^2 \alpha. \tag{7.33}$$

We can again use the positive P-representation[96] to analyse this model. If we include in the term in $\hat{H}_I$ corresponding to the Duffing nonlinearity (see Eq. (7.11)), we find that the P-representation model is equivalent to the pure Duffing model with the replacement $\chi \to \tilde{\chi} = \chi - i\Gamma_2/2$. The steady-state solution is also given by Eq. (7.25) with this replacement.

7.4 SELF-PULSING AND LIMIT CYCLES

In the preceding sections we have been concerned with the fixed points of dissipative nonlinear systems. In these cases, the long-time solution is attracted onto a zero-dimensional set in the phase space of the system. However, this is not the only kind of stable long-time solution that is possible. A fixed point can itself become unstable (typically via a Hopf bifurcation; see below) giving rise to a one-dimensional attractor, a closed curve, in the phase space. This is called a *limit cycle*, as the long-time solution relaxes onto a stable oscillation [268]. This is sometimes referred to as *self-pulsing*. In optics, the canonical example of a limit cycle as a stable attractor is found in the semiclassical model of a laser in terms of the van der Pol oscillator [240]. As the laser pumping is driven above threshold, a fixed point at the origin in the phase space of the laser field goes unstable, leading to the emergence of a stable oscillation with nonzero amplitude. In fact, the word lasing is often used as a synonym for limit cycle in optics.

Self-sustained oscillations, in the form of limit cycles, can arise in nonlinear systems in which amplification competes with nonlinear damping or nonlinear detuning. One of the earliest observations of a limit cycle oscillation in an optomehanical system was by Kippenberg et al. [159] using a parametric instability. A heating induced (light absorption) forcing mechanism was seen to lead to limit cycles in the work of Metzger et al. [198].

A quantum theory of limit cycles in an optomechanical system was given by Lörch et al. [183] in a close analogy to the theory of the laser. This necessarily requires us to take into account quantum noise, which leads to a slow diffusion of the phase of the limit cycle oscillation and nonclassical effects in the phonon statistics.

We will discuss, here, a quantum model based on an oscillator with nonlinear damping as in Section 7.3, which is also subjected to phase-insensitive amplification. Including amplification at rate A, the master equation of Eq. (7.31) becomes

$$\dot{\rho} = -i\Omega[b^\dagger b, \rho] + \Gamma \mathcal{D}[b]\rho + \Gamma_2 \mathcal{D}[b^2]\rho + A\mathcal{D}[b^\dagger]\rho. \tag{7.34}$$

The amplification could, for example, be implemented by driving on the blue sideband (see Section 4.4.1) of an optomechanical element and adiabatically eliminating the cavity field. Equations of motion for the first-order moments can be obtained using Eq. (5.11). Factorising moments that are second-order, or higher, in these equations we arrive at the semiclassical equations

$$\dot{\beta}_r = \Omega\beta_i - \Gamma\beta_r/2 - \Gamma_2(\beta_r^2 + \beta_i^2 - A/2\Gamma_2)\beta_r \tag{7.35}$$
$$\dot{\beta}_i = -\Omega\beta_r - \Gamma\beta_i/2 - \Gamma_2(\beta_r^2 + \beta_i^2 - A/2\Gamma_2)\beta_i, \tag{7.36}$$

where $\langle b \rangle = \beta = \beta_r + i\beta_i$.

Exercise 7.5 *Derive these equations of motion.*

In polar coordinates, with $\beta \equiv re^{i\theta}$, the equations take the form

$$\dot{r} = r(C - \Gamma_2 r^2) \tag{7.37}$$
$$\dot{\theta} = \Omega \tag{7.38}$$

with $C \equiv (A - \Gamma)/2$.

The fixed points (critical points) of this equation are $\beta_{r0} = \beta_{i0} = 0$. Linearising around these fixed points, we see that they are stable provided $A < \Gamma$ ($C < 0$). For A above the critical amplification value $A_c = \Gamma$ we have $C > 0$ and the origin becomes unstable, with the system exhibiting a Hopf bifurcation. In that case, the long-time solution is attracted onto the curve $r^2 = C/\Gamma_2$. This is the limit cycle [10]. In the optomechanical literature, this limit cycle bifurcation is often called a lasing transition, as it has some similarity to the limit cycle transition in the semiclassical theory of the laser.

7.4.1 Hopf bifurcation from a quadratic optomechanical interaction

Another example of an exactly solvable quantum model with limit cycles was presented in [140]. This model is based on the quadratic optomechancial interaction discussed later in Section 7.5. Thompson et al. [283] describe an optomechanical system comprising an optical cavity containing a thin dielectric membrane of SiN placed between rigid high-finesse mirrors: the membrane-in-the-middle model. Including coherent driving of the cavity with amplitude ϵ_c and frequency Ω_L to the basic Hamiltonian for such systems given in Eq. (7.63), and moving into the interaction picture for the light we arrive at the driven Hamiltonian

$$\hat{H} = \hbar\Delta a^\dagger a + \hbar\Omega b^\dagger b + \hbar(\epsilon_c^* a e^{i\Omega_L t} + \epsilon_c a^\dagger e^{-i\Omega_L t}) + \frac{\hbar}{2}g_2 a^\dagger a(b+b^\dagger)^2, \quad (7.39)$$

where, as usual, $a, a^\dagger$ are the lowering and raising operators for the cavity field, $b, b^\dagger$ are the lowering and raising operators for the mechanical oscillator, the optical detuning $\Delta \equiv \Omega_c - \Omega_L$, with Ω_c the cavity optical frequency, Ω is the mechanical resonance frequency, and g_2 is the vacuum quadratic optomechanical coupling rate defined later in Eq. (7.64).

The model can be simplified by linearising around the steady-state field inside the cavity, $\langle a\rangle_{ss} \equiv \alpha_0$, in a similar manner to that performed for the usual optomechanical interaction in Section 2.7. Assuming that the mechanical frequency is much larger than the cavity linewidth, $\Omega \gg \kappa$, the optical cavity can be driven on resolved sidebands. This is achieved when the parametric resonance condition

$$\Delta = \pm 2\Omega \quad (7.40)$$

is met. Dropping nonresonant terms and neglecting terms that are fourth order in the operators – a form of linearisation approximation – results in two different kinds of resonant interactions:

$$\hat{H}_r = \hbar\chi(a^\dagger b^2 + a b^{\dagger 2}) + \hbar g_2|\alpha_0|^2 b^\dagger b \quad (7.41)$$

when the laser driving the cavity is tuned to the red (lower frequency) of the cavity frequency $\Omega_L = \Omega_c - 2\Omega$, and

$$\hat{H}_b = \hbar\chi(a^\dagger b^{\dagger 2} + a b^2) + \hbar g_2|\alpha_0|^2 b^\dagger b \quad (7.42)$$

when the laser driving frequency is tuned to the blue (higher frequency) of the cavity frequency $\Omega_L = \Omega_c + 2\Omega$. In both cases $\chi \equiv g_2\alpha_0/2$.

Exercise 7.6 *Derive these results from Eq. (7.39).*

These Hamiltonians describe effective two-phonon Raman processes in which a photon is destroyed and two phones are either created or destroyed. The optomechanically induced mechanical frequency shift $(g_2|\alpha_0|^2)$ will be absorbed into the definition of the mechanical frequency. Henceforth, we will only consider the red sideband case, $\Delta = 2\Omega$. Including a mechanical driving force, ϵ (in frequency units), in an interaction picture for both light and mechanics, the Hamiltonian is

$$\hat{H}_I(t) = \hbar\chi\left(a^\dagger b^2 + a(b^\dagger)^2\right) - \hbar\epsilon(b+b^\dagger). \quad (7.43)$$

Including dissipation, we thus have the master equation,

$$\dot{\rho} = -\frac{i}{\hbar}[\hat{H}_I(t), \rho] + \kappa \mathcal{D}[a]\rho + \Gamma(\bar{n} + 1)\mathcal{D}[b]\rho + \Gamma\bar{n}\mathcal{D}[b^\dagger], \tag{7.44}$$

where, as usual, κ and Γ are the amplitude damping rates of the optical and mechanical oscillators, respectively, and $\bar{n}$ is the mean bosonic occupation number for the mechanical bath. This model is now fully equivalent to the well-known quantum optical model of sub/second harmonic generation for two quantised fields interacting in a medium with a significant second-order optical nonlinearity [305, 95].

The study of the limit cycles in this model is given in [140] and we summarise the main results as typical of how one finds the limit cycles of a dynamical system. The classical nonlinear equations of motion for the red sideband interaction are

$$\begin{cases} \dot{\alpha} = \imath\chi\beta^2 - \frac{\kappa}{2}\alpha, \\ \dot{\beta} = 2\imath\chi\beta^*\alpha - \imath\epsilon - \frac{\Gamma}{2}\beta, \end{cases} \tag{7.45}$$

together with their complex conjugate counterparts. Here α and β are the complex amplitudes of the cavity field and the mechanical oscillator, respectively. We choose to measure time in units of χ^{-1}, which is equivalent to setting $\chi = 1$ and all other constants with units of frequency are then given as ratios with respect to χ. We can then write the equations of motion in terms of the real and imaginary parts of each variable,

$$\dot{\beta}_r = 2(\beta_i\alpha_r - \beta_r\alpha_i) - \frac{\Gamma}{2}\beta_r \tag{7.46}$$

$$\dot{\beta}_i = 2(\beta_r\alpha_r + \beta_i\alpha_i) - \frac{\Gamma}{2}\beta_i - \epsilon \tag{7.47}$$

$$\dot{\alpha}_r = -2\beta_r\beta_i - \frac{\kappa}{2}\alpha_r \tag{7.48}$$

$$\dot{\alpha}_i = \beta_r^2 - \beta_i^2 - \frac{\kappa}{2}\alpha_i, \tag{7.49}$$

where the subscript r's and i's, respectively label the real and imaginary parts of the complex amplitudes. This system has one critical point (where time derivatives vanish) for all values of ϵ. What are the long time solutions as ϵ is increased beyond the region of stability of these critical points?

A key concept here is the notion of a *centre manifold* [123]. This is the subspace of the full four-dimensional phase space in which the limit cycle lies for parameter values close to its creation. As the critical point becomes unstable due to a Hopf bifurcation (see Section 7.4), points will tend to be attracted onto the limit cycle in the centre manifold. In the case considered here, the limit cycle grows smoothly from the critical point and we thus have a *supercritical Hopf bifurcation*. To determine the stability of the critical points, we find the eigenvalues and eigenvectors of the linearised dynamics near those critical points. If the eigenvalue has a vanishing real part as ϵ is varied, the corresponding eigenvectors span the centre subspace. The centre manifold of the nonlinear system is then tangent to this centre subspace.

The critical points α_0 and β_0 have zero real parts ($\alpha_{r0} = \beta_{r0} = 0$), with the imaginary part of β_0 then given by the solutions to the cubic

$$\frac{4}{\kappa}\beta_{i0}^3 + \frac{\Gamma}{2}\beta_{i0} + \epsilon = 0. \tag{7.50}$$

If ϵ is positive, then β_{i0} is negative.[2] The imaginary part of the critical point for the field, α_0, is given in terms of the solution of the cubic as $\alpha_{i0} = -\frac{2\beta_{i0}^2}{\kappa}$.

Here, the centre eigenspace is the (β_r, α_r) space and the centre manifold is tangent to this space. Let $\epsilon = \epsilon_h + \Delta\epsilon$ where ϵ_h is the critical driving amplitude at which the Hopf bifurcation occurs. There is a supercritical Hopf bifurcation creating a stable limit cycle, which exists for $\Delta\epsilon$ small and positive. The detailed analysis of the dynamics on the centre manifold is given in [140]. We can summarise the results as follows for the case $\kappa \gg \Gamma$. The amplitude and frequency on the limit cycle are given approximately by

$$A = \frac{1}{\kappa}\sqrt{\frac{136\sqrt{2}\Delta\epsilon}{99}} \tag{7.51}$$

$$\omega_h = \frac{\kappa}{2}. \tag{7.52}$$

The dynamics on the centre manifold to first order in $\sqrt{\Delta\epsilon}$ is given by

$$\beta_r = 2\beta_{i0h}A\cos(\omega_h t + \phi) \tag{7.53}$$

$$\alpha_r = \omega A\sin(\omega_h t + \phi) - \frac{\kappa A}{2}\cos(\omega_h t + \phi) \tag{7.54}$$

$$\beta_i = -\sqrt{\frac{\kappa(\kappa+\Gamma)}{8}} - \frac{2\Delta\epsilon}{3\kappa+4\Gamma} \tag{7.55}$$

$$\alpha_i = -\frac{\kappa+\Gamma}{4} - \frac{2\sqrt{2\kappa(\kappa+\Gamma)}\Delta\epsilon}{\kappa(3\kappa+4\Gamma)}. \tag{7.56}$$

In Fig. 7.4 the limit cycles obtained by direct numerical integration are shown in solid lines and the approximations from centre manifold theory in dashed lines. The limit cycles are shown for various values of the bifurcation parameter $\Delta\epsilon$ and specific values of κ and Γ.

The semiclasical dynamical system discussed above gives us information on fixed points and limit cycles together with the conditions for their stability. A full quantum solution of the dynamics is not possible due to the nonlienar nature of the dynamics however we can sue our knowledge of the semiclassical dynamical structure to liveries the quantum dynamics around the semiclasscial

[2]The other case ($\epsilon < 0$, $\beta_{i0} > 0$) can be obtained from symmetry.

fixed points. This will fail for certain parameter values (i.e. at the fixed points themselves) as the linear terms in the equations of motion are zero at these points, and the eigenvalues of the linearised analysis go to zero. Despite this limitation, the linearised approach is effective in understanding quantum nose close to fixed points.

We can use the stochastic methods pioneered by Drummond et al. [95], to derive quantum stochastic differential equations in the generalised P-representation (see Section 7.2.1). This effectively replaces the quantum operators $a, a^\dagger, b, b^\dagger$ by independent complex variables, $\alpha, \alpha^\dagger, \beta, \beta^\dagger$. The semiclassical equations of motion then result when we neglect noise and restrict the stochastic dynamics to the subspace defined by $\alpha^\dagger \to \alpha^*, \beta^\dagger \to \beta^*$. As we are only interested here in dissipative quantum effects, we restrict the treatment

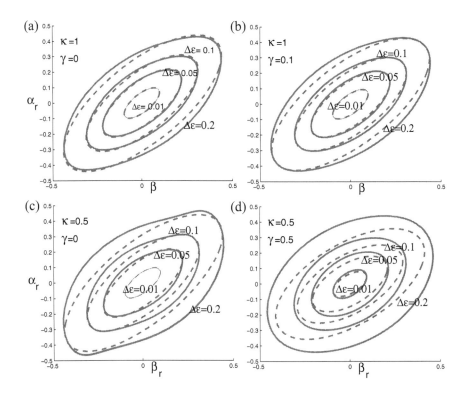

FIGURE 7.4 A comparison of the approximate limit cycles (dashed) with the numerical solutions (solid) for various values of the parameters, (a) $\kappa = 1.0, \Gamma = 0.0$, (b) $\kappa = 1.0, \Gamma = 0.1$, (c) $\kappa = 0.5, \Gamma = 0.0$, (d) $\kappa = 0.5, \Gamma = 0.5$. Reproduced with permission from [140]. Copyright WILEY-VCH Verlag GmbH & Co. KGaA, Weinheim (2009).

to zero temperature $\bar{n} = 0$. We then find that, [95]

$$\frac{\partial}{\partial t}\begin{pmatrix} \beta \\ \beta^\dagger \end{pmatrix} = \begin{pmatrix} 2\imath\chi\beta^\dagger\alpha - \imath\epsilon - \frac{\Gamma}{2}\beta \\ -2\imath\chi\beta\alpha^\dagger + \imath\epsilon - \frac{\Gamma}{2}\beta^\dagger \end{pmatrix}$$

$$+ \begin{pmatrix} 2\imath\chi\alpha & 0 \\ 0 & -2\imath\chi\alpha^\dagger \end{pmatrix}^{1/2} \begin{pmatrix} \eta_1(t) \\ \eta_1^\dagger(t) \end{pmatrix}$$

$$\frac{\partial}{\partial t}\begin{pmatrix} \alpha \\ \alpha^\dagger \end{pmatrix} = \begin{pmatrix} \imath\chi\beta^2 - \frac{\kappa}{2}\alpha \\ -\imath\chi(\beta^\dagger)^2 - \frac{\kappa}{2}\alpha^\dagger \end{pmatrix}.$$

Here $\eta_1(t), \eta_1^\dagger(t)$ are independent delta-correlated Langevin noise terms so that so that $\beta, \beta^\dagger$ are complex-conjugates in the mean. If we keep the driving amplitude to be less than that required for the Hopf bifurcation ($\epsilon < \epsilon_h$), linearisation around the semiclassical fixed points gives,

$$\frac{\partial}{\partial t}\begin{pmatrix} \delta\beta \\ \delta\beta^\dagger \\ \delta\alpha \\ \delta\alpha^\dagger \end{pmatrix} = \begin{pmatrix} -\frac{\Gamma}{2} & 2\imath\chi\alpha_0 & 2\imath\chi\beta_0^* & 0 \\ -2\imath\chi\alpha_0^* & -\frac{\Gamma}{2} & 0 & -2\imath\chi\beta_0 \\ 2\imath\chi\beta_0 & 0 & -\frac{\kappa}{2} & 0 \\ 0 & -2\imath\chi\beta_0^* & 0 & -\frac{\kappa}{2} \end{pmatrix}\begin{pmatrix} \delta\beta \\ \delta\beta^\dagger \\ \delta\alpha \\ \delta\alpha^\dagger \end{pmatrix}$$

$$+ \begin{pmatrix} 2\imath\chi\alpha_0 & 0 & 0 & 0 \\ 0 & -2\imath\chi\alpha_0^* & 0 & 0 \\ 0 & 0 & 0 & 0 \\ 0 & 0 & 0 & 0 \end{pmatrix}^{1/2}\begin{pmatrix} \eta_1(t) \\ \eta_1^\dagger(t) \\ \eta_2(t) \\ \eta_2^\dagger(t) \end{pmatrix},$$

which we can write as

$$\frac{\partial}{\partial t}[\delta\boldsymbol{\alpha}] = -\mathbf{A}[\delta\boldsymbol{\alpha}] + \mathbf{D}^{1/2}[\boldsymbol{\eta}(t)], \tag{7.57}$$

where $\mathbf{A}$ and $\mathbf{D}$ are the drift and diffusion matrices, respectively. The steady-state power spectral density for each normally ordered moment is given by (see Section 1.2.2.2):

$$S_{ij}(\omega) = \int_{-\infty}^{\infty} e^{\imath\omega\tau}\langle\alpha_i(t)\alpha_j(t+\tau)\rangle_{t\to\infty}\,d\tau, \tag{7.58}$$

We thus find the matrix of power spectral densities, $\mathbf{S}(\omega)$, in terms of the linearised drift and diffusion matrices as [305]

$$\mathbf{S}(\omega) = (-\imath\omega\mathbb{1} + \mathbf{A})^{-1}\mathbf{D}\left(\imath\omega\mathbb{1} + \mathbf{A}^T\right)^{-1}, \tag{7.59}$$

where $\mathbb{1}$ is the identity matrix and the superscript T denotes the transpose.

Signatures of the Hopf bifurcation will appear in the spectrum of the output field from the cavity. Fig. 7.5 shows the power spectral density of the optical field amplitude as the mechanical drive strength ϵ approaches the Hopf

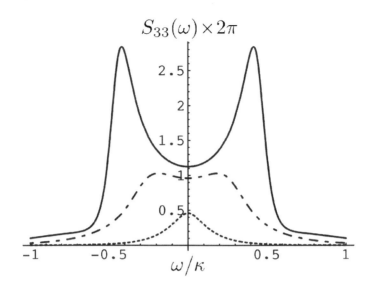

FIGURE 7.5 Linearized spectrum for $S_{33}(\omega)$, for $\Gamma/\kappa = 0.1$, with three different drive strengths: $\epsilon = 0.01/\kappa$ (dotted curve), $\epsilon/\kappa = 0.05$ (dot-dashed curve), and $\epsilon/\kappa = 0.13$ (solid curve). Reproduced with permission from [140]. Copyright WILEY-VCH Verlag GmbH & Co. KGaA, Weinheim (2009).

bifurcation at ϵ_h. As can be seen, two sharp peaks appear as the drive strength increases. As shown in [140], these peaks occur at the characteristic frequencies $\pm\omega_h = \pm\kappa/2$ associated with the Hopf bifurcation and given in Eq. (7.52). For driving beyond the critical value at the Hopf bifurcation, the real parts of α, β are no longer locked to zero but oscillate according to Eq. (7.53). The quantum noise will cause phase diffusion around the limit cycle. In the limit $\kappa \gg \Gamma$, the phase dynamics can be approximated by

$$\frac{d\phi}{dt} = \frac{1}{A}\sqrt{\frac{1}{2\kappa}}dW(t), \tag{7.60}$$

where dW is a Wiener process (see Section 5.1.1), and A is the amplitude of the limit cycle given in Eq. (7.51). Noting that, by definition, the coefficient of the Wiener increment is the square root of the phase diffusion constant, $\sqrt{D_\phi}$, the phase diffusion constant, for $\kappa \gg \Gamma$, therefore depends on the cavity linewidth κ and the excess drive amplitude $\Delta\epsilon$ above the critical drive strength ϵ_h as

$$D_\phi \propto \frac{\kappa}{\Delta\epsilon}. \tag{7.61}$$

We see that the phase diffusion decreases both as the cavity linewidth is reduced, and as the drive strength is increased. This behaviour is analogous to that of a laser above threshold.

7.5 NONLINEAR MEASUREMENT OF A MECHANICAL RES-ONATOR

For much of this book the optomechanical coupling has been taken as the standard radiation pressure interaction which is linear in the mechanical displacement. In this section we will consider an optomechanical interaction which is quadratic in the mechanical displacement.[3] Such an interaction enables measurement of the energy of the mechanical resonator or, equivalently, the phonon number.

The derivation of the radiation pressure interaction in Section 2.3 proceeds from the observation that the frequency of the optical resonator depends on the displacement of the mechanical element, $\Omega_c(q)$. If it is possible to engineer an optomechanical system so that $\Omega_c(q)$ is an extremum at a point $q = q_0$, the linear contribution to the interaction vanishes and the lowest significant order is quadratic in q. The nonlinear optomechanical coupling constant at some mechanical oscillator position q_0 then corresponds to the second-order term in the Taylor series for $\Omega_c(q)$,

$$\Omega_c(q) = \Omega_c(q_0) + \frac{1}{2}\frac{d^2\Omega_c(q)}{dq^2}\bigg|_{q_0} q^2 + ... \tag{7.62}$$

One way to achieve this quadratic coupling is via the *membrane-in-the-middle* model of [283, 248], depicted in Fig. 7.6. The resonance frequencies

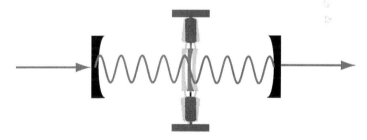

FIGURE 7.6 A schematic representation of the membrane-in-the-middle optomechanical system. The membrane is a thin dielectric membrane located at a point x_0 along the cavity axis between the two mirrors of a Fabry–Pérot cavity.

of the cavity may now be changed by moving the position of the membrane

---

[3]We briefly looked at the same interaction previously in Section 7.4.1.

along the cavity axis. Typically these frequencies vary periodically with q_0. Around a minimum, or a maximum, it is possible to find situations in which the variation around q_0 is quadratic in small displacements.

Exercise 7.7 *Substitute Eq. (7.62) for the optical cavity frequency in the presence of quadratic coupling to a mechanical resonator into the Hamiltonian in Eq. (2.14) to derive the quadratic coupling Hamiltonian*

$$\hat{H} = \hbar\Omega_c a^\dagger a + \hbar\Omega b^\dagger b + \frac{\hbar g_2}{2} a^\dagger a (b + b^\dagger)^2, \tag{7.63}$$

where

$$g_2 = x_{zp}^2 \left. \frac{\partial^2 \Omega_c(q)}{\partial q^2} \right|_{q=q_0} \tag{7.64}$$

and we have made the substitution $\hbar\Omega_c(q_0) \to \hbar\Omega_c$. You should neglect terms in Eq. (7.62) that are higher than quadratic in order.

Thompson et al. [283] achieved a value for g_2 of approximately $10^{-4}\mathrm{s}^{-1}$. In a rather different scheme, Brawley et al. [50] achieved a value of about $10^{-3}\mathrm{s}^{-1}$. In an interaction picture rotating at the mechanical frequency, we can neglect rapidly oscillating terms provided $g_2/\Omega \ll 1$ to give the approximate interaction of the form

$$\hat{H}_I = \hbar g_2 a^\dagger a b^\dagger b. \tag{7.65}$$

We will assume that the cavity is driven coherently near the appropriate cavity resonance and, as usual, linearise the optical field dynamics around the stable steady-state field in the absence of the optomechanical coupling (see Section 2.7). We then approximate the interaction Hamiltonian by

$$\hat{H}_I = \hbar\chi(a + a^\dagger) b^\dagger b, \tag{7.66}$$

where, analogously to the linear case treated in Section 2.7, we define $\chi \equiv g_2\alpha_0$. Including the cavity and mechanical dissipation, with rates of κ and Γ, respectively, we then arrive at the master equation

$$\dot{\rho} = -\frac{i}{\hbar}[\hat{H}_I, \rho] + \kappa\mathcal{D}[a]\rho + \Gamma(\bar{n} + 1)\mathcal{D}[b]\rho + \Gamma\bar{n}\mathcal{D}[b^\dagger]\rho, \tag{7.67}$$

where the superoperator, $\mathcal{D}$, is defined in Eq. (5.10a) and, as usual, $\bar{n}$ is the mean thermal occupation of the mechanical bath at frequency Ω.

The fact that the interaction Hamiltonian in Eq. (7.66) commutes with the canonical phonon number operator indicates that this coupling could realise a quantum nondemolition measurement of the phonon number of the mechanics. In the linearised approximation, the phonon number leads to a displacement of the cavity field away from its steady-state value, α_0, in the absence of the coupling. In principle this can be detected by performing homodyne measurements on the output field from the cavity. Of course, the phonon number will not be a constant of the motion due to the coupling to the mechanical environment. Is it possible to monitor these transitions in the phonon number by a sufficiently careful measurement of the changes in the cavity field amplitude as revealed by homodyne detection? In other words, can we see *quantum jumps* in the mechanical phonon number in the homodyne signal?

To answer this question we will follow the presentation of [109]. If real-time monitoring of the phonon Fock state transitions can be resolved, it is clear that the measurement bandwidth must be greater than the rate of thermal transitions, $\Gamma\bar{n}$ [249]. The master equation, Eq. (7.67), implies

$$\frac{d\langle a\rangle}{dt} = -i\frac{\chi}{2}\bar{n}_b - \frac{\kappa}{2}\langle a\rangle \tag{7.68}$$

$$\frac{d\bar{n}_b}{dt} = -\Gamma\bar{n}_b + \Gamma\bar{n} \tag{7.69}$$

with solutions

$$\bar{n}_b(t) = \bar{n}_b(0)e^{-\Gamma t} + \bar{n}(1 - e^{-\Gamma t}) \tag{7.70}$$

$$\langle a\rangle(t) = \langle a\rangle(0)e^{-\kappa t/2} - i\frac{\chi}{2}\left[(\bar{n}_b(0) - \bar{n})\frac{(e^{-\Gamma t} - e^{-\kappa t/2})}{\kappa/2 - \Gamma} + \bar{n}\frac{(1 - e^{-\kappa t/2})}{\kappa/2}\right]. \tag{7.71}$$

In the adiabatic limit, $\kappa \gg \Gamma$, we find $\langle a\rangle(t) \approx -i\frac{\chi}{\kappa}\bar{n}_b(t)$. Information about the average phonon number can be obtained by monitoring the quadrature phase amplitude of the steady-state amplitude α_0 (which has been chosen as real). The mechanical resonator will be in thermal equilibrium with its environment before the optomechanical coupling is turned on, so $\bar{n}_b(0) = \bar{n}$. In the long time limit, the cavity amplitude changes from the steady background amplitude by the amount $\Delta\alpha = -i\chi\bar{n}/\kappa$. We then define χ/κ as the *gain* of the measurement.

Extending our treatment of continuous measurement upon a mechanical oscillator in Section 5.1, the conditional dynamics of a cavity optomechanical system subject to continuous homodyne measurement of the output optical field is given by the stochastic master equation (SME) [318]

$$d\rho = -\frac{i}{\hbar}[H_I, \rho]dt + \Gamma(\bar{n} + 1)\mathcal{D}[b]\rho dt + \Gamma\bar{n}\mathcal{D}[b^\dagger]\rho dt + \kappa\mathcal{D}[a]\rho dt + \sqrt{\eta\kappa}dW\mathcal{H}[ia]\rho. \tag{7.72}$$

where η is the detector efficiency. The homodyne photocurrent is proportional to the cavity phase quadrature amplitude with stochastic noise due to the local oscillator and intrinsic quantum noise of the cavity field:

$$i_h(t)dt = i\eta\kappa\langle a^\dagger - a\rangle dt + \sqrt{\eta\kappa}dW. \tag{7.73}$$

Here, we need the stochastic master equation after the cavity field has been adiabatically eliminated to give a stochastic master equation for the mechanical degree of freedom alone. This is [109]:

$$d\rho_b = \Gamma(\bar{n}+1)\mathcal{D}[b]\rho_b dt + \Gamma\bar{n}\mathcal{D}[b^\dagger]\rho_b dt + \mu\mathcal{D}[b^\dagger b]\rho_b dt - \sqrt{\eta\mu}\mathcal{H}[b^\dagger b]\rho_b dW, \quad (7.74)$$

where $\mu \equiv \chi^2/\kappa$ is the measurement rate, as defined in Chapter 5. In the adiabatic limit the homodyne photocurrent becomes

$$i_h(t)dt = -2\eta\chi\langle b^\dagger b\rangle_c dt + \sqrt{\eta\kappa}dW(t). \quad (7.75)$$

The conditional Pauli master equation for the diagonal elements of ρ_b is

$$
\begin{aligned}
dp_n &= \Gamma\bar{n}[np_{n-1} - (n+1)p_n]dt \quad &(7.76)\\
&+\Gamma(\bar{n}+1)[(n+1)p_{n+1} - np_n]dt - 2\sqrt{\eta\mu}(n - \langle n\rangle)p_n dW.
\end{aligned}
$$

Henceforth, we will set the detection efficiency $\eta = 1$ for ease of presentation.

The stochastic conditional evolution tends to lead to a reduced variance of the mechanical state in the number state basis. In competition with this, thermal transitions have the opposite effect. To see transitions in the phonon number, the measurement rate must dominate the thermalisation rate. This requires the *good measurement* limit as follows.

The conditional Pauli master equation, Eq. (7.76), indicates that, in the absence of measurement, the thermalisation rate of a mechanical Fock state $|n\rangle$ is $\Gamma[\bar{n}(n + 1) + (\bar{n} + 1)n]$ [249] (see Exercise 2.1). We thus require the *adiabatic condition* $\kappa \gg \Gamma[\bar{n}(n_{max}+1)+(\bar{n}+1)n_{max}]$, where n_{max} is the likely largest phonon number state. Even in the adiabatic limit, the measurement will extract information on the phonon number at a finite rate. The *good measurement condition* is then $\mu \gg \Gamma[\bar{n}(n_{max} + 1) + (\bar{n} + 1)n_{max}]$, together with the *adiabatic condition*. The good measurement limit is valid for arbitrary values of χ/κ; the strong coupling regime $(\chi/\kappa \gtrsim 1)$ is not required.

In the good measurement limit, the cavity adiabatically follows the number state of the mechanical mode, and the measurement collapses the state of the mechanical mode in the number state basis very rapidly. We thus expect that the cavity amplitude will reach its steady-state value, $-i\frac{\chi}{\kappa}n$, between consecutive jump times t_j, and the conditional average of the phase quadrature amplitude will exhibit a random telegraph signal.

We can demonstrate this by a numerical integration of Eq. (7.72), choosing the parameters $\bar{n} = 0.5$, $\chi/\kappa = 1.5$, $\kappa = 100\Gamma\bar{n}$, and $\chi^2/\kappa = 225\Gamma\bar{n}$ to ensure that the system is in the good measurement limit for low phonon numbers. We assume the mechanics begins in the ground state perhaps due to a prior period of resolved sideband cooling (see Section 4.2). The resulting trajectories are shown in Fig. 7.7, exhibiting well-resolved quantum jumps in the conditional mean phonon number. We also plot the homodyne photocurrent of Eq. (7.73) convoluted with a low-pass filter.

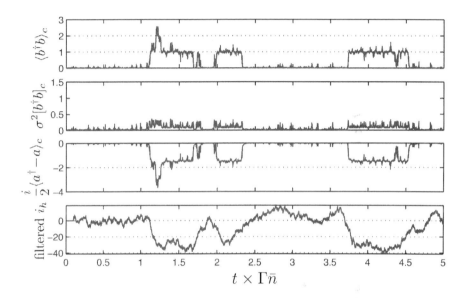

FIGURE 7.7 Evolution of the conditional (a) phonon number, (b) phonon number variance, and (c) cavity field phase quadrature in the good measurement limit for low phonon numbers ($n \sim 1$). Quantum jumps in the phonon number are tracked by the conditional cavity phase quadrature amplitude. (d) The filtered homodyne current gives a noisy version of the cavity phase quadrature trajectory. From Gangat et al. [109]. IOP Publishing & Deutsche Physikalische Gesellschaft. CC BY-NC-SA.

CHAPTER 8

Hybrid optomechanical systems

CONTENTS

8.1	Electromechanical systems	....	256
	8.1.1	Superconducting circuit coupled to a mechanical element	256
	8.1.2	Superconducting junction coupled to a mechanical element	258
8.2	Coupling a mechanical resonator and a two-level system	..	261
	8.2.1	Mechanical cooling via a two-level system	263
	8.2.2	Single phonon control via a superconducting qubit	268
8.3	Microwave to optical interface.	....	270
	8.3.1	A frequency convertor MOM interface	271
	8.3.2	A teleportation MOM interface	274

In this chapter we will explore some directions in the study of optomechanical systems that incorporate additional quantum degrees of freedom, beyond the cavity field and a mechanical resonator. A typical example includes a single two-level system coupled, either, to the field, the mechanics, or both. We will, further, introduce optomechanical systems within which the mechanical resonator is simultaneously coupled to both an optical cavity and a microwave resonator. In other examples not considered in detail here, the optomechanical system can be coupled to a gas of cold atoms, for example, a Bose-Einstein condensate [285, 266, 56, 186, 146, 30], or a superfluid [84, 135, 5].

There are many examples of the inclusion of a two-level system. Usually this takes the form of an electric dipole coupled to the light in the cavity. The dipole could be a single atom [25] or a quantum dot[326]. The two-level system is often directly coupled to the mechanical degree of freedom through strain induced effects[315] in an embedded two-level system such as an nitrogen vacancy (NV) centre [153, 279] or quantum dot [247, 326]. In magnetic force resonance microscopy, a mechanical resonator is coupled to one or more nuclear or electronic spins in a solid or in a molecule [242]. In other cases, the two-level system is coupled to the mechanical resonator and can be used as a transducer for the mechanical motion [17].

8.1 ELECTROMECHANICAL SYSTEMS

As we mentioned in Chapter 2, the physical effects seen in optomechanical systems can also be realised in the microwave domain by a capacitive coupling of a mechanical resonator to a superconducting microwave circuit. In this section we will give a brief introduction to this field. We give three examples. In the superconducting electromechanics group in JILA/NIST, Boulder [216], a lumped LC superconducting circuit with a suspended membrane capacitor – the mechanical element – is coupled to the microwave field in the LC circuit. As the second example, we discuss the experiment of LaHaye et al. [169] which involves a small superconducting island (a Cooper pair box) with an electrostatic energy controlled by a capacitive coupling to a mechanical element. Finally, we discuss the landmark experiment of O'Connell et al [213] in which a very high frequency mechanical resonator was cooled to its ground state using a dilution refrigerator and coherently controlled at the single-phonon level by coupling it to a superconducting qubit.

8.1.1 Superconducting circuit coupled to a mechanical element

The new field of superconducting quantum circuits, also known as circuit quantum electrodynamics (cQED) is, like quantum optomechanics, an example of how systems can be engineered to show quantum behaviour in macroscopic collective degrees of freedom. In the case of optomechanics, this is the quantised motion of a bulk mechanical mode. In the case of a quantum circuit, it is the quantum dynamics of charge and flux on capacitors and inductors. Their quantum state can be controlled by standard capacitive and inductive control, and a variety of quantum limited measurement schemes have been implemented. A key role is played by Josephson junctions, which appear as a nonlinear circuit element. For a review of quantum circuits we refer the reader to the review by Clarke and Wilhelm [72].

We consider a superconducting LC circuit made up of lumped elements: an inductive loop and a capacitor. The mechanical resonator is the fundamental drumhead mode of the top plate of a thin (100 nm), 15 mm diameter superconducting aluminium membrane forming a 50 nm vacuum gap capacitor [216]. See Fig. 8.1. The lumped LC resonator is itself coupled to a superconducting coplanar transmission line that is used to drive the LC resonator.

The LC circuit frequency is Ω_c and the mechanical resonator frequency is Ω. The capacitive coupling is a function of the mechanical resonators displacement from its equilibrium position, q. To lowest order in the displacement the capacitance is $C_0(q) = C_0(1 - q/d)$ where C_0 represents an equilibrium capacitance and d is the equilibrium plate separation. Thus the LC has an equivalent

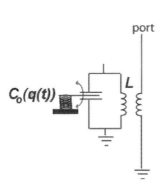
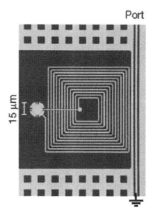

FIGURE 8.1 A co-planar superconducting circuit composed of an inductive loop and a capacitor with a mechanically compliant suspended top plate; and the equivalent circuit for the system. Adapted by permission from Macmillan Publishers Ltd. from [216], copyright (2011).

capacitance $C_\Sigma = C + C_0$, and equivalent inductance L, such that the coupled resonance frequency is $\Omega_c \approx 1/\sqrt{LC_\Sigma}$. The capacitive energy of the system is to first order given by $Q^2/2C_\Sigma + (\beta/2dC_\Sigma)qQ^2$, where $\beta = C_0/C_\Sigma$ and Q is the charge on the equivalent capacitor C_Σ. [1] The system is described by the classical Hamiltonian,

$$
\begin{aligned}
H &= \frac{p^2}{2m} + \frac{1}{2}m\Omega^2 q^2 + \frac{\Phi^2}{2L} + \frac{Q^2}{2C_\Sigma} \\
&\quad + \frac{\beta}{2dC_\Sigma}qQ^2 + \frac{1}{2}E(t)\Phi,
\end{aligned}
\tag{8.1}
$$

where (q,p) are the canonical position and momentum coordinates of the mechanical resonator, (Q, Φ) are the canonical coordinates for the cavity (representing the charge on the capacitor C_Σ and flux through the inductor L), and the explicit time dependence in our notation has been retained only for the driving field $E(t)$ entering via the transmission line.

We may quantize the Hamiltonian by imposing the commutation relations $[\hat{q}, \hat{p}] = i\hbar$ and $[\hat{Q}, \hat{\Phi}] = i\hbar$ in a similar manner to what we did for optomechanics in Sections 2.2 and 2.3. It is not immediately obvious that this procedure does give a valid quantum description of a quantum circuit. The reason it works is that – just like for bulk mechanical modes – for superconducting circuits, the collective degrees of freedom described by current and flux largely factor out of the underlying microscopic dynamics. Of course, the residual

---

[1] This should not be confused with the dimensionless mechanical position position, or the mechanical quality factor, for which we use the same symbol in other parts of the book.

interactions with microscopic degrees of freedom are still present as a source of noise and dissipation. Effective quantisation such as this is characteristic of quantum engineered systems.

In the Schrödinger picture, the Hamiltonian is then

$$\hat{H} = \hbar\Omega_c a^\dagger a + \hbar\Omega b^\dagger b + \frac{1}{2}\hbar g_0 (b + b^\dagger)(a + a^\dagger)^2$$
$$+\hbar(\mathcal{E}_2^* e^{i\omega_d t} + \mathcal{E}_2 e^{-i\omega_d t})(a + a^\dagger), \qquad (8.2)$$

where

$$a = \sqrt{\frac{\Omega_c L}{2\hbar}}\hat{Q} + \frac{i}{\sqrt{2\hbar\Omega_c L}}\hat{\Phi} \qquad (8.3a)$$

$$b = \sqrt{\frac{m\Omega}{2\hbar}}\hat{q} + \frac{i}{\sqrt{2\hbar m\Omega}}\hat{p}, \qquad (8.3b)$$

and the coupling constant is

$$g_0 \equiv \frac{\beta\Omega_c}{2}\frac{x_{zp}}{d}, \qquad (8.4)$$

with $x_{zp} = (\hbar/2m\Omega)^{1/2}$ being, as usual, the standard deviation of the mechanical ground state wavefunction, i.e., the zero-point length.

Transforming the Hamiltonian of Eq. (8.2) into an interaction picture with respect to the bare Hamiltonian $\hat{H}_0 = \hbar\Omega_L a^\dagger a + \hbar\Omega b^\dagger b$, where Ω_L is the frequency of the drive "laser" – in this case a microwave field – and making the rotating wave approximation leads to

$$\hat{H}_I = \hbar\Delta a^\dagger a + \sqrt{2}\hbar g_0 \hat{X}_M(t) a^\dagger a + \hbar\left(\mathcal{E}_1^* a + \mathcal{E}_1 a^\dagger\right), \qquad (8.5)$$

where $\Delta \equiv \Omega_c - \Omega_L$ is the detuning between the cavity resonance and the drive, and, as usual, the mechanical X-quadrature is $\hat{X}_M(t) = (be^{-i\Omega t} + b^\dagger e^{i\Omega t})/\sqrt{2}$. This is exactly equivalent to the optomechanical radiation pressure interaction, and our previous discussion can be applied as long as we keep an eye on the quite different parameter values. For example, in the experiment of [216], the parameters are $\Omega_c \approx 2\pi \times 7.5$ GHz, $\Omega = 2\pi \times 10.5$ MHz, $m = 48$ pg, $x_{zp} = 4.1$ fm, and $g_0 = 2\pi \times 200$ Hz.

8.1.2 Superconducting junction coupled to a mechanical element

In the second example, LaHaye et al. [169] use a Copper pair box (CPB). This is a superconducting metal island coupled by a split Josephson junction to a superconducting circuit. The junction is split so that magnetic flux through the hole, so formed, can be used to tune the Josephson energy of the junction. A DC voltage gate can be used to tune the charging energy of the island. The coupling between the CPB and the microwave field is given by [121]

$$\hat{H} = 4E_C \sum_N (N - n_g(t))^2 |N\rangle\langle N| - \frac{E_J(\phi)}{2}\sum_{N=0}(|N\rangle\langle N+1| + |N+1\rangle\langle N|),$$
$$(8.6)$$

where N is the number of Cooper pairs on the junction and

$$
\begin{aligned}
E_C &= \frac{e^2}{2C_\Sigma} \\
E_J(\phi) &= E_J \cos(2\pi\phi/\phi_0) \\
n_g(t) &= \frac{C_g V_g(t)}{2e}
\end{aligned}
$$

with C_Σ being the capacitance between the island and the rest of the circuit, C_g the capacitance between the CPB island and the bias gate for the island, $e = 1.6 \times 10^{-19}$ C the elementary charge, and $V_g(t)$ the total voltage applied to the island by the bias gate composed of a DC field, $V_g^{(0)}$, and the microwave field in the cavity, $\hat{v}(t)$. Thus we can write $V_g(t) = V_g^{(0)} + \hat{v}(t)$, where the hat indicates a quantisation of the cavity field. The Hamiltonian in Eq. (8.6) is written in the Cooper pair number basis. In this basis the electrostatic energy of the first term is quite clear. The Josephson energy term describes tunneling of single Cooper pairs across the junction.

We now allow for the possibility that the CPB (or its bias gate) is mounted on a mechanical oscillator. This leads to a periodic modulation of the gate capacitance that, to lowest order in the displacement, is

$$
C_g(t) = C_g(1 - \hat{q}(t)/d) \tag{8.7}
$$

where $\hat{q}$ is the displacement operator for the mechanical oscillator and d is a typical length scale. Including both the time-dependent cavity field and the modulated gate capacitance, we can write

$$
n_g(t) = n_g^{(0)} + \delta\hat{n}_g(t), \tag{8.8}
$$

where

$$
\delta\hat{n}_g(t) = \frac{C_g}{2e}\hat{v}(t) - \frac{n_g^{(0)}}{d}\hat{q}(t) - \frac{C_g}{2ed}\hat{q}(t)\hat{v}(t). \tag{8.9}
$$

Restricting the Hilbert space of the CPB to $N = \{0, 1\}$ (i.e., a qubit), we can write the electronic Hamiltonian as

$$
\hat{H}_{el} = \hat{H}_q - 4E_C\delta\hat{n}_g(t)(1 - 2n_g^{(0)} - \bar{\sigma}_z) \tag{8.10}
$$

with the bare qubit Hamiltonian defined by

$$
\hat{H}_q = \frac{\hbar\epsilon}{2}\bar{\sigma}_z + \frac{\hbar\delta}{2}\bar{\sigma}_x \tag{8.11}
$$

with $\bar{\sigma}_z = |0\rangle\langle 0| - |1\rangle\langle 1|, \bar{\sigma}_x = |1\rangle\langle 0| + |0\rangle\langle 1|$, and the parameters δ and ϵ given by

$$
\begin{aligned}
\hbar\epsilon &= -4E_c(1 - 2n_g^{(0)}) \tag{8.12} \\
\hbar\delta &= -E_J \cos(2\pi\phi/\phi_0). \tag{8.13}
\end{aligned}
$$

It is important to note that *both* of these parameters may be tuned over a wide range by varying a DC voltage gate and a flux bias on the split junction CPB. For example, when the CPB is biased at the charge degeneracy, $n_g = 1/2$, ϵ goes to zero.

If there is no AC bias voltage, we may neglect the terms involving $\hat{v}(t)$ in Eq. (8.9) for the quantum fluctuations in the number of Cooper pairs on the junction. Assuming that the DC bias voltage is set by the voltage drop between the CPB and the mechanical resonator, so that $V_g^{(0)} \equiv V_m$ and $C_g \equiv C_m$, the interaction term between the junction and the mechanical resonator in Eq. (8.10) is then given by

$$\hat{H}_I = \hbar\lambda(b + b^\dagger)\bar{\sigma}_z, \tag{8.14}$$

where we have used the standard relation $\hat{q} = x_{zp}(b + b^\dagger)$, and the junction-resonator coupling rate is defined as

$$\lambda \equiv -\frac{eV_m}{\hbar}\frac{C_m}{C_\Sigma}\frac{x_{zp}}{d} \tag{8.15}$$

with x_{zp} being the usual zero-point length scale (the standard deviation in the ground state of the mechanical resonator). We have ignored a DC bias force on the mechanical resonator, as this can be removed by changing the definition of its equilibrium position. In the experiment of LaHaye et al. [169], λ ranged from 0.3 to 3 MHz. The mechanical frequency was $\Omega \sim 60$ MHz and the qubit parameters had maximum values of $\epsilon \sim \delta \sim 14$ MHz. More recently, Pirkkalainen et al. [227] have reported junction-resonator coupling rates as large as $\lambda = 160$ MHz. This substantial coupling rate allowed the demonstration of a six order of magnitude effective enhancement in the optomechanical coupling strength g_0 between the field and the resonator. As proposed in [137], this was achieved by taking advantage of a direct qubit-field interaction which is present in the form of a qubit-state-dependent dispersive Stark shift of the frequency of the cavity field. This interaction allowed the resonator-qubit coupling to be converted into coupling between the resonator and the field.

If the two-level system is a spin-half magnetic dipole, we need to consider a magnetic dipole coupling to the mechanical resonator, perhaps by placing a small permanent magnet on the mechanical resonator. The interaction Hamiltonian takes a similar form to Eq. (8.14) [235]. This is the basis for the study of magnetic force resonant microscopy (MRFM)[242]. In this case, the spin exerts a force on the mechanical displacement that changes sign if the spin is flipped (much like in the Stern-Gerlach experiment).

If there is an AC microwave field present, there are four extra terms corresponding to (1) an AC driving of the microwave resonator proportional to $\hat{v}(t)$; (2) a direct coupling between the mechanical system and the microwave proportional to $\hat{q}\hat{v}(t)$, (3) a coupling between the microwave driving field and the qubit proportional to $\hat{v}(t)\bar{\sigma}_z$, and (4) a small cubic coupling term proportional to $\hat{q}\hat{v}(t)\bar{\sigma}_z$. In what follows we will assume that the AC microwave field is a strong classical coherent field at frequency ω_d and only concern ourselves with how the qubit responds to this field. This is described by the interaction Hamiltonian

$$\hat{H}_d = \hbar E_0 \cos(\omega_d t) \bar{\sigma}_z, \tag{8.16}$$

where

$$E_0 = \frac{eV_0}{\hbar} \left(\frac{C_g}{C_\Sigma} \right) \tag{8.17}$$

with V_0 being the amplitude of the microwave field voltage at the qubit.

8.2 COUPLING A MECHANICAL RESONATOR AND A TWO-LEVEL SYSTEM

In the previous section we showed how the coupling between a two-level system and a mechanical resonator is described by a linear potential, providing a mechanical force controlled by the two-level system. We now consider two applications of this kind of coupling. In the first example the qubit is rapidly damped to provide a mechanism to cool the mechanical resonator. In the second example, based on the experiment of O'Connell et al. [213], we will show how the two-level system can be used to implement quantum control of the mechanics at the single phonon level.

We first discuss the coupling of a mechanical resonator to a single two-level system, a qubit. Such systems have been proposed as an alternative approach to mechanical cooling [144] than the resolved sideband and feedback cooling treated, respectively, in Sections 4.2 and 5.2. If the mechanical resonator is coupled to two or more two-level systems, it can provide a means to directly couple remote qubits [267]. Using the results from the previous section, a general theory for the model may be given in terms of the Hamiltonian for the qubit-mechanical resonator

$$\hat{H}_{qm} = \hbar\Omega b^\dagger b + \frac{\hbar\epsilon}{2}\bar{\sigma}_z + \frac{\hbar\delta}{2}\bar{\sigma}_x + \hbar\lambda(b + b^\dagger)\bar{\sigma}_z + \hbar E_0 \cos(\omega_d t)\bar{\sigma}_z, \tag{8.18}$$

where $b, b^\dagger$ are the raising and lowering operator for mechanical excitations, Ω is the mechanical frequency, and the coupling between the mechanical resonator and the CPB qubit is described by the coupling constant, λ, and E_0 is the amplitude of the AC microwave driving field. Note that we do not assume that the coupling of the qubit to the mechanics is diagonal in the energy eigenstates of the bare qubit so that in general $\delta \neq 0$. In the example of capacitive coupling to a superconducting qubit [169], the parameters δ and ϵ can be independently controlled by a voltage gate bias and a magnetic flux bias. As mentioned previously, the experiments of LaHaye et al. [169] and Pirkkalainen et al. [227] achieved qubit-mechanical resonator coupling rates, λ, as high as 3 MHz and 160 MHz, respectively. In the experiment of Yeo et al. [326] on a strained quantum dot in a GaAs nanowire, the coupling constant was of the order of 500 kHz.

For now, let us neglect the AC driving field, setting $E_0 = 0$. The first step is to make a canonical transformation to diagonalise the bare Hamiltonian of the qubit. To this end we make a rotation of the qubit defined by

$$\begin{pmatrix} \sigma_x \\ \sigma_z \end{pmatrix} = \begin{pmatrix} \epsilon/\Omega_q & -\delta/\Omega_q \\ \delta/\Omega_q & \epsilon/\Omega_q \end{pmatrix} \begin{pmatrix} \bar\sigma_x \\ \bar\sigma_z \end{pmatrix}, \tag{8.19}$$

where the qubit splitting is $\Omega_q = \sqrt{\epsilon^2 + \delta^2}$. In the new variables the Hamiltonian becomes

$$\hat H = \hbar\Omega b^\dagger b + \frac{\hbar\Omega_q}{2}\sigma_z + \hbar\lambda(b + b^\dagger)\left(\frac{\epsilon}{\Omega_q}\sigma_z - \frac{\delta}{\Omega_q}\sigma_x\right). \tag{8.20}$$

In an interaction picture at the qubit frequency the Hamiltonian is

$$\hat H_I = \hbar\Omega b^\dagger b + \hbar\lambda(b + b^\dagger)\left(\frac{\epsilon}{\Omega_q}\sigma_z - \frac{\delta}{\Omega_q}(\sigma_+ e^{i\Omega_q t} + \sigma_- e^{-i\Omega_q t})\right), \tag{8.21}$$

where $\sigma_\pm = \sigma_x \pm i\sigma_y$.

It would be natural, at this point, to make the usual rotating wave approximation of quantum optics by moving to an interaction picture for the mechanical resonator in a frame rotating at the qubit frequency and neglecting counter-rotating terms. This gives the Jaynes–Cummings Hamiltonian,

$$\hat H_I = \hbar\Delta b^\dagger b + \hbar g_0(b\sigma_+ + b^\dagger\sigma_-), \tag{8.22}$$

where $g_0 = -\frac{\lambda\delta}{\Omega_q}$ and $\Delta = \Omega - \Omega_q$. We will return to a discussion of this model in Section 8.2.2.

The rotating wave approximation is problematic if either of the parameters δ, ϵ is be varied through zero, as in [169]. Furthermore, the mechanical frequency is typically very much lower than the qubit frequency. For example, if the qubit is realised as a Josephson junction, then $\Omega_q \sim 1 - 10$ GHz [169]. A similar order of magnitude holds if the qubit is realised as an electron spin in a nitrogen vacancy (NV) defect in diamond.

Returning to Eq. (8.21), we now make use of the fact that the mechanical frequency is much lower than the qubit frequency to obtain an effective dispersive interaction Hamiltonian. Substituting Eq. (8.21) into the Dyson expansion of the unitary evolution operator over a time $t \gg 1/\Omega$, we find that, on this time scale, the dynamics is well described to second order in λ by the time-independent Hamiltonian

$$\tilde H_I(t) = \hbar\Omega b^\dagger b + \hbar\lambda(b + b^\dagger)\frac{\epsilon}{\Omega_q}\sigma_z + \frac{\hbar\lambda^2\delta^2}{\Omega_q^3}(b + b^\dagger)^2\sigma_z. \tag{8.23}$$

In a further interaction picture rotating at the mechanical frequency, we can neglect terms proportional to $b^2, b^{\dagger 2}$, thus giving a final approximate interaction Hamiltonian

$$\tilde H_I \approx \hbar\Omega b^\dagger b + 2\hbar\frac{\lambda^2\delta^2}{\Omega_q^3}b^\dagger b\sigma_z. \tag{8.24}$$

We see that qubit acts as a conditional frequency shift for the mechanical system.

When the microwave drive is included and the rotating wave approximation is made, the interaction Hamiltonian in the interaction picture at the frequency of the driving field is

$$\hat{H}_d = -\frac{\hbar\delta_q}{2}\sigma_z + \hbar\Omega b^\dagger b + \hbar g(b+b^\dagger)\sigma_z + \frac{\hbar\mathcal{E}_0}{2}\sigma_x, \qquad (8.25)$$

where

$$\mathcal{E}_0 = \frac{E_0\delta}{\Omega_q} \qquad (8.26)$$

$$g = \frac{\lambda\epsilon}{\Omega_q} \qquad (8.27)$$

and the detuning between the qubit and the microwave driving frequency is defined by $\delta_q \equiv \omega_d - \Omega_q$.

8.2.1 Mechanical cooling via a two-level system

Thus far we have ignored dissipation of the mechanical resonator and the qubit. The full dynamics is given by the master equation,

$$\frac{d\rho}{dt} = -\frac{i}{\hbar}[\hat{H}_d, \rho] + \Gamma(\bar{n}+1)\mathcal{D}[b]\rho + \Gamma\bar{n}\mathcal{D}[b^\dagger]\rho \qquad (8.28)$$
$$+\Gamma_q(\bar{n}_q+1)\mathcal{D}[\sigma_-]\rho + \Gamma_q\bar{n}_q\mathcal{D}[\sigma_+]\rho + \Gamma_p\mathcal{D}[\sigma_z]\rho ,$$

where the interaction picture Hamiltonian is given in Eq. (8.25), and Γ, Γ_q are the dissipation rates for the mechanical resonator and qubit, respectively, Γ_p is a dephasing rate for the qubit, while $\bar{n}, \bar{n}_q$ are the mean thermal occupation numbers of the mechanical and qubit baths. The rate of irreversible dynamics in the qubit is typically much greater than the reversible and irreversible dynamics for the mechanical system. For example, in the experiment of La-Haye et. al [169], $\Gamma = 1$ kHz and $\Gamma_q = 0.3$ GHz. Under these circumstances we will assume that the qubit relaxes to a steady state. This enables us to adiabatically eliminate the qubit dynamics, obtaining an effective master equation for the mechanics alone.

From the point of view of the mechanical resonator, the qubit looks like an effective bath coupled via the time-dependent terms in Eq. (8.21). As we will see below, this can lead to heating or cooling of the mechanics, as well as resonance frequency shifts. Before we perform the adiabatic elimination discussed above, we can estimate the size of the frequency shift due to the interaction with the two-level system. Assuming that there is no microwave driving field, so that $\mathcal{E}_0 = 0$, we can replace the operator σ_z in Eq. (8.24) with the steady-state value it would take in the absence of the interaction with the mechanical system. This is given by

$$\langle\sigma_z\rangle_{ss} = -\frac{1}{2\bar{n}_q+1} \qquad (8.29)$$

with $\langle \sigma_+ \rangle_{ss} = 0$.

Exercise 8.1 *Using the master equation, Eq. (8.28), verify that these are the correct steady-state values of the qubit.*

Within this rather crude approximation, the mechanical frequency shift would be

$$\delta\Omega = \frac{2g^2}{\Omega_q} \langle \sigma_z \rangle_{ss} = -\frac{2g^2}{\Omega_q} \frac{1}{2\bar{n}_q + 1} \qquad (8.30)$$

with $\bar{n}_q$ given by Eq. (1.7). We will give a better estimate of the frequency shift below, using a more accurate adiabatic elimination of the qubit. LaHaye et al. investigated this frequency shift experimentally as a function of the qubit parameters in [169].

The qubit may be adiabatically eliminated using the Zwanzig projection operator method (see Gardiner and collaborators [114, 111] and also [144]). This results in the effective master equation for the oscillator,

$$\dot{\rho} = -\frac{i}{\hbar}[\hat{H}', \rho] + \Gamma(\bar{n} + 1)\mathcal{D}[b]\rho + \Gamma\bar{n}\mathcal{D}[b^\dagger]\rho \qquad (8.31)$$
$$+ \Gamma_\sigma(\bar{n}_\sigma + 1)\mathcal{D}[b]\rho + \Gamma_\sigma\bar{n}_\sigma\mathcal{D}[b^\dagger]\rho,$$

where

$$\hat{H}' = \hbar(\Omega + \delta\Omega)b^\dagger b. \qquad (8.32)$$

This shows that the coupling to the qubit does indeed caused a resonance frequency shift $\delta\Omega$ to the mechanical resonator, as well as effectively introducing a second thermal bath with a new damping rate Γ_σ and thermal occupation $\bar{n}_\sigma$.

Both Γ_σ and $\bar{n}_\sigma$ may be determined from the power spectral density $S_{FF}(\pm\Omega)$ of the force exerted on the mechanical oscillator by the qubit, via the same approach as was used to quantify the effective optical bath in cavity cooling of a mechanical resonator in Section 4.2.1. From Eq. (8.25), in an interaction picture for both the qubit and the mechanical resonator, the qubit-resonator interaction Hamiltonian can be seen to be

$$\hat{H}_I = \frac{\hbar g}{x_{zp}} \hat{q}\sigma_z, \qquad (8.33)$$

where we have used the definition of the momentum position in Eq. (1.9a). The force applied to the mechanical resonator by the qubit is then

$$\hat{F} = \frac{\partial \hat{H}_I}{\partial \hat{q}} = \frac{\hbar g}{x_{zp}}\sigma_z. \qquad (8.34)$$

Using the definition of the power spectral density in Eq. (1.42), we then directly find the force power spectral density in terms of the power spectral density of σ_z:

$$S_{FF}(\omega) = \left(\frac{\hbar g}{x_{zp}}\right)^2 S_{\sigma_z\sigma_z}(\omega). \qquad (8.35)$$

Eqs. (1.62) and (1.54b) then give the damping rate Γ_σ and effective bath occupancy $\bar{n}_\sigma$:

$$\Gamma_\sigma = g^2 \left(S_{\sigma_z \sigma_z}(\Omega) - S_{\sigma_z \sigma_z}(-\Omega) \right) \tag{8.36a}$$

$$\bar{n}_\sigma = \frac{g^2}{\Gamma_\sigma} S_{\sigma_z \sigma_z}(-\Omega), \tag{8.36b}$$

which, together with the intrinsic mechanical damping rate Γ and bath occupancy $\bar{n}$, determine the final mechanical occupancy as dictated by Eq. (4.23). So long as the effective occupancy of the qubit bath $\bar{n}_\sigma$ is much smaller than $\bar{n}$, and the coupling rate to the qubit bath is comparable, or larger, than the Γ, the mechanical resonator will experience substantial cooling.

To determine the qubit-induced resonance frequency shift $\delta\Omega$ we define the one-sided power spectral density:

$$S_{\mathcal{OO}}^{\text{one-sided}}(\omega) \equiv \int_0^\infty dt \, e^{i\omega t} \langle \mathcal{O}(t)\mathcal{O}(0) \rangle, \tag{8.37}$$

which – as is also the case for the two-sided power spectral density when defined in terms of an autocorrelation function – is only valid in the long-time limit for time-stationary statistics in the steady-state.

Exercise 8.2 *By using the property that, in the long-time limit and with stationary statistics, the power spectral density is invariant under time translation, show that the two- and one-sided power spectral densities are related via*

$$S(\omega) = 2 \, \text{Re}[S^{\text{one-sided}}(\omega)]. \tag{8.38}$$

The qubit-induced frequency shift is given in terms of the one-sided power spectral density of σ_z as [144]

$$\delta\Omega = g^2 \, \text{Im} \left[S_{\Delta\sigma_z \Delta\sigma_z}^{\text{one-sided}}(\Omega) + S_{\Delta\sigma_z \Delta\sigma_z}^{\text{one-sided}}(-\Omega) \right], \tag{8.39}$$

where $\Delta\sigma_z \equiv \sigma_z - \langle\sigma_z\rangle$.

We now understand that, in this limit where the qubit dynamics have been adiabatically eliminated, the effect of the qubit on the mechanical resonator is to appear as a second effective bath. To quantitatively determine the effect of this bath, we must calculate the power spectral density of $\Delta\sigma_z$ or, equivalently, the correlation function $\langle\Delta\sigma_z(t)\Delta\sigma_z(0)\rangle$ in the long-time steady-state limit. This correlation function can be calculated using the quantum regression theorem for the interaction picture qubit dynamics. To do this, we first find the equations of motion for the qubit moments. The equations of motion for the qubit, including damping, are determined by the interaction picture master equation

$$\dot{\rho} = -i\frac{1}{\hbar}[\hat{H}_q, \rho] + \Gamma_q(\bar{n}_q + 1)\mathcal{D}[\sigma_-]\rho + \Gamma_q\bar{n}_q\mathcal{D}[\sigma_+]\rho + \Gamma_p\mathcal{D}[\sigma_z]\rho, \tag{8.40}$$

where the qubit Hamiltonian is

$$\hat{H}_q = -\frac{\hbar \delta_q}{2}\sigma_z + \frac{\hbar \mathcal{E}_0}{2}\sigma_x, \tag{8.41}$$

with $\bar{n}_q$ being the mean thermal occupation of the qubit bath at the qubit frequency, Γ_q the qubits spontaneous emission rate, and Γ_p its dephasing rate.

Writing the moments in vector form using $\vec{\sigma} = (\langle \sigma_- \rangle, \langle \sigma_+ \rangle, \langle \sigma_z \rangle)^T$, we find that

$$\frac{d\vec{\sigma}}{dt} = \mathbf{A}\vec{\sigma} + \vec{\Gamma}, \tag{8.42}$$

where $\vec{\Gamma} = (0, 0, -\Gamma_q)$ and

$$\mathbf{A} = \begin{pmatrix} -\Gamma_2/2 - i\delta_q & 0 & -i\mathcal{E}_0/2 \\ 0 & -\Gamma_2/2 + i\delta_q & i\mathcal{E}_0/2 \\ -i\mathcal{E}_0/2 & i\mathcal{E}_0/2 & -\Gamma(2\bar{n}_q + 1) \end{pmatrix} \tag{8.43}$$

with $\Gamma_2 = \Gamma_q(2\bar{n}_q + 1) + 8\Gamma_p$. The eigenvalues of $\mathbf{A}$ have negative real parts, and the solution to this differential equation is

$$\vec{\sigma}(t) = e^{\mathbf{A}t}\vec{\sigma}(0) + e^{\mathbf{A}t} \int_0^t dt' e^{-\mathbf{A}t'} \vec{\Gamma}. \tag{8.44}$$

The steady-state solution is

$$\vec{\sigma}_{ss} = \lim_{t\to\infty} e^{\mathbf{A}t} \int_0^t dt' e^{-\mathbf{A}t'} \vec{\Gamma} = \lim_{t\to\infty} \int_0^t dt'' e^{\mathbf{A}t''} \vec{\Gamma} = -\mathbf{A}^{-1}\vec{\Gamma}. \tag{8.45}$$

Now consider the differential equation for $\vec{\Delta\sigma} = \vec{\sigma} - \vec{\sigma}_{ss}$:

$$\frac{d}{dt}\vec{\Delta\sigma} = \frac{d}{dt}\vec{\sigma} = \mathbf{A}\vec{\sigma} + \vec{\Gamma} = \mathbf{A}\vec{\Delta\sigma} + \mathbf{A}\vec{\sigma}_{ss} + \vec{\Gamma} = \mathbf{A}\vec{\Delta\sigma}. \tag{8.46}$$

The quantum regression theorem states that the differential equations for the time-ordered second-order moments like $\langle \Delta\sigma_z(\tau)\Delta\sigma_z(0)\rangle_{ss}$ obey the same differential equations as the mean values, so that if we define the matrix

$$\mathbf{G}(\tau) = \begin{pmatrix} \langle \Delta\sigma_-(\tau)\Delta\sigma_-(0)\rangle_{ss} & \langle \Delta\sigma_-(\tau)\Delta\sigma_+(0)\rangle_{ss} & \langle \Delta\sigma_-(\tau)\Delta\sigma_z(0)\rangle_{ss} \\ \langle \Delta\sigma_+(\tau)\Delta\sigma_-(0)\rangle_{ss} & \langle \Delta\sigma_+(\tau)\Delta\sigma_+(0)\rangle_{ss} & \langle \Delta\sigma_+(\tau)\Delta\sigma_z(0)\rangle_{ss} \\ \langle \Delta\sigma_z(\tau)\Delta\sigma_-(0)\rangle_{ss} & \langle \Delta\sigma_z(\tau)\Delta\sigma_+(0)\rangle_{ss} & \langle \Delta\sigma_z(\tau)\Delta\sigma_z(0)\rangle_{ss} \end{pmatrix} \tag{8.47}$$

then

$$\frac{d\mathbf{G}}{d\tau} = \mathbf{A}\mathbf{G}, \tag{8.48}$$

so that

$$\mathbf{G}(\tau) = e^{\mathbf{A}\tau}\mathbf{G}(0). \tag{8.49}$$

Finally, we are interested in the one-sided spectra

$$\mathbf{S}(\omega) = \int_0^\infty d\tau \, e^{i\omega\tau} \mathbf{G}(\tau) = \int_0^\infty d\tau \, e^{i\omega\tau} e^{\mathbf{A}\tau} \mathbf{G}(0) = -(i\omega\mathbb{1} + \mathbf{A})^{-1}\mathbf{G}(0).$$

(8.50)

The initial condition can be expressed in terms of steady-state expectation values

$$\mathbf{G}(0) = \begin{pmatrix} -\langle\sigma_-(0)\rangle_{\text{ss}}^2 & \frac{1}{2}(1-\langle\sigma_z\rangle_{\text{ss}})-|\langle\sigma_-\rangle_{\text{ss}}|^2 & \langle\sigma_-\rangle_{\text{ss}}(1-\langle\sigma_z\rangle_{\text{ss}}) \\ \frac{1}{2}(1+\langle\sigma_z\rangle_{\text{ss}})-|\langle\sigma_-\rangle_{\text{ss}}|^2 & -\langle\sigma_+(0)\rangle_{\text{ss}}^2 & -\langle\sigma_+\rangle_{\text{ss}}(1+\langle\sigma_z\rangle_{\text{ss}}) \\ -\langle\sigma_-\rangle_{\text{ss}}(1+\langle\sigma_z\rangle_{\text{ss}}) & \langle\sigma_+\rangle_{\text{ss}}(1-\langle\sigma_z\rangle_{\text{ss}}) & 1-\langle\sigma_z\rangle_{\text{ss}}^2 \end{pmatrix}$$

(8.51)

Exercise 8.3 *In the undriven case, $(\mathcal{E}_0 \to 0)$, show that for $t \geq 0$*

$$\langle\Delta\sigma_z(t)\Delta\sigma_z(0)\rangle_{\text{ss}} = \left[1 - \frac{1}{(2\bar{n}_q+1)^2}\right] e^{-\Gamma_q(2\bar{n}_q+1)t} \tag{8.52}$$

$$\langle\Delta\sigma_-(t)\Delta\sigma_+(0)\rangle_{\text{ss}} = \frac{\bar{n}_q+1}{2\bar{n}_q+1} e^{-\Gamma_2 t/2} e^{-i\Omega_q t} \tag{8.53}$$

$$\langle\Delta\sigma_+(t)\Delta\sigma_-(0)\rangle_{\text{ss}} = \frac{\bar{n}_q}{2\bar{n}_q+1} e^{-\Gamma_2 t/2} e^{i\Omega_q t}, \tag{8.54}$$

where $\Gamma_2 = \Gamma_q(2\bar{n}_q+1)+8\Gamma_p$. Then, using these results along with Eq. (8.37), show that full one-sided power spectral density of $\Delta\sigma_z$ is

$$
\begin{aligned}
S_{\Delta\sigma_z\Delta\sigma_z}^{\text{one-sided}}(\omega) &= \frac{\epsilon^2\lambda^2}{\Omega_q^2}\left(1 - \frac{1}{(2\bar{n}_q+1)^2}\right)\left(\frac{1}{\Gamma_q(2\bar{n}_q+1)-i\omega}\right) \\
&\quad + \frac{\delta^2\lambda^2}{\Omega_q^2}\left[\frac{\bar{n}_q+1}{2\bar{n}_q+1}\left(\frac{1}{\Gamma_2/2-i(\omega-\Omega_q)}\right)\right. \\
&\quad \left. + \frac{\bar{n}_q}{2\bar{n}_q+1}\left(\frac{1}{\Gamma_2/2-i(\omega+\Omega_q)}\right)\right].
\end{aligned}
$$

(8.55)

The oscillator frequency shift is given by substituting the power spectral density of Eq. (8.55) into Eq. (8.39), and can be approximated as

$$\delta\Omega \simeq -\frac{1}{2\bar{n}_q+1}\left(\frac{\epsilon}{\Omega_q}\right)^2 \frac{2\lambda^2\Omega_q}{(\Gamma_2/2)^2+\Omega_q^2}, \tag{8.56}$$

where ϵ is defined in Eq. (8.27). This should be compared to the frequency shift $\delta\Omega$ obtained in Eq. (8.30) by ignoring dissipation and simply using the dispersive Hamiltonian. The two results agree when $\Gamma_2 \ll \Omega_q$, i.e., when the qubit frequency Ω_q is much greater than both its thermal decoherence rate $\Gamma_q(2\bar{n}_q+1)$, and its dephasing rate Γ_p. The frequency shift given in Eq. (8.56), which explicitly includes dissipation, avoids a divergence in the expression for the frequency shift given by the simpler dispersive analysis when Ω_q is tuned through zero.

The explicit form of the spectrum including driving ($\mathcal{E}_0 \neq 0$) is given in [144]. In general, it has three peaks, one at zero frequency and two equally spaced on either side at $\pm\sqrt{\mathcal{E}_0^2 + \delta_q^2}$. The widths of the central peak Γ_0 and side-peaks $\Gamma_\pm$ are given, respectively, by:

$$\Gamma_0 = \Gamma_q \left(\frac{2\delta_q^2 + \mathcal{E}_0^2}{\delta_q^2 + \mathcal{E}_0^2} \right) \left(1 + \frac{4\Gamma_p}{\Gamma_q} \left(\frac{\mathcal{E}_0^2}{2\delta_q^2 + \mathcal{E}_0^2} \right) \right) \tag{8.57}$$

$$\Gamma_\pm = \frac{\Gamma_q}{2} \left(\frac{\delta_q^2 + 3\mathcal{E}_0^2}{\delta_q^2 + \mathcal{E}_0^2} \right) \left(1 + \frac{4\Gamma_p}{\Gamma_q} \left(\frac{2\delta_q^2 + \mathcal{E}_0^2}{2\delta_q^2 + 3\mathcal{E}_0^2} \right) \right). \tag{8.58}$$

We can now return to the question of the steady-state mean vibrational quantum number as given by the effective qubit-bath coupling rate and occupancy in Eqs. (8.36), combined with Eq. (4.23). As we saw for the general case in Section 1.2, $S_{\sigma_z \sigma_z}(\Omega)$ is responsible for driving cooling transitions, while $S_{\sigma_z \sigma_z}(-\Omega)$ is drives heating transitions. Net cooling will occur so long as $S_{\sigma_z \sigma_z}(\Omega) > S_{\sigma_z \sigma_z}(-\Omega)$ and the effective bath occupancy of the qubit $\bar{n}_\sigma < \bar{n}$. Similar to the case for resolved sideband cooling in Section 4.2.2, where red-detuning of the optical field is required for cooling, the condition $S_{\sigma_z \sigma_z}(\Omega) > S_{\sigma_z \sigma_z}(-\Omega)$ can only be satisfied here for negative detuning, $\delta_q < 0$.

In order to clearly see the effect of cooling, we will assume that the qubit is strongly coupled to a zero temperature heat bath $\bar{n}_q = 0$ and there is no dephasing $\Gamma_p = 0$. Then, using the full nose power spectrum in [144], we can approximate $\bar{n}_\sigma = 0$ and for weak driving $\mathcal{E}_0 \ll |\delta_q|$,

$$\bar{n}_{b,ss} = \frac{\bar{n}}{1 + \beta(\delta/\Omega_q)^2}, \tag{8.59}$$

where

$$\beta = \frac{2g^2}{\Gamma\Gamma_q}. \tag{8.60}$$

Thus, the steady-state average vibrational phonon number is less than the equilibrium steady-state value of $\bar{n}$. In the experiment of LaHaye et al. [169], $\beta \sim 20$.

8.2.2   Single phonon control via a superconducting qubit

In this section we will discuss the landmark experiment of O'Connell et al. [213], which used a superconducting qubit in a microwave cavity to control the quantum state of a microwave-frequency mechanical oscillator. This was the first experiment to show ground state cooling and single-phonon control of a bulk mechanical resonator.

The key to the experiment was a very high frequency (6 GHz) mechanical resonator whose ground state is reached for temperatures $T < 0.1$ K which can be achieved using passive cooling in a dilution refrigerator run at

25 mK. The mechanical resonator was fabricated from a piezoelectric material (aluminium nitride) with a fundamental dilational mode, sandwiched between two aluminium electrodes. By applying a voltage to the electrodes, the piezoelectric aluminium nitride resonator dilates and compresses in the direction perpendicular to the metal electrodes. The mechanical element is suspended and clamped at one end; see Fig. 8.2.

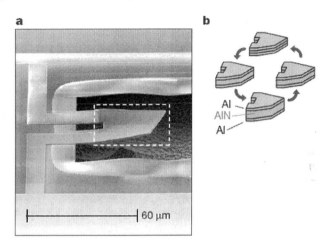

FIGURE 8.2 (a) A scanning electron micrograph of a suspended aluminium nitride film bulk acoustic resonator. (b) The fundamental dilational resonant mechanical mode in which the thickness oscillates. Adapted by permission from Macmillan Publishers Ltd., from [213], copyright 2010.

The mechanical resonator is coupled to a Josephson phase qubit made from a Josephson junction shunted in parallel by a capacitor and an inductor. The resonance frequency of the qubit was varied by a current bias so as to take values in the range 5 to 10 GHz. The state of the qubit can be measured using a single-shot readout. This is done using a flux bias current to cause the flux to selectively tunnel from the qubit excited state, $|e\rangle$, into a macroscopically distinguishable flux state of the junction while leaving the ground state, $|g\rangle$, of the qubit unchanged. This realises a good approximation to a projective measurement of the qubit excited state. Repeating this procedure many times enables one to sample the probability of finding the qubit in the excited state.

The qubit and the mechanical resonator are capacitively coupled. The coupling is described using the Jaynes–Cummings Hamiltonian given in Eq. (8.22). On resonance, $\Omega_q = \Omega$, the eigenstates of this coupled system are given by the dressed states

$$|\pm, n\rangle = \frac{1}{\sqrt{2}}(|g, n+1\rangle \pm |e, n\rangle) \tag{8.61}$$

with corresponding eigenvalues

$$\lambda_n = \pm g_0 \sqrt{n+1}. \tag{8.62}$$

The splitting between the two lowest eigenvalues ($n = 0$) is $2g_0$, the vacuum Rabi splitting.

A microwave drive can be applied to the electronic circuit, thus exciting the qubit. This enables a measurement of the spectrum of the coupled mechanical-qubit system. The measurement gives the excited-state probability, P_e, as a function of the qubit frequency and the microwave excitation frequency. At resonance $\Omega_q = \Omega$, the Jaynes–Cummings coupling leads to an avoided crossing of the qubit levels and this gives a vacuum level splitting of $\delta\omega = 2g_0$, which gives an experimental value of $g_0/2\pi = 124$ MHz for the coupling rate.

To estimate the temperature of the mechanical resonator, the qubit is initially detuned from the resonator (with a frequency of $\Omega_q = 5.44$ GHz) and prepared in the ground state. A flux bias pulse is then applied to bring the qubit into resonance with the mechanical resonator for a chosen interaction time and then detuned. Thermal excitations in the mechanical resonator can then excite the qubit during the time of interaction. A subsequent readout of the qubit can be used to estimate the average occupancy of the mechanical resonator. A value of less that $\bar{n}_b < 0.007$ was found, indicating that the mechanical resonator was indeed very close to its ground state. This is not surprising given the frequency of the mechanical resonator and its temperature.

Exercise 8.4 *Using the result*

$$\bar{n}_b = \frac{1}{e^{\beta \hbar \Omega} - 1} \tag{8.63}$$

where $\beta = (k_B T)^{-1}$. Show that for $\Omega = 6$ GHz and $T = 15$ mK, the experimental result for the average phonon number is reasonable.

In a final experiment, the qubit is detuned from resonance with the mechanics and prepared in the *excited state*. If the interaction time is now chosen to be one half a Rabi cycle, this excitation should be transferred into a single-phonon excitation of the mechanical resonator. This, of course, assumes that the qubit is not rapidly damped. O'Connell et al. estimate that the qubit decay time is less that 17 ns while the time needed for an exchange of excitation is estimated to be < 3.8 ns. By sampling the probability of finding the qubit in the excited state at the end of the interaction time as a function of the interaction time, one sees clear evidence of Rabi oscillations between the photonic and phononic resonators.

8.3 MICROWAVE TO OPTICAL INTERFACE.

As we have seen in this chapter, the mechanical interaction of radiation and bulk mechanical motion has an equivalent description in both the microwave and optical domains. This suggests that a mechanical resonator could serve as a quantum frequency convertor between microwaves and optics if a means can be found to couple to a common mechanical element. We will refer to this as a micro-optomechanical (MOM) quantum interface. This would enable a new approach to hybrid systems coupling optically active systems such as atoms and molecules to superconducting solid-state devices. An example of a MOM quantum interface has been developed by the group at JILA [236, 13, 14]. This approach uses a flexible silicon nitride membrane coupled via radiation pressure to both a microwave circuit and a Fabry–Pérot optical cavity. Other examples of cavity opto-electromechanical systems that can be applied as a MOM interface include whispering gallery mode [174, 323] and photonic crystal geometries [33, 316].

It is worthwhile to note that other applications exist for cavity opto-electromechanical systems beyond the quantum interfaces discussed above, and further detailed below. In particular, we would draw the readers attention to their use in the optical transduction of radio- or microwave-frequency signals[276] which, for example, has been shown to enable low noise radio frequency receivers[21].

8.3.1   A frequency convertor MOM interface

The phenomenon of frequency conversion is a typical phenomenon in the non-linear response of a dielectric and was the subject of a quantum description at the very beginning of quantum optics [287]. In the case of an MOM interface, the nonlinear response of the electromagnetic fields is mediated by an effective nonlinear susceptibility provided by the mechanical element. We will follow the discussions in [13, 196].

The Hamiltonian describing the interactions of the optical and microwave fields with a common mechanical resonator is

$$
\begin{aligned}
\hat{H} \;=\; & \hbar\Omega_c a^\dagger a + \hbar\Omega b^\dagger b + \hbar\Omega_\mu c^\dagger c + \hbar(g_0 a^\dagger a + g_{0,\mu} c^\dagger c)(b + b^\dagger) \quad (8.64)\\
& + (E_o^* a e^{i\omega_o t} + E_o a^\dagger e^{-i\omega_o t}) + (E_\mu^* c e^{i\omega_\mu t} + E_\mu c^\dagger e^{-i\omega_\mu t}),
\end{aligned}
$$

where Ω_c, Ω_μ, and Ω are the resonant frequencies of the optical, microwave, and mechanical resonators, respectively, g_0 and $g_{0,\mu}$ are the radiation pressure coupling constants for the optical and microwave fields, and E_0 and E_μ are the classical driving field amplitudes for the optical and microwave resonators with carrier frequencies ω_o and ω_μ, respectively. The optical (microwave) field annihilation operator is $a(c)$, with b, as usual, representing the annihilation operator for the mechanical resonator.

We now assume that both the optical and microwave cavities are driven on their respective red sidebands, so that $\Omega_c - \omega_o = \Omega_\mu - \omega_\mu = \Omega$. As usual we linearise the optomechanical interaction around the steady-state fields that each cavity would have in the absence of the radiation pressure interaction to give the following Hamiltonian in the interaction picture,

$$\hat{H}_I = \hbar g(ab^\dagger + a^\dagger b) + \hbar g_\mu(cb^\dagger + c^\dagger b), \tag{8.65}$$

where $g \equiv g_0\langle a\rangle_{ss}, g_\mu \equiv g_{0,\mu}\langle c\rangle_{ss}$. In the experiment [13] the linearised optomechanical coupling constants are of the order of 10% of the mechanical frequency. Dissipation is included in the usual way using the master equation

$$\frac{d\rho}{dt} = -\frac{i}{\hbar}[\hat{H}_I, \rho] + \kappa_o \mathcal{D}[a]\rho + \kappa_\mu \mathcal{D}[c]\rho \tag{8.66}$$
$$+\Gamma(\bar{n}+1)\mathcal{D}[b]\rho + \Gamma\bar{n}\mathcal{D}[b^\dagger]\rho,$$

where κ_o and κ_μ are the optical and microwave cavity decay rates.

Exercise 8.5 *If $\{\kappa_o, \kappa_\mu\} \gg \{\chi_o, \chi_\mu, \Gamma\}$ so that the optical and microwave cavity fields are rapidly damped, the optical and microwave fields become slaved to the mechanical system dynamics. Show that, in that case, the mechanical dynamics can be approximated by the quantum Langevin equation*

$$\frac{db}{dt} = -\frac{(\Gamma_o + \Gamma_\mu + \Gamma)}{2}b - i\sqrt{\Gamma_o}a_{in} - i\sqrt{\Gamma_\mu}c_{in} + \sqrt{\Gamma}b_{in}, \tag{8.67}$$

where, we see that mechanical heating and damping has been introduced by the optical and microwave fields, with respective rates give by $\Gamma_o \equiv 4g^2/\kappa_o, \Gamma_\mu \equiv 4g_\mu^2/\kappa_o$. Thus, as we saw in Section 3.2, the mechanical system sees each field cavity mode as a additional dissipation channel.

Following Exercise 8.5, we adiabatically eliminate the optical and microwave cavity field modes and, using the input-output relations (see Section 1.4.3), obtain the following expressions for the coupling between the output fields and input fields in the frequency domain:

$$a_{out}(\omega) = -\frac{(\Gamma_o - A/2 + i\omega)}{A/2 - i\omega}a_{in}(\omega) - \frac{\sqrt{\Gamma_o\Gamma_\mu}}{A/2 - i\omega}c_{in}(\omega) \tag{8.68}$$
$$-\frac{i\sqrt{\Gamma_o\Gamma}}{A/2 - i\omega}b_{in}(\omega)$$

$$c_{out}(\omega) = -\frac{(\Gamma_\mu - A/2 + i\omega)}{A/2 - i\omega}c_{in}(\omega) - \frac{\sqrt{\Gamma_o\Gamma_\mu}}{A/2 - i\omega}a_{in}(\omega) \tag{8.69}$$
$$-\frac{i\sqrt{\Gamma_\mu\Gamma}}{A/2 - i\omega}b_{in}(\omega)$$

where Γ_o and Γ_μ quantify the rates at which phonons are lost via the red sideband Raman coupling to the optical and microwave cavity fields, respectively, and $A \equiv \Gamma + \Gamma_o + \Gamma_\mu$ is the total effective decay rate of the mechanical system, including loss of phonons through up-conversion to optical and microwave photons. The correlation functions for the mechanical input operators,

$b_{\text{in}}(\omega), b_{\text{in}}^{\dagger}(\omega)$, are given by the thermal bath to which the mechanical oscillator is coupled as defined in Section 1.4.1. In an ideal convertor we would expect $a_{\text{out}} \sim c_{\text{in}}$ and $c_{\text{out}} \sim a_{\text{in}}$.

We can define a transfer function for microwaves to optical modes as

$$t(\omega) = \frac{\langle a_{\text{out}}(\omega) \rangle}{\langle c_{\text{in}}(\omega) \rangle}, \tag{8.70}$$

which, in this ideal case, is given by

$$t(\omega) = \frac{\sqrt{\Gamma_o \Gamma_\mu}}{A/2 - i\omega}. \tag{8.71}$$

Recall that we are working in an interaction picture rotating at the mechanical frequency, so $\omega = 0$ in this expression corresponds to $\omega = \Omega$, the mechanical frequency, in the lab frame. In the experiment of Andrews et al. [13] this ideal transfer function is reduced by about 17% as measured due to various out-coupling inefficiencies and departures from the perfect resolved sideband limit.

Equation (8.68) indicates how the input microwave field can be converted to an optical output field, and vice versa. As an interesting application we consider the case in which the microwave input field is a single-photon state (see Chapter 6) defined by

$$|1\rangle_\mu = \int d\omega \xi(\omega) c_{\text{in}}^{\dagger}(\omega) |0\rangle. \tag{8.72}$$

As there is only one photon in the field, the success probability is given by

$$p_s = \int dt \langle a_{\text{out}}^{\dagger}(t) a_{\text{out}}(t) \rangle. \tag{8.73}$$

The one-photon state has a zero average amplitude, $\langle c_{\text{in}}(t) \rangle = 0$, but a non-zero intensity, $\langle c_{\text{in}}^{\dagger}(t) c_{\text{in}}(t) \rangle = |\xi(t)|^2$, where $\xi(t)$ is the time-domain Fourier transform of $\xi(\omega)$. If we now assume that the input optical field is in a vacuum state, we see that the output optical field also has zero amplitude, but an intensity given by

$$\langle a_{\text{out}}^{\dagger}(t) a_{\text{out}}(t) \rangle = |\nu(t)|^2 + \frac{\Gamma_o \Gamma N}{\Gamma_o + \Gamma_\mu + \Gamma}, \tag{8.74}$$

where $\nu(t)$ is the time-domain Fourier transform of

$$\tilde{\nu}(\omega) = \frac{\sqrt{\Gamma_o \Gamma_\mu}}{A/2 - i\omega} \xi(\omega) \tag{8.75}$$

and the last term is a thermal background due to mechanical bath phonons being converted into optical photons. If this term is neglected, the success probability is given by

$$p_s = \frac{4\Gamma_o\Gamma_\mu}{(\Gamma_o + \Gamma_\mu + \Gamma)^2}\eta \tag{8.76}$$

where

$$\eta = \int d\omega \frac{|\xi(\omega)|^2}{1 + (\omega/A)^2}. \tag{8.77}$$

If the optical and microwave damping rates are equal ($\Gamma_o = \Gamma_\mu$) and both much larger than the mechanical loss rate Γ,[2] the success probability is given only by η. For this to be unity, we require that the bandwidth of the single microwave photon pulse is much less than the conversion rates Γ_o and Γ_μ. As it is relatively easier to create single-photon states in the microwave domain than in the optical, this might be a viable path to deterministic optical single-photon sources provided the mechanics is cooled.

8.3.2 A teleportation MOM interface

A nonlinear optical susceptibility can also lead to nondegenerate parametric amplification of two optical modes. Likewise, the nonlinear susceptibility provided by a common mechanical element can be used to implement parametric amplification of the optical and mechanical cavity fields. In this case, the optical and microwave fields can become entangled via the mechanism of two-mode squeezing.[3] This entanglement can be used as the basis of a teleportation scheme between optical and microwave degrees of freedom. We will follow the treatment of Barzanjeh et al. [24].

Unlike the case of frequency conversion, we now drive one of the electromagnetic field resonators on the blue side and and the other on the red. The linearised Hamiltonian in the interaction picture of the mechanical and microwave driving frequencies is

$$\begin{aligned} \hat{H} &= \hbar\Delta_o\hat{a}^\dagger\hat{a} + \hbar\Delta_\mu\hat{c}^\dagger\hat{c} + \hbar\Omega\hat{b}^\dagger\hat{b} \\ &\quad -\hbar g(\hat{a}^\dagger + \hat{a})(\hat{b} + \hat{b}^\dagger) - \hbar g_\mu(\hat{c}^\dagger + \hat{c})(\hat{b} + \hat{b}^\dagger), \end{aligned} \tag{8.78}$$

where $\Delta_o = \Omega_o - \omega_o$ and $\Delta_\mu = \Omega_\mu - \omega_\mu$. Unlike the simple transfer scheme discussed previously, here we choose opposite detunings $\Delta_o = -\Delta_\mu = \Omega$, and assume the regime of fast mechanical oscillations, $\Omega \gg \{\chi_o, \chi_\mu, \kappa_o, \kappa_\mu\}$, so that we are in the resolved sideband regime for both cavities, with red sideband driving on the optical cavity and blue sideband driving on the microwave cavity. If we move to an interaction picture at the frequency Ω and make the rotating wave approximation, we find the approximate interaction

---

[2]This corresponds to a regime where both the optical and microwave optomechanical cooperativities are much greater than one (see Section 3.2).

[3]The reader is referred to Section 4.4 for an analysis of optomechanical entanglement produced in this manner.

Hamiltonian as

$$\hat{H}_a = \hbar g(\hat{a}^\dagger \hat{b} + \hat{a}\hat{b}^\dagger) + \hbar g_\mu(\hat{c}\hat{b} + \hat{c}^\dagger \hat{b}^\dagger). \tag{8.79}$$

The second term, arising from blue detuning, describes nondegenerate parametric amplification and is responsible for entangling the microwave resonator with the mechanical resonator as discussed in Section 4.4. The first term arises from red detuning and, as in the previous example, corresponds to frequency conversion. We now look for a regime in which the optical and microwave resonators become entangled, thus providing a resource for a continuous variable teleportation protocol. Dissipation is included using the same irreversible terms as in the previous master equation.

Exercise 8.6 *If $\{\kappa_o, \kappa_\mu\} \gg \{g, g_\mu, \Gamma\}$, so that the optical and microwave cavity fields are rapidly damped, the optical and microwave fields are slaved to the mechanical system dynamics. Show that, in that case, the mechanical dynamics can be approximated by*

$$\frac{db}{dt} = -\frac{1}{2}(\Gamma_o + \Gamma)b + \frac{\Gamma_\mu}{2}b - i\sqrt{\Gamma_o}a_{\rm in} - i\sqrt{\Gamma_\mu}c_{\rm in}^\dagger + \sqrt{\Gamma}b_{\rm in}, \tag{8.80}$$

where, as before, $\Gamma_o = 4g^2/\kappa_o$ and $\Gamma_\mu = 4g_\mu^2/\kappa_\mu$. Thus the mechanical system sees the optical mode as a damping channel while it sees the microwave mode as an amplifier channel.

Again, using adiabatic elimination of the optical and microwave modes, we now find that the effective input-output relations are

$$a_{\rm out}(\omega) = \frac{(\Gamma_o - B/2 + i\omega)}{B/2 - i\omega}a_{\rm in}(\omega) - \frac{\sqrt{\Gamma_o \Gamma_\mu}}{B/2 - i\omega}c_{\rm in}^\dagger(-\omega) \tag{8.81}$$

$$-i\frac{\sqrt{\Gamma_o \Gamma}}{B/2 - i\omega}b_{\rm in}(\omega)$$

$$c_{\rm out}^\dagger(-\omega) = \frac{(\Gamma_\mu + B/2 - i\omega)}{B/2 - i\omega}c_{\rm in}^\dagger(-\omega) - \frac{\sqrt{\Gamma_o \Gamma_\mu}}{B/2 - i\omega}a_{\rm in}(\omega) \tag{8.82}$$

$$+i\frac{\sqrt{\Gamma_\mu \Gamma}}{B/2 - i\omega}b_{\rm in}(\omega),$$

where $B \equiv \Gamma_o - \Gamma_\mu + \Gamma$. The key difference between these equations and Eqs. (8.68) is the change in the sign of Γ_μ in the expression for B indicating linear amplification. Corresponding to this is the substitution of $c_{in}^\dagger(-\omega)$ for $c_{in}(\omega)$, also indicative of the noise added by a linear amplifier. The physics of this amplification and noise addition has been experimentally studied in microwave optomechanical systems in [172]. In the case of no mechanical damping, $\Gamma = 0$, the input-output relation for the fields reflects a multimode squeezing transformation of the optical input fields $a_{\rm in}$ and the microwave input fields $c_{\rm in}$ (see Section 4.4.3). Provided that the mechanical system is coupled to a zero temperature bath, the optical and microwave output fields will be entangled. Barzanjeh et al. [24] show how this can be used to implement a continuous variable teleportation protocol between the optical and microwave fields. In so far as teleportation is a quantum communication protocol with no classical analog, this scheme can be considered a true quantum interface.

CHAPTER **9**

Arrays of optomechanical systems

CONTENTS

9.1   Synchronisation in optomechanical arrays 277
      9.1.1   Reversibly coupled arrays with a common cavity
              mode 279
      9.1.2   Quantum noise and synchronisation 288
9.2   Irreversibly coupled arrays of optomechanical systems 293

In this chapter we consider optomechanical systems that comprise more than one bulk mechanical element. This could include the case of more than one mechanical element coupled to a single intracavity mode, or even coupled arrays of distinct optomechanical systems. The generic phenomenon we will be interested in is *synchronisation*. Some examples of optomechanical arrays are illustrated in Fig. 9.1.

9.1   SYNCHRONISATION IN OPTOMECHANICAL ARRAYS

Synchronisation in networks of nonlinear oscillators [225] is the unifying explanation behind diverse phenomenon, from biological systems such as networks of neural cells [262], to engineered systems such as MEMS (microelectromechanical systems) [332], with applications to sensing [306]. Recent developments in optomechanical resonators now give access to optically actuated and optically transduced synchronisation in microfabricated devices [335, 22].

     The theoretical study of optomechanical synchronisation is at an early stage, as are experimental demonstrations [335]. Marquardt investigated the possibility of Kuramoto-type synchronisation [167] in two coupled optomechanical systems [138]. Lee and Cross [175] have considered the synchronisation of two coupled nonlinear optical cavities in the quantum domain. While their study does not explicitly propose a mechanical element, it bears some similarity to the parametrically driven models of interest in optomechanics. Heinrich et al. [138] consider mechanical resonators directly coupled by elastic

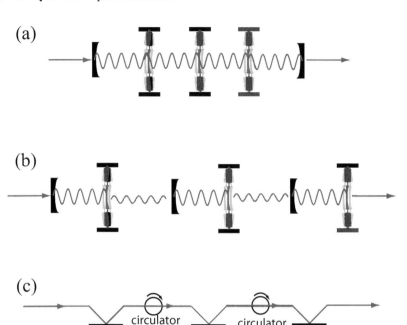

FIGURE 9.1 Different schemes for optomechanical arrays. In (a) multiple mechanical resonators interact reversibly with a common optical cavity field. In (b) different optomechanical cavities are coupled reversibly. In (c) the optical cavities are coupled irreversibly using circulators.

forces and each locally coupled to an electromagnetic field mode. This leads to synchronisation through a similar mechanism to Kuramoto's model.

We will not include thermal noise in our analysis, as it does not enter the semiclassical stability analysis that establishes the conditions for synchronisation. Of course, the inclusion of thermal nose is very relevant for the experimental realisation, and can be included using classical stochastic differential equations. The case of quantum noise at zero temperature, however, cannot be treated using this approach, as we have already seen for the different scenario of quantum tunnelling in a Duffing oscillator (Section 7.2.1). We will include a brief discussion of the effect of quantum noise on synchronisation to indicate the way it might be treated using the positive P-function from quantum optics. Classical thermal noise can also be treated using this approach.

9.1.1 Reversibly coupled arrays with a common cavity mode

We will begin with an example of a type-1 array, in which two or more mechanical resonators are *reversibly* coupled to a single cavity mode (see Fig. 9.1a). Holmes et al. [139] gave a detailed discussion of the synchronisation of two or more nanomechanical resonators coupled to a common electromagnetic resonator, revealing synchronisation in an amplitude-phase model that can exhibit oscillator death, unlike the phase models of Kuromoto [167]. An experimental realisation of a similar system was implemented by Zhang et al. [335].

We will define the semiclassical cavity field amplitude as $\alpha = \langle a \rangle$, and the conjugate classical position and momentum variables $q_j = \langle \hat{q}_j \rangle$ and $p_j = \langle \hat{p}_j \rangle$ for each mechanical resonator j, and assume that the cavity field is driven by a coherent source. We will factorise all other-order moments. A more rigorous justification of the semiclassical equations of motion, thus derived, is given in Section 9.1.2 using the positive P-function. If the uncoupled mechanical resonators are identical, then the oscillators synchronise. This is a natural consequence of linear damping and the fact that each oscillator experiences the same forcing (see [139]). For simplicity, we also assume, for now, that all the mechanical resonators have the same mass (m), frequency (Ω), and energy decay rate (Γ), but we do not assume identical optomechanical couplings. If the mechanical resonators naturally oscillate at different frequencies, desynchronisation can occur if the variation in frequencies is too great. The case of a distribution of mechanical frequencies has been considered in Cudmore and Holmes [82]. It is important to note that synchronisation does not simply mean that each resonator oscillates with the same frequency; rather it means that the collective dynamics is attracted onto a stable limit cycle with a single frequency. That is, the oscillations of the resonators have well defined phase relationships.

It is convenient to work in terms of dimensionless canonical variables for the mechanical resonators defined, similarly to Eqs. (1.13), as

$$\hat{Q}_j \;=\; \left(\frac{\hbar}{m\Omega}\right)^{-1/2} \hat{q}_j \tag{9.1}$$

$$\hat{P}_j \;=\; (\hbar m\Omega)^{-1/2}\,\hat{p}_j. \tag{9.2}$$

Assuming the usual optomechanical coupling between the cavity field and each mechanical resonator, the interaction Hamiltonian is of the form

$$\hat{H} = \hbar\Delta a^\dagger a + \hbar\Omega_j \sum_i b_j^\dagger b_j + \hbar a^\dagger a \sum_j g_j \hat{Q}_j, \tag{9.3}$$

where g_j is the vacuum optomechanical coupling rate between the field and resonator j and, as usual, $\Delta \equiv \Omega_c - \Omega_L$ is the detuning of the optical drive field from the cavity resonance frequency.

Exercise 9.1 *(a) Using the quantum Langevin equation of Eq. (1.112), and* *introducing the classical optical driving term $\hbar(\epsilon^* a + \epsilon a^\dagger)$ into the Hamiltonian of Eq. (9.3), where ϵ represents the amplitude of the driving, derive equations of motion for the cavity field a, and the mechanical variables $\hat{Q}_i$ and $\hat{P}_i$.*

(b) Defining the collective variables

$$X \equiv \sum_j g_j Q_j \tag{9.4a}$$

$$Y \equiv \sum_j g_j P_j \tag{9.4b}$$

$$G \equiv 2 \sum_j g_j^2, \tag{9.4c}$$

where $Q_j \equiv \langle \hat{Q}_j \rangle$ and $P_j \equiv \langle \hat{P}_j \rangle$, show that – factorising moments higher than first-order in the operators – the semiclassical equations of motion of the system may be expressed succinctly as

$$\frac{d\alpha}{dt} = -i\Delta\alpha - i\epsilon - i\alpha X - \frac{\kappa}{2}\alpha$$

$$\frac{dX}{dt} = \Omega Y - \frac{\Gamma}{2}X \tag{9.5}$$

$$\frac{dY}{dt} = -\Omega X - \frac{G}{2}|\alpha|^2 - \frac{\Gamma}{2}Y,$$

where κ and Γ are the usual energy decay rates for the electromagnetic and mechanical energy, respectively.

We see from the result of the above exercise that, in this special case, the full semiclassical dynamics of the system can be reduced to the dynamics of two coupled oscillators, with the contributions from each mechanical resonator in the ensemble weighted by its respective optomechanical coupling rate g_j, and the influence of the field on the collective momentum Y of the resonators enhanced above the single resonator coupling rate as quantified by Eq. (9.4c).

There are two time scales in the system: the amplitude decay rate κ of the common cavity mode, and the decay rate of the resonators, which is generally significantly smaller and will be important for the derivation of the amplitude equations. The amplitude decay rate κ of the common cavity mode provides a natural time scale and we introduce a new time parameter $t' = \kappa t/2$; rescaled optomechanical variables $Q_i' = 2\frac{Q_i}{\kappa}$ and $P_i' = 2\frac{P_i}{\kappa}$; and dimensionless coupling constants $\delta' = 2\frac{\Delta}{\kappa}$, $\epsilon' = \frac{2\epsilon}{\kappa}$, $\Omega' = \frac{2\Omega}{\kappa}$, $\Gamma' = \frac{2\Gamma}{\kappa}$, $G' = \frac{4G}{\kappa^2}$, and $\bar{\Omega}' = \sqrt{\Omega'^2 + (\Gamma')^2/4}$. The synchronised equations of motion of Eqs. (9.5) can then be represented in collective variables which, suppressing the use of primes and rewriting the collective resonator dynamics as a single second-order differential equation, gives

$$\frac{d\alpha}{dt} = -(1+i\delta)\,\alpha - i\alpha X - i\epsilon$$

$$\frac{d^2 X}{dt^2} = -\bar{\Omega}^2 X - \frac{G\Omega}{2}\,|\alpha|^2 - \Gamma\frac{dX}{dt}. \tag{9.6}$$

We will henceforth suppress the uses of primes in the notation but keep in mind that all couplings are now dimensionless, with the cavity decay rate determining the natural time scale of the system.

The cavity driving ϵ can be treated as the bifurcation parameter. The position, X_0, of the critical point is given by the single real root of the cubic

$$2\bar{\Omega}^2 X_0 \left(1 + (\delta + X_0)^2\right) + G\Omega\epsilon^2 = 0, \tag{9.7}$$

where

$$\alpha_0 = -\frac{i\epsilon}{1 + i\,(\delta + X_0)}. \tag{9.8}$$

It can be seen from this expression that, the critical point moves away from the origin as the drive strength ϵ increases from zero. However, the system loses stability on a Hopf bifurcation[123], creating a periodic orbit, for both $\delta > 0$ (red detuning) and $\delta < 0$ (blue detuning) provided $\Gamma > 0$ and much less than unity. Periodic orbits and multiple periodic orbits can exist for weak damping and weakly forced oscillators. This multistable behaviour, resulting from the play off between weak damping and cavity forcing, has been noted elsewhere [192, 138, 205].

For $\delta > 0$ (red detuning) the Hopf curve (the line in parameter space defined by the critical value for ϵ) is a perturbation of that from the $\Gamma = 0$ case where $\sqrt{G}\,\epsilon = \sqrt{2\delta}\,\bar{\Omega}$. To first order in Γ it is given by

$$\epsilon = \epsilon_H\,(\Omega, \delta, \Gamma, G) = \sqrt{\frac{2\Omega}{G}\left(\delta + \Gamma\frac{(1+\Omega^2)^2}{4\delta\,\Omega^2}\right)}. \tag{9.9}$$

For $\delta < 0$ (blue detuning) ϵ is of order $\sqrt{\Gamma}$:

$$\epsilon = \epsilon_H\,(\Omega, \delta, \Gamma, G) = \sqrt{\frac{\Gamma\,(1+\delta^2)\left((\delta^2 - \Omega^2 + 1)^2 + 4\Omega^2\right)}{-\delta G\Omega}}. \tag{9.10}$$

The Hopf bifurcation is subcritical for $\delta < -\sqrt{\frac{8\Omega^2 + 3}{5}}$ (blue detuning), where periodic orbits can exist for $\epsilon < \epsilon_H\,(\Omega, \delta, \Gamma, G)$. In fact, many stable limit cycles can exist for some parameter values because of the presence of saddle node bifurcations of limit cycles, each creating a stable and unstable pair of limit cycles. This leads to multistable behaviour that has been noticed elsewhere for similar systems [192, 94]. These bifurcations are shown in Fig. 9.2 for $\Omega = 2$

and $\Gamma = 0.001$. The limit cycle bifurcation diagrams were produced using the amplitude equations described below. We will restrict the discussion to the case $\delta < 0$ (blue detuning), for which eight of the saddle node bifurcations of limit cycles are shown, indicating regions where there are 1 to 8 pairs of stable and unstable periodic orbits. The dynamics are more complicated for $\delta > 0$ (red detuning), involving period doubling and regions of chaos; see [139] for more details.

The optomechanical bifurcation diagram in Fig. 9.2 is quite complex. In the shaded region there are no periodic orbits and there is one stable critical point. The Hopf bifurcation curve (indicated by supercritical Hopf) provides a partial boundary of this region. At the generalized Hopf (GH), which is at $\delta = \sqrt{7}$ for $\Omega = 2$, the Hopf bifurcation changes from super to subcritical. For $\delta < \sqrt{7}$ the Hopf bifurcation is subcritical and periodic orbits exist to the left of the Hopf curve. Also for $0 < \delta < \sqrt{3}$ there are regions where periodic orbits exist to the left of the Hopf curve. The curves A-GH, BGK, CFK, DEK, KHCusp, KMcusp, etc., are saddle node bifurcations of periodic orbits creating a stable and an unstable periodic orbit, existing to their right. The lozenge-like dashed curves are also saddle node bifurcations of periodic orbits, this time destroying a stable and an unstable periodic orbit. In region ABG(GH) and HM Cusp there is one stable critical point and a pair of periodic orbits with opposite stability. In region G(GH)HK there is one unstable critical point and one stable periodic orbit. In region BCFG and the region to the left of M Cusp there is one stable critical point and two pairs of periodic orbits with opposite stability. In region FGK there is one unstable critical point and two stable periodic orbits and one unstable periodic orbit. In region CDEF there is one stable critical point and three stable and three pairs of periodic orbits with opposite stability etc..

For the experiments described in [192, 138] we need to consider the case $\Omega = 2$ (that is to say, the mechanical frequency is twice the optical decay rate). However, for $\Omega > 2$, as in [281, 271], there is no qualitative change in the bifurcation diagram, although the generalized Hopf bifurcation ($\delta = -\sqrt{\frac{8\Omega^2+3}{5}}$) occurs for larger values of $|\delta|$. In Fig. 9.3 we see the corresponding situation for (a) $\Omega = 5$ and (b) $\Omega = 10$ and $\Gamma = 0.001$ with $\delta < 0$ (blue detuning). Multistable behaviour, due to the presence of limit cycles stacked above each other, remains an important feature.

We will now consider in more detail the periodic motion that occurs above the Hopf bifurcation using the method of *amplitude equations*. The method relies on defining a slow time which is proportional to the damping rate of the resonators, ($\tau = \Gamma t$), and on assuming that the forcing is on the order of the square root of the damping, so we define $\epsilon \equiv \sqrt{\Gamma}\,\bar{\epsilon}$. Then the cavity amplitude is naturally of the same order as the forcing and we can obtain equations for the slowly varying amplitude $A(\tau)$. Let

$$X = X_0 + \left(A(\tau)e^{i\bar{\Omega}t} + \text{c.c.}\right) = X_0 + 2\,|A(\tau)|\cos\left(\bar{\Omega}t + \theta\right), \qquad (9.11)$$

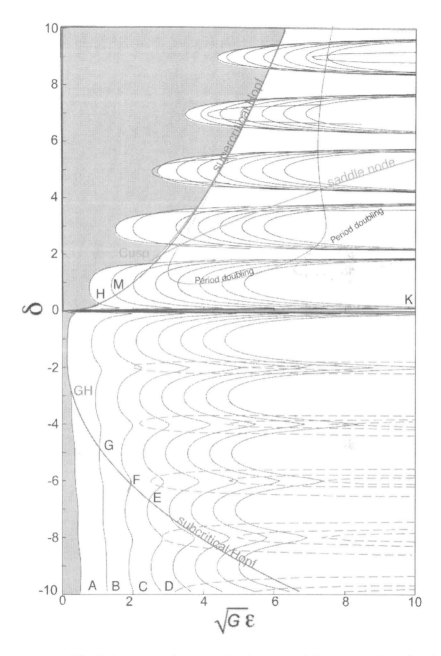

FIGURE 9.2 The bifurcation diagram for $\Omega = 2$ and $\Gamma = 0.001$ from [139]. See text for details. Reprinted with permission [139] Copyright 2012 by the American Physical Society.

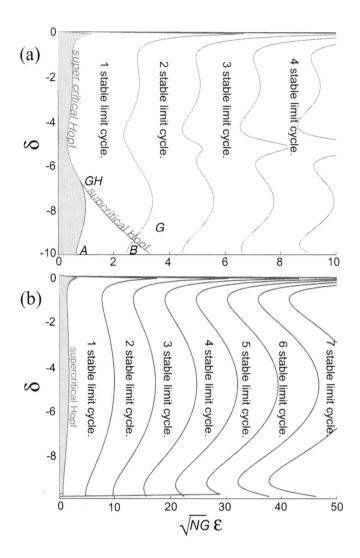

FIGURE 9.3 Bifurcation diagrams: (a) $\Omega = 5$ and (b) $\Omega = 10$ and $\Gamma = 0.001$ with $\delta < 0$ (blue detuning) showing the Hopf bifurcation and saddle node bifurcations of periodic orbits. The labelling in (a) is similar to 9.2. For instance, in the region ABG(GH) there is one stable critical point and a pair of periodic orbits with opposite stability. Reprinted with permission [139] Copyright 2012 by the American Physical Society.

where X_0 is the critical point of the system undergoing the Hopf bifurcation (see Eq. (9.7)). We substitute this expression into the field equation (the top equation in Eqs. (9.6)) to obtain

$$\frac{\mathrm{d}\alpha}{\mathrm{d}t} = -\left[1 + i\left(\delta + X_0 + 2\,|A|\cos\left(\bar{\Omega}t + \theta\right)\right)\right]\alpha - i\epsilon. \tag{9.12}$$

Using the ansatz

$$\alpha = \mathrm{e}^{i\psi(t)}\sum_m B_m \mathrm{e}^{im\Omega t}, \tag{9.13}$$

it follows that

$$\dot{\alpha} = i\dot{\psi}(t)\alpha + \mathrm{e}^{i\psi(t)}\sum_m im\Omega B_m \mathrm{e}^{im\Omega t}. \tag{9.14}$$

Substituting these expressions into Eq. (9.12), we find that

$$\dot{\psi}(t) = -\frac{2\,|A|}{\Omega}\cos\left(\Omega t + \theta\right) \quad \text{and} \quad B_m = -\frac{i^{m+1}\epsilon J_m\left(\frac{2|A|}{\Omega}\right)}{\bar{\kappa} + im\Omega}, \tag{9.15}$$

where $\bar{\kappa} \equiv 1 + i\left(\delta + X_0\right)$ and $J_m(x)$ are Bessel functions of the first kind. Substituting this back into the equation for X, Eq. (9.11), gives an amplitude equation for the oscillation in terms of sums of pairs of Bessel functions,

$$\frac{\mathrm{d}A}{\mathrm{d}\tau} = -A - \frac{iG\epsilon^2 \mathrm{e}^{i\theta}}{4}\sum_{m=-\infty}^{\infty}\frac{J_m\left(\frac{2|A|}{\Omega}\right)J_{m+1}\left(\frac{2|A|}{\Omega}\right)}{\left(\bar{\kappa} + i\left(m+1\right)\Omega\right)\left(\bar{\kappa}^* - im\Omega\right)}. \tag{9.16}$$

Identical mechanical resonators synchronise to oscillate with amplitude $A(\tau)$ given by this equation.

Since each term in the sum in Eq. (9.16) has $|A|$ as a factor, the amplitude equation may be rewritten as

$$\frac{\mathrm{d}A}{\mathrm{d}\tau} = -A + G\bar{\epsilon}^2 AF(|A|,\,\Omega,\,\delta), \tag{9.17}$$

where $F\left(r, \Omega, \delta\right)$ is a complex function.

Using circular polar coordinates, $A = r\mathrm{e}^{i\theta}$, the conditions for the Hopf bifurcation can be obtained by setting $\frac{\mathrm{d}r}{\mathrm{d}\tau} = 0$ in the linearised radial equation. We then find that θ does not appear in the equation for r, so the periodic orbits of the system are given by

$$F_r\left(r, \Omega, \delta\right) = \frac{1}{r}\sum_{m=0,\infty} a_{mr}\left(\bar{\delta}, \Omega\right) J_m\left(\frac{2r}{\Omega}\right) J_{m+1}\left(\frac{2r}{\Omega}\right) = \frac{1}{G\bar{\epsilon}^2}. \tag{9.18}$$

These curves are plotted in Fig. 9.4 for $\Omega = 2$ and $\Gamma = 0.01, 0.001, 0.0001$, and various values of δ. Corresponding to these oscillations, the cavity field amplitude oscillates with frequency $\bar{\Omega} + F_i\left(r, \Omega, \delta\right)$ and amplitude $\epsilon\sqrt{2r\,|F|}$:

$$\left(\text{Leading oscillatory term in } |\alpha|^2\right) = \tag{9.19}$$
$$2r\epsilon^2\,|F\left(r, \Omega, \delta\right)|\cos\left(\left(\bar{\Omega} + F_i\left(r, \Omega, \delta\right)\right)t + \zeta\right),$$

where ζ is a constant.

If each individual mechanical resonator has a different resonant frequency and/or damping, reduction to a single collective variable is no longer possible. However, the results of the previous section can be generalised to give a set of coupled amplitude equations. We will consider two groups of mechanical resonators with the case of approximately equal resonant frequencies $\bar{\Omega}_i = \Omega + \gamma \Delta\Omega_i$. In this case the amplitude equations are

$$\frac{\mathrm{d}A_1}{\mathrm{d}\tau} = -\left(1 + i\Delta\Omega_1\right)A_1 + GN_1\bar{\epsilon}^2\left(A_1 + A_2\right)F\left(|A_1 + A_2|\right)$$
$$\frac{\mathrm{d}A_2}{\mathrm{d}\tau} = -\left(1 + i\Delta\Omega_2\right)A_2 + GN_2\bar{\epsilon}^2\left(A_1 + A_2\right)F\left(|A_1 + A_2|\right),$$
(9.20)

where N_i is the number of resonators in each group. The radial motion is

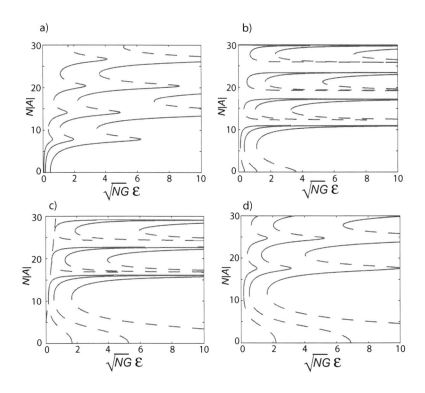

FIGURE 9.4 The amplitudes ($|A| = r$) of the periodic orbits of the system calculated from the amplitude equations as a function of $\sqrt{G}\,\epsilon$ for $\Omega = 2$ and $\Gamma = 0.01, 0.001, 0.0001$ and various values of δ. (a) $\delta = -2$, (b) $\delta = -5$, (c) $\delta = -9$, (d) $\delta = -10$. The unstable periodic orbits are given by dashed lines and the stable ones are given by solid lines.

unchanged if the $\Delta\Omega_i$ are equal and we still have

$$\frac{\mathrm{d}r}{\mathrm{d}\tau} = -r + G\bar{\epsilon}^2 N r F_r \left(Nr, \Omega, \delta\right), \qquad (9.21)$$

which implies that $N^2 r^2 = r_1^2 + r_2^2 + 2r_1 r_2 \cos(\theta_2 - \theta_1)$ is a constant of the motion. Substituting this into the equations for A_i results in a linear system whose symmetrical solution $N_1 A_2 = N_2 A_1$ is stable. The difference $N_1 A_2 - N_2 A_1$ obeys the equation of motion $\frac{\mathrm{d}(N_1 A_2 - N_2 A_1)}{\mathrm{d}t} = -(\gamma + i\Delta\omega)(N_1 A_2 - N_2 A_1)$. Therefore, as time progresses $|N_1 A_2 - N_2 A_1| \to 0$ and, after initial transients, the individual oscillators synchronise. If $\Delta\Omega_i$ are not equal,

$$\frac{\mathrm{d}r_1}{\mathrm{d}\tau} = -r_1 + \bar{\epsilon}^2 G N_1 \left(r_1 F_r \left(Nr\right) + r_2 \left(F_r \left(Nr\right) \cos\phi - F_i \left(Nr\right) \sin\phi\right)\right)$$

$$\frac{\mathrm{d}r_2}{\mathrm{d}\tau} = -r_2 + \bar{\epsilon}^2 G N_2 \left(r_2 F_r \left(Nr\right) + r_1 \left(F_r \left(Nr\right) \cos\phi + F_i \left(Nr\right) \sin\phi\right)\right)$$

$$\frac{\mathrm{d}\phi}{\mathrm{d}\tau} = \Delta\Omega_{21} + \bar{\epsilon}^2 G F_i \left(Nr\right) \left(\left(N_2 - N_1\right) + \left(\frac{N_2 r_1}{r_2} - \frac{N_1 r_2}{r_1}\right) \cos\phi\right) \qquad (9.22)$$

$$+ \bar{\epsilon}^2 G F_r \left(Nr\right) \left(\frac{N_2 r_1}{r_2} + \frac{N_1 r_2}{r_1}\right) \sin\phi,$$

where $F_{i,r}\left(Nr\right) = F_{i,r}\left(Nr, \omega, \delta\right)$, $Nr = |A_1 + A_2| = \sqrt{r_1^2 + r_2^2 + 2r_1 r_2 \cos\phi}$, and $\Delta\Omega_{21} = \Delta\Omega_2 - \Delta\Omega_1$ and where $N = N_1 + N_2$. For $N_1 = N_2$ we can assume that $\Delta\Omega_{21} > 0$ as the transformation ($\Delta\Omega_{21} \to -\Delta\Omega_{21}$, $\phi \to -\phi$) and ($r_1 \to r_2$ and vice versa) leaves the equations unchanged. The coupling, however, is strong rather than weak and the system cannot be reduced to a Kuramoto phase model. But it is, nevertheless, useful to compare the dynamics with those of similar phase and phase amplitude models, such as [225, 167, 179, 18, 268].

In the simplest two-oscillator phase model ($\dot{\phi} = \Delta\Omega_{21} - K \sin\phi$ with $\phi = \theta_2 - \theta_1$) there are two critical points, approximately an in-phase and an out-of-phase solution. One of the critical points is stable, for $|\Delta\Omega_{21}|$ sufficiently small ($|\Delta\Omega_{21}| < K$). Unsynchronised oscillation arises when the critical points are lost via a saddle node bifurcation ($|\Delta\Omega_{21}| > K$). The model, here, can also be discussed in terms of the stabilities of in-phase and out-of-phase solutions. However, the unsynchronised behaviour occurs as a transient state, resembling the transient rotational motion of a damped nonlinear pendulum started close to the separatrix of the undamped system. Similar motion has been noted for other systems with multistability [205]. Nonzero $\Delta\Omega_{21}$ breaks the symmetry and the in-phase critical points, which are still stable for $\Delta\Omega_{21}$ very small, only exist with $r_1 \neq r_2$. Their relative sizes as $|\Delta\Omega_{21}|$ is varied are shown in Fig. 9.5b. As $|\Delta\Omega_{21}|$ is increased, they loose stability via a Hopf bifurcation, Fig. 9.5a. This creates a stable periodic orbit which does not initially envelope the origin. However, in a bifurcation scenario, typical of large amplitude coupling [225], it grows rapidly to enclose the origin (In (r_1, r_2, ϕ) space this transition is a heteroclinic bifurcation with saddles at $r_{1 \text{ or } 2} = 0$, $\phi = \pm\frac{\pi}{2}$). Transient unsynchronised motion results for solutions started near the (unstable) in-phase solution, where solutions appear unbounded in-phase, but eventually become trapped by a stable out-of-phase solution. (In fact, the out-of-phase solutions are only unstable for $\Delta\Omega_{21}$ very small, where they exist at large amplitude, see Fig. 9.5c).

9.1.2 Quantum noise and synchronisation

There has been little work on the effect of quantum dynamics on a semiclassical limit cycle. Lörch et al. [183] consider the phonon statistics on a stable limit cycle, extended in [182]. Wang et al. [307] also consider second-order coherence of the mechanical phonons, in a model with two optical modes with a coupling modulated by the mechanical degree of freedom (in fact, the model is similar to that discussed in Section 6.4). Suchoi et al. [269] consider the case of two field modes, one optical and one microwave, coupled to a single mechanical mode and show how highly nonclassical superposition states might arise in the quantum regime of a limit cycle.

One might expect the role of quantum noise on the limit cycle describing synchronisation to be similar to the role of quantum noise in producing diffusion on a limit cycle that we discussed in Section 7.4. Here we will show how the question may be addressed using stochastic methods that were developed for dissipative nonlinear systems in quantum optics.

The quantum Hamiltonian corresponding to the semiclassical dynamics is given in the interaction picture for the field by

$$\hat{H} = \hbar\Delta a^\dagger a + \sum_{i=1}^{N} \hbar\Omega_i b_i^\dagger b_i + \hbar(\epsilon^* a + \epsilon a^\dagger) + \sum_{i=1}^{N} \hbar g_i a^\dagger a(b_i + b_i^\dagger). \quad (9.23)$$

We describe the dissipative dynamics with the master equation in the limit of weak damping, zero temperature bath, and in the rotating wave approximation for the system-environment couplings (see Section 5.1.3):

$$\frac{d\rho}{dt} = -\frac{i}{\hbar}[\hat{H}, \rho] + \kappa \mathcal{D}[a]\rho + \sum_{i=1}^{N} \Gamma_i \mathcal{D}[b_i] . \quad (9.24)$$

Corresponding to the classical description, we are interested in the collective quantities $\hat{X}$ and $\hat{Y}$ defined by replacing the classical canonical variables in Eq. (9.4a) with the canonical operators. We can define creation and annihilation operators for these collective mechanical modes,

$$\hat{B} = \sum_j b_j \quad (9.25)$$

$$\hat{B}^\dagger = \sum_j b_j \quad (9.26)$$

for which

$$[\hat{B}, \hat{B}^\dagger] = \frac{G}{2}\hat{I} \quad (9.27)$$

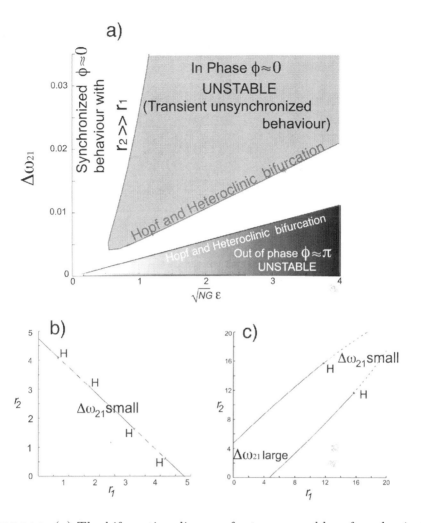

FIGURE 9.5 (a) The bifurcation diagram for two ensembles of mechanical resonators for $N_1 = N_2$, $\Gamma = 0.0001$, $\Omega = 2$, and $\delta = -1.5$. The in-phase solutions are stable outside the shaded regions. The out-of-phase solutions are stable outside the slice near the horizontal axis. They are singular at $\Delta\Omega_{21} = 0$ and unstable for $|\Delta\Omega_{21}|$ small, where they occur for very large values of r_i. In the unshaded regions both in-phase and out-of-phase solutions are stable, but have different basins of attraction. (b) shows the in-phase solution in (r_1, r_2) space as $\Delta\Omega_{21}$ is varied. $r_1 + r_2$ remains approximately constant. (c) shows the out of phase solution in (r_1, r_2) space as $\Delta\Omega_{21}$ is varied. Reprinted with permission from [139], Copyright, 2012, by the American Physical Society.

where G is defined in Eq. (9.4c).

From the master equation, we proceed by deriving a Fokker–Planck-like equation describing the evolution of the optomechanical system, which is the equation of motion of the Positive P-function, $P(\chi)$. Using methods in [305] we arrive at the Fokker–Planck-like equation

$$\frac{dP(\chi)}{dt} = -\sum_i \frac{\partial}{\partial \chi_i} \left[\mathbf{A}(\chi)\right]_i P(\chi) + \frac{1}{2}\sum_{ij} \frac{\partial}{\partial \chi_i} \frac{\partial}{\partial \chi_j} \left[\mathbf{B}(\chi)\mathbf{B}(\chi)^T\right]_{ij} P(\chi),$$

$$(9.28)$$

where

$$\chi = \begin{bmatrix} \alpha & \beta & \mu & \nu \end{bmatrix}^T. \qquad (9.29)$$

The correspondence between these variables and the corresponding quantum operators is $a \leftrightarrow \alpha$, $a^\dagger \leftrightarrow \beta$, $b \leftrightarrow \mu$, $b^\dagger \leftrightarrow \nu$. The drift term vector $\mathbf{A}(\chi)$ is

$$\mathbf{A}(\chi) = \begin{bmatrix} -(1+i\delta)\alpha - i\frac{1}{2}\alpha(\mu+\nu) - i\epsilon \\ -(1-i\delta)\beta + i\frac{1}{2}\beta(\mu+\nu) + i\epsilon \\ -(\Gamma+i\omega)\mu - i\frac{G}{2}\alpha\beta \\ -(\Gamma-i\omega)\nu + i\frac{G}{2}\alpha\beta \end{bmatrix} \qquad (9.30)$$

and the diffusion term matrix $\mathbf{B}(\chi)\mathbf{B}(\chi)^T$ is

$$\mathbf{B}(\chi)\mathbf{B}(\chi)^T = \begin{bmatrix} 0 & 0 & -i\frac{G}{2}\alpha & 0 \\ 0 & 0 & 0 & i\frac{G}{2}\beta \\ -i\frac{G}{2}\alpha & 0 & 0 & 0 \\ 0 & i\frac{G}{2}\beta & 0 & 0 \end{bmatrix}. \qquad (9.31)$$

Considering only the drift term of the Fokker–Planck equation, and make the mappings $\beta \mapsto \alpha^*$ and $u \mapsto v^*$, to reduce the phase space dimension by half onto the semiclassical phase space,[1] we obtain the semiclassical equations of motion.

$$\frac{d\alpha}{dt} = -i\Delta\alpha - i\epsilon - i\alpha X - \frac{\kappa}{2}\alpha$$

$$\frac{dX}{dt} = \Omega Y - \frac{\Gamma}{2}X \qquad (9.32)$$

$$\frac{dY}{dt} = -\Omega X - \frac{G}{2}|\alpha|^2 - \frac{\Gamma}{2}Y.$$

These equations match the equations of motion in Section 9.1.1 obtained by factoring moments $\langle \hat{a}\hat{X}\rangle = \langle \hat{a}\rangle\langle \hat{X}\rangle$ in the quantum equations of motion of operator expectations. With this factorisation, the mappings from expectation values of quantum operators to semiclassical dynamic variables is then $\langle \hat{a}\rangle \mapsto \alpha$, $\langle \hat{X}\rangle \mapsto X$, $\langle \hat{Y}\rangle \mapsto Y$.

We can now determine the effect of quantum noise on the synchronisation.

---

[1]The positive P-function has twice the dimensionality of the classical phase space.

To do this, we calculate the linearised spectrum as the driving strength is increased to approach the first Hopf bifurcation at the supercritical Hopf line for blue detuning ($\delta < 0$) in Fig. 9.2 and Fig. 9.3. We do this calculation for the case of a single group of optomechanical resonators following the procedure of Ref. [140]. For a single group, using the dimensionless notation where the coupling coefficients and time are rescaled by the cavity dissipation rate κ, the stochastic differential equations of motion corresponding to the Fokker–Planck equation (9.28) are found to be

$$\frac{d\chi}{dt} = \mathbf{A}(\chi) + \mathbf{B}(\chi)\mathbf{E}(t), \qquad (9.33)$$

where $\mathbf{E}(t)$ is the noise process. The principal matrix square root of the diffusion matrix $\mathbf{B}(\chi)\mathbf{B}(\chi)^T$ is

$$\mathbf{B}(\chi) = \mathbf{B}(\chi)^{\mathrm{T}} = \frac{\sqrt{G}}{2} \begin{bmatrix} \sqrt{\alpha} & 0 & -i\sqrt{\alpha} & 0 \\ 0 & \sqrt{\beta} & 0 & i\sqrt{\beta} \\ -i\sqrt{\alpha} & 0 & \sqrt{\alpha} & 0 \\ 0 & i\sqrt{\beta} & 0 & \sqrt{\beta} \end{bmatrix}. \qquad (9.34)$$

The diffusion matrix and its square root have determinants

$$\det\left\{\mathbf{B}(\chi)\mathbf{B}(\chi)^{\mathrm{T}}\right\} = \frac{1}{16}G^4\alpha^2\beta^2$$

$$\det\left\{\mathbf{B}(\chi)\right\} = \frac{1}{4}G^2\alpha\beta, \qquad (9.35)$$

and the two matrices are positive definite on the semiclassical manifold where $\beta = \alpha^*$, so do, indeed, describe a valid stochastic process. However, the off-diagonal terms with the factors of i in the matrix square root $\mathbf{B}(\chi)$ will take the solution off the semi-classical manifold. This is a direct manifestation of quantum correlations in the dynamics.

We can linearise these equations of motion about the semiclassical fixed points obtained in the classical analysis in Section 9.1. The linearised stochastic differential equations are

$$\frac{d\chi}{dt} \approx \mathbf{M}\left(\chi - \chi_0\right) + \mathbf{D}^{\frac{1}{2}}\mathbf{E}(t), \qquad (9.36)$$

where the Jacobian, $\mathbf{M}$, is

$$\begin{bmatrix} -(1+i\delta) - i\frac{1}{2}(\mu_0 + \mu_0^*) & 0 & -i\frac{1}{2}\alpha_0 & -i\frac{1}{2}\alpha_0 \\ 0 & -(1-i\delta) + i\frac{1}{2}(\mu_0 + \mu_0^*) & i\frac{1}{2}\alpha_0^* & i\frac{1}{2}\alpha_0^* \\ -i\frac{G}{2}\alpha_0^* & -i\frac{G}{2}\alpha_0 & -(\Gamma+i\omega) & 0 \\ i\frac{G}{2}\alpha_0^* & i\frac{G}{2}\alpha_0 & 0 & -(\Gamma-i\omega) \end{bmatrix},$$

$$(9.37)$$

and the diffusion matrix about the semiclassical fixed points $\mathbf{D} = \mathbf{B}(\boldsymbol{\chi}_0)\mathbf{B}(\boldsymbol{\chi}_0)^T$ is

$$
\mathbf{D} = \begin{bmatrix} 0 & 0 & -i\frac{G}{2}\alpha_0 & 0 \\ 0 & 0 & 0 & i\frac{G}{2}\alpha_0^* \\ -i\frac{G}{2}\alpha_0 & 0 & 0 & 0 \\ 0 & i\frac{G}{2}\alpha_0^* & 0 & 0 \end{bmatrix}, \tag{9.38}
$$

with α_0 given by Eq. (9.8). The linearised normally ordered moments at steady-state can be expressed in terms of these matrices [305],

$$
\begin{aligned}
\mathbf{S}(\omega) &= \frac{1}{2\pi} \int_{-\infty}^{\infty} e^{-i\omega\tau} \left\langle \boldsymbol{\chi}(t)\boldsymbol{\chi}(t+\tau)^T \right\rangle_{t\to\infty} d\tau \tag{9.39} \\
&= \frac{1}{2\pi} (i\omega\mathbb{1} - \mathbf{M})^{-1} \mathbf{D} \left(-i\omega\mathbb{1} - \mathbf{M}^T\right)^{-1},
\end{aligned}
$$

where $\mathbb{1}$ is the identity matrix. We plot the cavity field component of these quantum noise spectra in Fig. 9.6a. As the Hopf bifurcation is approached, the spectrum becomes more sharply peaked at two frequencies. The frequency corresponding to the Hopf bifurcation is the peak at the mechanical frequency Ω. The second, shorter but broader peak, is at the detuning Δ. If the laser driving the cavity is detuned to a sideband, these two peaks coincide. Beyond the supercritical Hopf bifurcation, the semiclassical fixed point is no longer stable, with the system entering the regime dominated by the first stable limit cycle, where oscillatory motion is present as analysed by semiclassical amplitude equations in Section 9.1. However, we can continue to linearise about this point, with the results shown in Fig. 9.6b. As can be seen, the two peaks begin to converge as the driving strength and coupling are increased.

The stationary noise power spectrum we have calculated here can be directly measured by homodyne detection. Below the Hopf bifurcation, the noise power spectrum of these fluctuations is peaked at a frequency associated with the decay of fluctuations back to the fixed point. The width of the peaks gives the time scale of this decay. Above the Hopf bifurcation, the fluctuations correspond to diffusion noise on the limit cycle. As we have neglected thermal fluctuations, this diffusion is entirely due to quantum noise.

Nonlinear dissipative dynamical systems can exhibit other quantum features, for example, dissipative switching between fixed points as we saw in Chapter 7. This is not equivalent to coherent quantum tunnelling in a double well, neither is it reducible to thermally activated switching, as it occurs in the presence of dissipation at zero temperature. It is possible that Fig. 9.6b is evidence for dissipative quantum switching between limit cycles. Such phenomena have been investigated in the case of driven damped parametric amplification in quantum optics [158], which has a similar linearised diffusion matrix to the model considered here.

An example of the strong effects possible when quantum tunnelling is

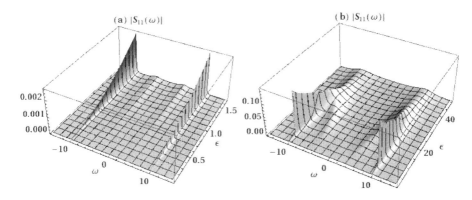

FIGURE 9.6 Linearised quantum noise spectrum of the optical cavity $S_{11}(\omega)$ (a) approaching the Hopf bifurcation and (b) beyond the Hopf bifurcation. The magnitude of the stationary normally ordered cavity spectrum $\frac{1}{2\pi} \int_{-\infty}^{\infty} e^{-i\omega\tau} \langle \alpha(t)\alpha(t+\tau)^T \rangle_{t\to\infty} \, d\tau$, the first diagonal element of $\mathbf{S}(\omega)$, is plotted at the frequency ω for varying driving amplitude ϵ. Here we have set $\Omega = 10$, $\delta = -4$, $\Gamma = 0.001$, $N = 1$, $G = 1$, for which the Hopf bifurcation occurs at a driving strength of $\epsilon_h \approx 1.76$. [Reprinted with permission [139], Copyright, 2012, by the American Physical Society.

achievable is give in Schmidt et al. [252]. They proposed a scheme in which an array of quantum optomechanical resonators could exhibit novel collective quantum behaviour including a Dirac-type tunable band structure.

9.2 IRREVERSIBLY COUPLED ARRAYS OF OPTOMECHANICAL SYSTEMS

We turn now to the case of irreversibly coupled optomechanical arrays as depicted in Fig. 9.1c, which we will refer to as type-2. In systems of type-2, many electromagnetic cavities, each containing a mechanical resonator, can be coupled optically in large arrays. This opens up the possibilities of networks of coupled cavities, where the network topology itself can be manipulated as desired, unlike the case of many oscillators in a single cavity which are necessarily coupled in an all-to-all fashion. Arrays of this type also open up the possibility of correlating the motion of two mechanical resonators separated by considerable distances. Cascaded cavity schemes of type-2 are a form of all-optical feedforward. Stannigel et al. have proposed a protocol for quantum information processing based on cascaded optomechanical cavities [267]. Shah et al. [257] demonstrated the coupling of two independent mechanical oscillators, situated far from each other (3.2 km), using a measurement-based feedforward. The light from the master cavity was detected by homodyne detection and the signal used to modulate the coherent input light to the slave optomechanical system.

The theory of cascaded quantum cavities was given in Section 6.2.2, using a master equation approach. For example, with three cascaded cavities we have that

$$\frac{d\rho}{dt} = -\frac{i}{\hbar}[\hat{H}, \rho] + \sum_{n=1}^{3} \kappa_n \mathcal{D}[a_n]\rho \tag{9.40}$$

$$+ \Gamma(\bar{n} + 1)\mathcal{D}[b]\rho + \Gamma\bar{n}\mathcal{D}[b^\dagger]\rho$$
$$+ \sqrt{\kappa_1\kappa_2}([a_1\rho, a_2^\dagger] + [a_2, \rho a_1^\dagger]) + \sqrt{\kappa_1\kappa_3}([a_1\rho, a_3^\dagger] + [a_3, \rho a_1^\dagger])$$
$$+ \sqrt{\kappa_2\kappa_3}([a_2\rho, a_3^\dagger] + [a_3, \rho a_2^\dagger]),$$

where κ_n is the decay rate for cavity-n and $\hat{H}$ contains a description of the coherent dynamics, including the optomechanical interaction between each cavity mode and the mechanical resonator to which it is coupled.

Ignoring the mechanical dynamics for the moment, the mean value equations of motion for the field amplitudes in each cavity are

$$\dot{\alpha}_1 = -\frac{\kappa}{2}\alpha_1$$

$$\dot{\alpha}_2 = -\kappa\alpha_1 - \frac{\kappa}{2}\alpha_2$$

$$\dot{\alpha}_3 = -\kappa\alpha_2 - \frac{\kappa}{2}\alpha_3,$$

where we have set all of the cavity decay rates to be equal ($\kappa_1 = \kappa_2 = \kappa_3 \equiv \kappa$). Note that this model already includes decay of the cavity amplitudes which arises simply from the field leaking out of one cavity and driving the next one downstream. In fact, it is determined by the same parameter as the coupling rate between the cavities. We can compare this with three coherently coupled cavities described by an interaction Hamiltonian of the form $\hat{H}_I = \chi \sum_k (a_k a_{k+1}^\dagger + a_{k+1} a_k^\dagger)$, which results in the equations of motion

$$\dot{\alpha}_1 = \frac{\chi}{2}\alpha_2 - \frac{\kappa}{2}\alpha_1$$

$$\dot{\alpha}_2 = -\frac{\chi}{2}\alpha_1 + \frac{\chi}{2}\alpha_3 - \frac{\kappa}{2}\alpha_2$$

$$\dot{\alpha}_3 = -\frac{\chi}{2}\alpha_2 - \frac{\kappa}{2}\alpha_3.$$

Exercise 9.2 *Check these two sets of equations using*

$$\dot{\alpha}_i = \text{tr}\left(a_i \frac{d\rho}{dt}\right). \tag{9.41}$$

One can easily check that, in the absence of dissipation ($\kappa = 0$), the dynamics conserves $\sum_k |\alpha_j|^2$. We can now see that, in the coherent case, the middle cavity is coupled to the cavity either side of it, while in the irreversible case the middle cavity is only coupled to the cavity upstream.

As an example of an irreversibly coupled cavity array, we will consider the model presented in Joshi et al. [147]. The objective, in this case, is to induce quantum correlations (entanglement) in the motion of the mechanical resonators. We will assume the optomechanical system is driven by strong coherent driving fields at the same carrier frequency on each cavity and tuned simultaneously to either the red, or the blue, sideband of each cavity. The linearised optomechanical interaction discussed in Section 2.7 may then be used. The dynamics is determined by the master equation of Eq. (9.40) with

$$\hat{H} = \sum_{k=1}^{3} \hbar \Delta_k a_k^\dagger a_k + \sum_{k=1}^{3} \hbar \Omega_k b_k^\dagger b_k + \hbar g_k (a_k^\dagger + a_k)(b_k^\dagger + b_k), \qquad (9.42)$$

where, as usual, the optical detuning of each cavity $\Delta_k \equiv \Omega_{c,k} - \Omega_L$, with $\Omega_{c,k}$ being the resonance frequency of cavity k and Ω_L the laser drive frequency, Ω_k is the frequency of the kth mechanical resonator, and a_k, b_k are the annihilation operators for photons in cavity-k and phonons in mechanical resonator-k, respectively. In writing this equation we have neglected the coherent field that cascades forward from each cavity. In practice, this may be filtered out in the circulator so that each cavity is driven by the same coherent amplitude (see [147]). We may simultaneously drive on the red or the blue sideband, by choosing $\Delta_k = \Omega_k$ or $\Delta_k = -\Omega_k$, respectively.

In Vitali et al. [299], it is shown that intracavity photon-phonon entanglement can arise in a single optomechanical unit (see Section 4.4). Can one induce entanglement between the optical and mechanical resonators in a network of cascaded optomechanical cavities? In the linearised approximation, all states would be Gaussian as we start from the vacuum, and the equations of motion are linear. We can then use the logarithmic negativity measure for Gaussian states formulated by Vidal [297] to quantify the entanglement (see Section 4.4.6). To determine the matrix elements of each of the 2 × 2 matrices $\mathbf{A}$, $\mathbf{B}$, and $\mathbf{C}$ that, combined, form the covariance matrix for the state, we calculate the second-order moments from the master equation of the composite system. Inserting into these matrices the relevant second-order moments, we can calculate the entanglement between any two modes of the system. We will restrict the discussion to the case of red sideband driving and zero temperature.

In Fig. 9.7a, we show the pairwise entanglement in the steady state between cavity modes and mechanical modes for a system consisting of three cascaded optomechanical cavities. We assume red sideband driving and zero temperature, but do not make the rotating wave approximation on the linearised interaction. In Fig. 9.7b, we plot the entanglement between all possible pairs of the optical and mechanical modes of different cavities, i.e., intercavity photon-phonon entanglement in the composite system. All six pairs of photon-phonon modes of the system are entangled with each other.

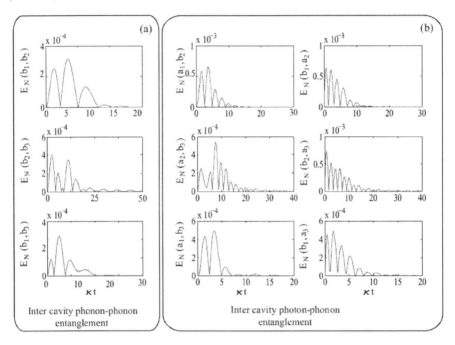

FIGURE 9.7 Entanglement between mechanical and optical elements in three cascaded optomechanical systems at zero temperature. The temporal evolution of the entanglement, quantified by the logarithmic negativity E_N (termed L in Chapter 4) between all possible intercavity modes of the optomechanical array: (a) phonon-phonon and (b) photon-phonon, where we have chosen $\kappa_{1,2,3} = \kappa; \Gamma_{1,2,3} = 0.01\kappa, \Delta_{1,2,3} = \Omega_{1,2,3} = 400\kappa, g_{1,2,3} = 0.5\kappa$. Reproduced with permission [8], Copyright, 2012 by the American Physical Society.

A, perhaps, more interesting case corresponds to driving some cavities on the red sideband, and others on the blue sideband. An experiment that is somewhat along these lines has been reported in [269], with one optical mode and one microwave mode simultaneously coupled to a single mechanical resonator. As we saw in Section 4.4.1, blue sideband driving generates quantum correlations between the optical and mechanical degrees of freedom. In the simple case of a system consisting of two optomechanical cavities, the objective is to use the red sideband driving on the second cavity to distribute entanglement across two spatially separated mechanical resonators using a cascaded interaction of the optical fields. In Fig. 9.8, we plot the steady-state entanglement between the two mechanical resonators as a function of different

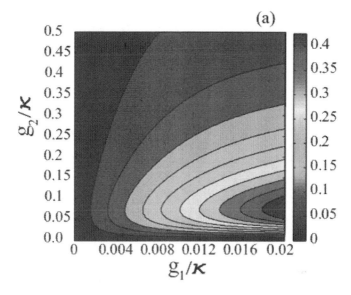

FIGURE 9.8 Irreversible coupling for a two element optomechanical array, with the source cavity blue detuned and a single receiver cavity red detuned. (a) The steady-state entanglement (quantified by the logarithmic negativity) between the intercavity phonons of two optomechanical units (b_1, b_2) versus g_1/κ and g_2/κ, with $\kappa_{1,2} = \kappa, \Gamma_{1,2} = 0.01\kappa$, and zero temperature. Reproduced with permission [8], Copyright, 2012, by the American Physical Society.

values of the optomechanical coupling strengths g_1 and g_2. The entanglement between the two mechanical resonators becomes larger in the weak-coupling regime of the receiver optomechanical system.

CHAPTER 10

Gravitational quantum physics and optomechanics

CONTENTS

10.1   What is gravitational decoherence?   300
10.2   Optomechanical tests of gravitational decoherence   308
        10.2.1 Single-photon optomechanics   308
        10.2.2 Tests of gravitational decoherence using entangled
                photons   311
10.3   Tests of nonstandard gravitational effects using geometric
        phase   315

Gravitational physics and optomechanics have had a long association through the development of laser interferometric gravitational observatories. In fact, these experiments are one of the historical drivers of the field of quantum optomechanics. The observation and routine study of gravitational radiation would initiate a revolution in our understanding of classical general relativity and cosmology. However, optomechanics has the potential to provide an even more profound impact on gravitational physics.

Until now most experiments in quantum physics have not needed to take gravitational interactions into account. This is because gravity is very weak compared to the electromagnetic force. There are, of course, some important exceptions, e.g. neutron and atom interferometry [223, 77]. In those cases, the gravitational field of the earth acts as a classical control of other quantum objects; we do not concern ourselves with the gravitational field of the atoms and neutrons themselves. In contrast, optomechanics, given its potential for controlling the quantum states of quite massive objects, provides a path to studying quantum systems in which the gravitational interactions between the objects themselves must necessarily be included [12]. This forces us to confront a major problem: what is the gravitational field of a massive object that is in a superposition of two distinct classical states of motion?

This problem has long been recognised [124, 85]. If we had a quantum theory of gravity, the solution would be given. As in quantum electrodynamics, a quantum theory of the gravitational field would provide a consistent description of both gravity and its sources and also provide the conditions under which gravity could be treated classically, providing a semiclassical approximation. That such a resolution has not been easy to find is due to the special character of gravity: it is described as space-time curvature and in principle cannot be screened, unlike the electromagnetic field.

The apparent difficult in finding a quantum theory of gravity has led some to suggest that perhaps it is unnecessary and we can live with a hybrid quantum-classical theory [57]. However there are well-known inconsistencies when one attempts to combine classical and quantum dynamics in the same framework [28, 222]. Diosi [88] has shown that apparent inconsistencies may be avoided if one includes a minimum amount of noise in both the classical and quantum dynamics. We will discuss another path to Diosi's result below, when we consider the possibility that gravitational interactions between massive objects are mediated by a classical measurement channel, suitably defined.

Penrose [220] has argued that the special nature of the gravitational field necessarily leads to a kind of spontaneous collapse of the quantum state, which he describes using an analogy with the spontaneous change for an excited electronic state to a ground state due to spontaneous emission. While Penrose does not use the word decoherence, it is possible to elaborate the analogy with spontaneous emission to make the connection. We will thus continue to refer to Penrose's argument as making a case for gravitational decoherence.

10.1   WHAT IS GRAVITATIONAL DECOHERENCE?

What is the gravitational field of a single massive object prepared in a superposition state of two well-localised Gaussian wavefunctions? The situation is depicted in Fig. 10.1. Gravitational fields are determined by measuring the trajectories of test particles. In Fig. 10.1 we indicate what one would expect for the case of a single massive object of mass M when test particles are directed, with a well-defined momentum, along the perpendicular line that bisects the line between the two possible positions and subsequently detected in the far field. If the large mass were on the left of the origin, the test particle would be found to be deflected to the left. If the large mass were on the right, a test particle would be found to be deflected to the right. The test particle thus performs a measurement of the position of the large mass, from which it follows that, if the large mass were in an equal superposition of two Gaussian wave functions localised on the left and the right, the test particle will be deflected either to the left or the right with equal probability from one

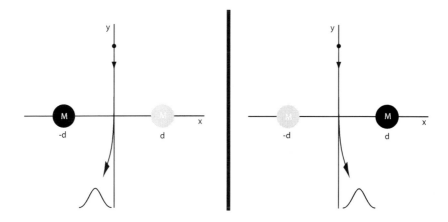

FIGURE 10.1 A single massive object of mass M is prepared in a superposition of two well localised Gaussian wave functions. Test particles (mass $m \ll M$) are used to determine the gravitational field.

trial to the next. It would appear that the gravitational field has an indefinite character.

One way to answer the question of what is the gravitational field for the situation depicted in Fig. 10.1 reverts to an old interoperation of the wavefunction first proposed by Schrödinger before Born's probability interpretation. In this *semiclassical* approach, the modulus squared of the wave function is interpreted as a kind of classical mass density functional $\rho(r) = M|\psi(r)|^2$ for which the gravitational potential field can be computed using

$$\nabla^2 \Phi(r) = -4\pi G \rho(r), \tag{10.1}$$

where G is the Newton gravitational constant. This is a kind of mean field theory. In this example, for sufficiently localised states, the mass distribution has two equal peaks, $\rho(r) \approx \rho_l(r) + \rho_r(r)$, made up of the two components of the superposition state. It should be intuitively clear that this will lead to a very different set of test particle trajectories: every test particle will pass undeflected directly between the two possible positions for the large mass. While a semiclassical approach of this kind is often valid, it is certainly not valid for the kind of macroscopic cat-like states we are considering in this example. Even classically, a double-peaked mass distribution is gravitationally unstable. In the real problem there is only one particle not two.

It should be noted that the problem highlighted here is not avoided by moving to a general relativistic formulation. The same problem of how to consistently combine quantum and classical dynamics arises in the Einstein field equations when one tries to replace the classical stress energy tensor density functions by the relevant operators derived from whatever quantum fields are under study: the left-hand side is a set of classical nonlinear differential equations for the metric field, while the right-hand side is an operator density. There is a long history of dealing with this by the same mean field approach that we just discussed [141]. The stress energy operator density is replaced by its average over some suitable semiclassical quantum state for the fields. This will fail for precisely the kind of highly nonclassical states that we are most interested in here.

Penrose's argument for gravitational decoherence can be best stated in the context of the experiment in Fig. 10.1. We need to focus on the mechanism that brings about the superposition state of the large mass in the first place. Let us assume that this is an entirely reversible process that begins at time t_i with a single large mass localised on the left (or right), creates the cat-like superposition state which persists for some finite time T and then restores the initial state at time t_f. This is depicted in Fig. 10.2. At the end of the process the position of the particle is probed using the deflection of test particles. If

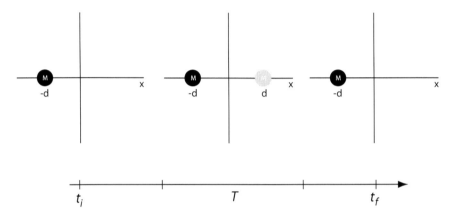

FIGURE 10.2 A single massive object of mass M is prepared in a Gaussian state localised on the left and then reversibly transformed to a superposition of two localised Gaussian wavefunctions on opposite sides of the origin and back again to the initial state. This is a kind of gravitational Mach–Zehnder interferometer.

this process is indeed unitary, the final probe by test particles will reveal that the large mass is located back at the initial position t_i *in every trial*. On the other hand, if Penrose's argument is correct, there will be a finite probability that the massive particle will be found on the opposite side at the end of the process to that in which it was prepared at time t_i. The reason for this is that, during the time T for which the gravitational field is indeterminate, the state of the massive particle will collapse, in Penrose's view, into one or the other of the two possible definite positions, either on the left or on the right. Penrose gives an estimate of the probability per unit time (rate) for this collapse to occur as

$$R_{grav} = \frac{\Delta_p}{\hbar}, \tag{10.2}$$

where

$$\Delta_p = -G \int d^3x \int d^3y \; \frac{(\rho_l(x) - \rho_r(x))(\rho_l(y) - \rho_r(y))}{|x - y|}, \tag{10.3}$$

which Penrose describes as the "the gravitational self-energy of the difference between the mass distributions of each of the two lump locations."

The concept of a spontaneous collapse is somewhat unclear in Penrose's argument. This is is not central to the argument, as we can see by formulating the problem as a measurement-decoherence problem. To see this, let us suppose that, during the ostensibly unitary process that creates the superposition, some process is weakly and continuously monitoring the position of the massive particle. For example, we could imagine a stream of very weak test particles as in Fig. 10.1. If this measurement can distinguish the two possible mass distributions that are involved, it must necessarily cause a gradual stochastic localisation of the conditional state of the massive particle to the left or the right consistent with the resulting measurement record that is doing the conditioning. If the measurement record is not known, we only need to compute the unconditional state of the massive particle. In that case we see that, for a good measurement, the coherence between the two components of the wavefunction rapidly decays. Repeated trials of the experiment will show that the massive particle is sometimes found on the opposite side to the initial localisation, just as Penrose has suggested and the relative fraction is determined by a fundamental rate of decoherence.

Of course, it is no surprise that a continually monitored system will show decoherence. It is well known that, to create cat-like states, one needs to keep all sources of decoherence to a minimum. For example, in the ion trap experiment of Monroe et al. [203] a single ion was placed in a superposition of two distinct motional states by reducing the effects of unknown stray electromagnetic fields to a minimum (a kind of electromagnetic shielding). It is at this point that we notice something quite special about the gravitational field: in principle there is always an open measurement channel. In Einstein's formulation, the gravitational field is manifest as space-time curvature which can be monitored in any way whatsoever: rulers, clocks, even test particles. We cannot shield the gravitational field.

If we had a quantum theory of gravity, presumably we could take this open measurement channel into account just as we do in the case of the electromagnetic field. The particle position becomes entangled with the quantum gravitational field and, if the field is not monitored, it will lead to decoherence of the source of the gravitational field. Clearly, this kind of decoherence must constrain possible theories of quantum gravity. It is for this reason that experimentally searching for evidence of gravitational decoherence is so important: it will give us hints on a future quantum theory of gravity.

An alternative, but equivalent, theory of gravitational decoherence has been proposed by Diosi [87, 88]. In this approach the role of weak continuous measurement plays a central role. The decoherence rate in Diosi's approach depends on a different functional integral of the mass density to that of Penrose. Diosi uses

$$\Delta_d = -G \int d^3x \int d^3y \; \frac{\rho_l(x)\rho_r(y)}{|x-y|} \tag{10.4}$$

and proceeds in a similar manner to the weak continuous measurement protocol discussed in Chapter 5. In the case of a spherical ball (as in the example discussed above) one can average over the internal coordinates to obtain an effective decoherence rate for the centre of mass displacement, $\hat{x}$, leading to a Markov master equation for the state of the centre of mass of the form

$$\frac{d\rho}{dt} = \mathcal{L}\rho - \Gamma[\hat{q}, [\hat{q}, \rho]], \tag{10.5}$$

where

$$\Gamma = \left(\frac{4\pi\kappa G\rho_m}{3\hbar}\right) M, \tag{10.6}$$

$\mathcal{L}$ is the Liouville operator for the rest of the particle dynamics, ρ_m is the density of the material, and κ is a dimensionless constant of the order of unity.

The double commutator terms in the second term in Eq. (10.5) leads to a decay of the off-diagonal matrix elements in the position basis,

$$\frac{d\langle y|\rho|x\rangle}{dt} = (\dots) - \Gamma(y-x)^2\langle y|\rho|x\rangle \tag{10.7}$$

and momentum diffusion (heating)

$$\frac{d\langle \hat{p}^2\rangle}{dt} = (\dots) + 2\Gamma\hbar^2. \tag{10.8}$$

Diosi's equation can also be obtained under the assumption that gravity is equivalent to a classical channel [149, 148]. An example of a classical channel between two point masses m_1, m_2 can be given by postulating simultaneous weak continuous measurement of the position of each mass. The two measurement records are then fed forward to apply exactly the right kind of force on the opposite mass to simulate the gravitational interaction. If the noise is minimised, the decoherence rate obtained is of the same order of magnitude as estimated by Diosi.

As an illustration of the classical channel model, we consider the situation depicted in Fig.10.3. Two masses harmonically bound interact purely gravitationally. As gravity is weak, we can expand the Newtonian force to second order in the displacements to give an effective Hamiltonian. The term linear in the displacement represents a constant force between the masses and simply modifies the equilibrium position of the masses and can be absorbed into the

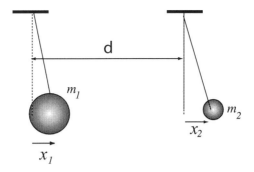

FIGURE 10.3   A gravitationally coupled system of two harmonic oscillators comprising two suspended masses m_1, m_2.

definition of the displacement coordinates. The quadratic terms proportional to $\hat{q}_k^2$ can be incorporated into the definition of the harmonic frequency of each mass. The total mechanical Hamiltonian is then given by

$$\hat{H}_{qm} = \hat{H}_0 + K\hat{q}_1\hat{q}_2, \tag{10.9}$$

where

$$\hat{H}_0 = \sum_{k=1}^{2} \frac{\hat{p}_k^2}{2m_k} + \frac{m_k\Omega_k^2}{2}\hat{q}_k^2. \tag{10.10}$$

Here, Ω_k is the gravitational interaction-modified mechanical resonance frequency of oscillator k:

$$\Omega_k^2 = \omega_k^2 - K/m_k, \tag{10.11}$$

where

$$K = \frac{2Gm_1m_2}{d^3}, \tag{10.12}$$

and ω_k is the bare frequency of each oscillator in the absence of the gravitational interaction. In a standard quantum description, the usual canonical commutation relations are $[\hat{x}_k, \hat{p}_j] = i\hbar\delta_{kj}$.

Exercise 10.1 *(a) Transform this system into the normal mode basis $\hat{q}_\pm \equiv (\hat{q}_1 \pm \hat{q}_2)/\sqrt{2}$, where $\hat{q}_+$ is the* centre-of-mass *mode and $\hat{q}_-$ is the* breathing *mode.*

(b) Show that the eigenfrequencies of the centre-of-mass and breathing modes are given, respectively, by

$$\omega_\pm^2 = \frac{\Omega_1^2 + \Omega_2^2}{2} \pm \frac{1}{2}\left[\left(\Omega_1^2 - \Omega_2^2\right)^2 + \frac{4K^2}{m_1m_2}\right]^{1/2}. \tag{10.13}$$

Let us now assume, for simplicity, that $m_1 = m_2 = m$ and $\Omega_1 = \Omega_2 = \Omega$. The normal mode frequencies then become

$$\omega_+ = \Omega \; ; \qquad \omega_- = \Omega \left[1 - \frac{2K}{m\Omega^2} \right]^{1/2} . \tag{10.14}$$

In the realistic limit where the gravitational coupling is sufficiently weak that $2K/m\Omega^2 \ll 1$, the difference in frequency between the two normal modes (the normal mode splitting) can be approximated as

$$\Delta \equiv \omega_+ - \omega_- \approx \frac{K}{m\Omega} . \tag{10.15}$$

In the classical channel model we introduce an effective weak continuous measurement of the displacement of each mass and feedforward control process. The measurement record is a continuous conditional stochastic process which is used to modulate a linear force acting on each mass in such a way that, when averaged over measurement results, the resulting systematic dynamics is equivalent to that given by the effective Hamiltonian in Eq. (10.9). However, noise must enter through two mechanisms. Firstly, the measurement itself leads to back-action noise, and, secondly, the noisy measurement record acts as a noisy force on each oscillator. These two sources of noise are complementary: when the back-action noise is large the control noise is small. We can set the measurement rate to minimise the overall effect of noise. We can then average over the measurement records[318] to obtain a master equation that depends only on the gravitational interaction strength K as[149],

$$\frac{d\rho}{dt} = -\frac{i}{\hbar}[\hat{H}_0, \rho] - \frac{i}{\hbar}K[\hat{q}_1\hat{q}_2, \rho] - \frac{K}{2\hbar} \sum_{k=1}^{2}[\hat{q}_k, [\hat{q}_k, \rho]] . \tag{10.16}$$

This is consistent with Diosi's model, which gives the same decoherence rate as obtained here under similar approximations.

For simplicity we will assume that the two mechanical resonators have the same mass ($m_1 = m_2 = m$) and frequency ($\omega_1 = \omega_2 = \omega$). The last term in Eq. (10.16) is responsible for two complementary effects: it drives a diffusion process in the momentum of each of the oscillators at the rate $\hbar K$, which we will call the gravitational heating rate

$$D_{grav} = \hbar K . \tag{10.17}$$

The momentum diffusion leads to heating of the mechanical resonators. Defining the rate of change of the phonon number as the rate of change of the average mechanical energy divided by $\hbar\Omega$, the heating rate is given by

$$R_{grav} = \frac{K}{2m\Omega} . \tag{10.18}$$

The decay of off-diagonal coherence in the position basis of each mechanical resonator is given by

$$\frac{d\langle q_k'|\rho|q_k\rangle}{dt} = (\ldots) - \frac{K}{2\hbar}(q_k' - q_k)^2 . \tag{10.19}$$

To obtain an estimate of the size of this effect, we need to scale the displacement and momentum to give dimensionless quantities. As we will discuss in the next section, most experimental proposals for testing gravitational decoherence in optomechanics confine the particle in a harmonic potential. The zero-point position uncertainty $x_{zp} = \sqrt{\frac{\hbar}{2m\Omega}}$ then provides a convenient length scale. In terms of this scale, we can write the gravitational decoherence rate as

$$\Lambda_{grav} = \frac{K}{2\hbar} x_{zp}^2 = \frac{K}{4m\Omega}. \tag{10.20}$$

Thus the gravitational decoherence rate for position is one half the gravitational heating rate.

These rates can equivalently be expressed in terms of the normal mode splitting when the gravitational interaction is weak, Eq. (10.15).

$$R_{grav} = \frac{\Delta}{2} \tag{10.21}$$

$$\Lambda_{grav} = \frac{\Delta}{4}. \tag{10.22}$$

We thus see that the key parameter responsible for gravitational decoherence is of the order of the normal mode splitting between the two mechanical resonators due to their gravitational coupling. This has significant consequences for observation. Most materials that might be used in an experimental test have density $\rho_m \sim 10^3$ kg m^{-3}. As the gravitational constant G is of the order of 10^{-11} in SI units, it is desirable to use a trap frequency as low as possible, for example, in the case of the suspended (pendulum) end mirrors in laser interferometric gravitational wave observatories, $\Omega \sim 1 - 100$ s^{-1}. Thus, an optimistic estimate of the gravitational decoherence/heating rate is $\Lambda_{grav} \sim 10^{-7}$ s^{-1}.

In reality, the heating due to gravitational decoherence will be masked by real thermal fluctuations. If thermal effects are included, the master equation of the two coupled oscillators in Eq. (10.16) includes the term [318],

$$\left.\frac{d\rho}{dt}\right|_{diss} = \sum_{j=1}^{2} -\frac{i\Gamma_j}{\hbar}[\hat{q}_j, \{\hat{p}_j, \rho\}] - \frac{2\Gamma_j k_B T m_j}{\hbar^2}[\hat{q}_j, [\hat{q}_j, \rho]], \tag{10.23}$$

where Γ_j is the dissipation rate for each of the mechanical resonators assumed to be interacting with a common thermal environment at temperature T.

Exercise 10.2 *Consider the case of a single harmonic oscillator of mass m, decay rate Γ, and frequency ω. Show that the thermal terms added to the master equation (Eq. (10.23)) lead to the average momentum decaying at the rate 2Γ, and also introduce a heating term that increases the momentum variance increases at the constant rate $D = 4\Gamma k_B T m$. Thus, show that the average vibrational occupancy increases at the constant rate $2\Lambda_{thermal}$ with*

$$\Lambda_{thermal} = \frac{k_B T}{\hbar Q}, \tag{10.24}$$

where $Q \equiv \Omega/\Gamma$ is the quality factor of the resonator.

We thus need to compare the heating rate due to gravitational decoherence to the thermal heating rate. For the two terms to be of the same order, we would need a thermal environment of around a nanokelvin and a quality factor of about one billion, a challenging experiment but within the reach of optomechanical technology. It is important to note, also, that the heating rate considered here is only a lower bound. It is possible that gravity is a nonoptimal classical control channel and therefore introduces a greater amount of heating.

From a phenomenological perspective, the effect of gravitational decoherence is analogous to a Browning heating effect. To see this, we note that the average vibrational quantum number increases diffusively

$$\left.\frac{d\langle b^\dagger b\rangle}{dt}\right|_{grav} = 2\Lambda_{grav}. \tag{10.25}$$

Indeed, one could simulate this effect by adding a stochastic driving force to the mechanical element via the stochastic Hamiltonian

$$\hat{H}_s = \frac{dP}{dt}(b + b^\dagger), \tag{10.26}$$

where $P(t)$ satisfies an Ito stochastic differential equation,

$$dP(t) = \sqrt{2\Lambda_{grav}}\, dW(t) \tag{10.27}$$

with $dW(t)$, as usual, being the Weiner increment (see Section 5.1.1). Averaging over all histories of the stochastic driving force gives the final term in Eq. (10.5).

10.2 OPTOMECHANICAL TESTS OF GRAVITATIONAL DECO-HERENCE

10.2.1 Single-photon optomechanics

One of the first proposals to search for gravitational decoherence was given by Bouwmeester [193, 221] based on the optomechanical interaction between a harmonically bound mirror and a single photon. The model is very similar to that described in Section 6.5. Consider the Mach–Zender interferometer in Fig. 10.3. Each arm contains an optomechanical cavity and these need to be as similar as possible to make the arms indistinguishable. A single photon injected at one input port means that the input to each optomechanical cavity is a superposition state of zero and one photon. As the optomechanical interaction is conditioned on the photon number, this leads to a superposition of different mechanical states for the mechanical elements. As shown by Akram et al. [6], the conditional state of the mechanics, conditioned by late detection of a photon at time t, is very close to a cat state of the form

$$|\psi(t)\rangle = \mathcal{N}(|\beta(t)\rangle_1|0\rangle_2 + |0\rangle_1|\beta(t)\rangle_2) \tag{10.28}$$

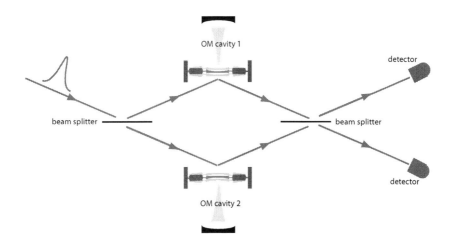

FIGURE 10.4 An optomechanical scheme for creating a superposition state of mechanical motion. using optomechanical cavities in each arm of a Mach–Zehnder interferometer. The two harmonically bound mirrors interact with each other gravitationally.

with $\beta = (e^{-i\Omega t} - 1)g_0/\Omega$ and $\mathcal{N}$ is a normalisation constant.

One of the difficulties in analysing the scheme in Fig. 10.3 is the stochastic nature of the detection times (see Section 6.5). This means that the values of $\beta(t)$ fluctuate due to the fluctuations in the time a photon spends in each optomechanical cavity. To avoid these fluctuations here, we will turn to a different model based on an intracavity single photon Raman source of the kind described in Nisbet-Jones et al. [211]. In this scheme (see Fig. 10.5) a control pulse can quickly and efficiently prepare a cavity mode in a single photon state by driving a Raman transition between two hyperfine levels that we label as $|g\rangle, |e\rangle$. Furthermore, by using a time-reversed control pulse, the single photon can be mapped back into the source and read out: an effective switchable single-photon detector. If the control pulse area is carefully controlled, we can prepare the entangled state of source and photon as $|g\rangle|0\rangle_a + |e\rangle|1\rangle_a$.

We will begin by neglecting photon and mechanical decay as well as gravitational decoherence. The initial total state, beginning with the Raman excitation pulse, is

$$|\psi(0)\rangle = \frac{1}{\sqrt{2}}(|g\rangle|0\rangle_a + |e\rangle|1\rangle_a)|0\rangle_b, \qquad (10.29)$$

where we have assumed that the mechanics is initially prepared in its ground state. The state at time $t > 0$ is determined by the optomechanical interaction alone, governed by the Hamiltonian in Eq. (2.18), where here we choose an interaction picture rotating at the cavity frequency so that the cavity detuning

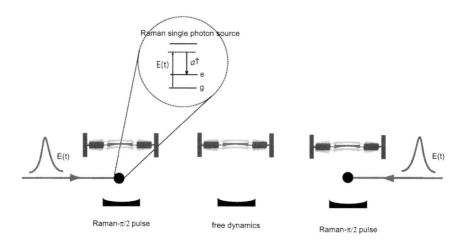

Raman-π/2 pulse   free dynamics

Raman-π/2 pulse

FIGURE 10.5 An optomechanical scheme for creating a superposition state of mechanical motion using a Raman single-photon excitation scheme. There are three steps: 1. Raman $\pi/2$ pulse to create and entangled state of atom and single photon, (2) free dynamics of the optomechanical system, (3) second Raman $\pi/2$ pulse followed by readout of the atomic state $|g\rangle$.

$\Delta = 0$. The state then evolves as

$$|\psi(t)\rangle = \frac{1}{\sqrt{2}}(|g\rangle|0\rangle_a|0\rangle_b + |e\rangle|1\rangle_a|\beta(t)\rangle_b \qquad (10.30)$$

with $\beta(t)$ the same as in Eq. (10.28).

Exercise 10.3 *Using the optomechanical Hamiltonian,* $H = \hbar\Omega b^\dagger b + \hbar g_0 a^\dagger a(b + b^\dagger)$, *derive Eq. (10.30) with* $\beta \equiv (e^{-i\Omega t} - 1)g_0/\Omega$.

We now consider a final application of the Raman pulse, which leaves the final state as

$$|\psi_f\rangle = \frac{1}{2}(|g\rangle|0\rangle_a(|0\rangle_b - |\beta(t)\rangle_b) + |e\rangle|1\rangle_a(|0\rangle_b - |\beta(t)\rangle_b). \qquad (10.31)$$

If, at this point, a projective readout of the atomic state is made, the probability to find the system in the ground state is

$$p_g(t) = \frac{1}{2}\left[1 - e^{-|\beta(t)|^2/2}\right]. \qquad (10.32)$$

If the dynamics is precisely unitary, as this assumes, then after one period the system can never be found in the ground state. On the other hand, if there is gravitational decoherence of the mechanical system over one period, there is a nonzero probability of finding the system in the ground state as we now show. This is the same kind of scheme that we discussed in relation to Penrose's argument in Section 10.1.

To determine the effect of gravitational decoherence, we use the equivalence noted above between the effect of gravitational decoherence and a fluctuating classical force acting on the mechanical system. This requires the replacement $g_0 \to g_0 + \frac{dP}{dt}$ with dP obeying the stochastic differential equation in Eq. (10.27). This leads to the addition of a stochastic component to the amplitude of the mechanical displacement $\beta(t) \to \beta(t) + \delta\beta(t)$. Including this in Eq. (10.32), averaging over the fluctuations, and evaluating after one period $T = 2\pi/\Omega$ gives to lowest order in the period

$$p_g(T) = \Lambda_{grav}T \ . \tag{10.33}$$

Given that the period is of the order of unity in our example, this means a very small probability for the system to be left in the ground state due to gravitational decoherence. Not surprisingly, this would be a challenging, but not impossible, experiment. Thermal effects would mimic the effect of gravitational decoherence, as they also look like a fluctuating classical force on the mechanical element. Of course, the dependence on the parameters is quite different for thermal effects, so they might be distinguishable in a sophisticated experiment.

10.2.2   Tests of gravitational decoherence using entangled photons

It is now routine to produce pairs of entangled photons. As entanglement is often lost rapidly to decoherence, perhaps these sources might be a more sensitive way to probe gravitational decoherence. A scheme to do this is shown in Fig. 10.6. A continuously driven spontaneous down conversion source with two distinguishable output modes (the signal and the idler) is used to excite two optomechancial systems, one on each of the output modes. The light is then directed onto photodetectors and the coincidence current is analysed. The essential idea is that gravitational heating will cause phase diffusion of the optical field emitted from the cavity and this will effect the indistinguishability of the two paths leading to coincidence detection (see [337] for a proposal of a purely optical implementation of this idea).

The state of the input field to each cavity for this source may be taken to be [288]

$$|\psi\rangle = \int_0^\infty d\omega_1 \int_0^\infty d\omega_2\, \rho(\omega_1,\omega_2)a_1^\dagger(\omega_1)a_2^\dagger(\omega_2)|0\rangle, \tag{10.34}$$

where a_1, a_2 designate the signal and idler modes, respectively. For spontaneous parametric down conversion with a classical pump field at frequency 2Ω, we can assume

$$\rho(\omega_1,\omega_2) = \alpha(\omega_1 + \omega_2)\phi(\omega_1,\omega_2), \tag{10.35}$$

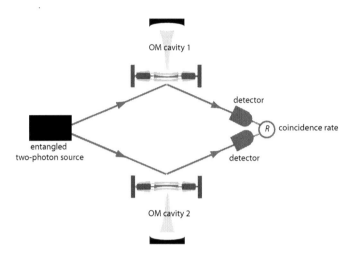

FIGURE 10.6 An optomechanical scheme to use an entangled photon pair to test for gravitational decoherence. The signal and idler beams of a spontaneous down conversion source are reflected off two optomechanical cavities and the coincidence rate at two detectors recorded. The two harmonically bound mirrors interact with each other gravitationally.

where $|\alpha(\omega)|^2$ is very sharply peaked at $\omega = 2\Omega$, twice the frequency of the pump field Ω. We now make the change of variable $\epsilon = \Omega - \omega_1$ and assume that the bandwidth, B, over which $\alpha(\omega_1)$ is significantly different from zero is such that $\Omega \gg B$ then we can write

$$|\psi\rangle = \lambda \int_{-\infty}^{\infty} d\epsilon\, \beta(\epsilon) a_1^\dagger(-\epsilon) a_2^\dagger(\epsilon)|0\rangle, \qquad (10.36)$$

where $\lambda = \alpha(0)$ and we have defined the amplitude function $\beta(\epsilon) \equiv \phi(\Omega - \epsilon, \Omega + \epsilon)$ and $a_1(-\epsilon) \equiv a_1(\Omega - \epsilon), a_2(\epsilon) \equiv a_2(\Omega + \epsilon)$. We will also assume that $\beta(-\epsilon) = \beta(\epsilon)$. In physical terms $|\lambda|^2$ is related to the temporal width of the pump field. Normalisation of $|\psi\rangle$ requires that

$$|\lambda|^2 \int_{-\infty}^{\infty} d\epsilon |\beta(\epsilon)|^2 = 1. \qquad (10.37)$$

Defining the positive frequency component of the field at carrier frequency ω_a,

$$a_1(t) = \frac{1}{\sqrt{2\pi}} \int_0^{\infty} d\omega\, a_1(\omega) e^{-i\omega t}, \qquad (10.38)$$

with $[a_1(\omega'), a_1^\dagger(\omega)] = \delta(\omega' - \omega)$. In terms of the new frequency variable ϵ this is

$$a_1(t) = \frac{e^{-i\Omega t}}{\sqrt{2\pi}} \int_{-\infty}^{\infty} d\epsilon\, a_1(\Omega + \epsilon) e^{-i\epsilon t} \qquad (10.39)$$

where we have extended the lower limit $-\Omega$ to negative infinity; an assumption that the carrier frequency Ω is much greater than the bandwidths at optical frequencies.

Putting aside, for the moment, the interactions with the optomechanical cavities, the probability per unit time to detect a photon from this field with a unit efficiency detector is

$$n_1(t) \;=\; \langle \psi | a_1^\dagger(t) a_1(t) | \psi \rangle \tag{10.40}$$

$$=\; \frac{1}{2\pi} \int_{-\infty}^{\infty} \int_{-\infty}^{\infty} d\epsilon_1 d\epsilon_2 \langle \psi | a_1^\dagger(\epsilon_1) a_1(\epsilon_2) | \psi \rangle e^{i(\epsilon_1 - \epsilon_2)t} \tag{10.41}$$

and we find that

$$n_1(t) = |\lambda|^2 \int_{-\infty}^{\infty} d\omega |\beta(\omega)|^2 = 1 \tag{10.42}$$

by normalisation. A similar result holds for mode 2. The probability per unit time to detect a photon from either mode is thus independent of time. This simply means that photons will be counted at randomly distributed times from each of the fields a_1 and a_2.

Let us now compute the coincidence rate,

$$C(t, t') = \langle \psi | a_1^\dagger(x, t') a_1(x, t') a_2^\dagger(x, t) a_2(x, t) | \psi \rangle. \tag{10.43}$$

This is given by

$$C(t, t') \;=\; |\lambda|^2 \left| \int_{-\infty}^{\infty} d\epsilon \beta(\epsilon) e^{-i\epsilon(t'-t)} \right|^2 \tag{10.44}$$

$$=\; |\lambda|^2 |\tilde{\beta}(\tau)|^2 \tag{10.45}$$

$$\equiv\; \mathcal{C}(\tau), \tag{10.46}$$

where $\tau = t' - t$ and $\tilde{\beta}(\tau)$ is the Fourier transform of $\beta(\epsilon)$ The coincidence rate is thus symmetrical about $t' - t = 0$ and takes a maximum value of unity when $t' = t$. Rohde and Ralph [241] give an example based on spontaneous parametric down conversion for which the amplitude function $\beta(\epsilon)$ is a Lorentzian.

We now consider what happens when the two correlated modes a_1 and a_2 are injected into a pair of optomechanical cavities, as shown in Fig. 10.6. We will assume that the optomechanical cavities are far from the resolved sideband regime, $\kappa \gg \Omega$. In this case, the time scale of the optical system is much faster than that of the mechanical system and especially much faster than the gravitational heating rate. We can, thus, only hope to be sensitive to the thermal steady-state of the mechanics that would result for a combination of gravitational and thermal heating. In this limit, the effect of the optomechanical cavities on the field emitted from each cavity is as described in Section 6.3 and we can treat the canonical displacements, $\hat{Q}_1, \hat{Q}_2$ of each mechanical element as classical random variables X_1, X_2 with a Gaussian (thermal) distribution.

The result is that the output field emitted from the cavities is in the mixed state

$$\rho_f = \int dX_1 dX_2 P(X_1, X_2) |\psi(X_1, X_2)\rangle \langle \psi(X_1, X_2)|, \tag{10.47}$$

where

$$|\psi(X_1, X_2)\rangle = \int_{-\infty}^{\infty} d\epsilon \beta(\epsilon)\xi(\epsilon|X_1)\xi(\epsilon|X_2)a_1^\dagger(-\epsilon)a_2^\dagger(\epsilon)|0\rangle \tag{10.48}$$

with (in the interaction picture)

$$\xi(\epsilon|X_1) = \left[\frac{\gamma_0/2 + i(\epsilon + g_0 X_1)}{\gamma_0/2 - i(\epsilon + g_0 X_1)}\right] \tag{10.49}$$

$$\xi(\epsilon|X_2) = \left[\frac{\gamma_0/2 - i(\epsilon - g_0 X_2)}{\gamma_0/2 + i(\epsilon - g_0 X_2)}\right] \tag{10.50}$$

and $P(X_1, X_2)$ is the joint probability distribution for the mechanical displacements in each cavity. This is assumed to be in the thermal state.

Effectively, the optomechanical cavities transform the spectral correlations of the photons as

$$\beta(\epsilon) \rightarrow \beta(\epsilon)\,\xi(\epsilon|X_1)\xi(\epsilon|X_2) \tag{10.51}$$

$$= \beta(\epsilon) \left[\frac{\gamma_0/2 + i(\epsilon + g_0 X_1)}{\gamma_0/2 - i(\epsilon + g_0 X_1)}\right] \left[\frac{\gamma_0/2 - i(\epsilon - g_0 X_2)}{\gamma_0/2 + i(\epsilon - g_0 X_2)}\right]. \tag{10.52}$$

The resulting coincidence rate is then given by

$$C_g(\tau) = |\lambda|^2 \left| \int_{-\infty}^{\infty} d\epsilon \beta(\epsilon)\xi(\epsilon|X_1)\xi(\epsilon|X_2)e^{-i\epsilon\tau} \right|^2. \tag{10.53}$$

Note that, if $g_0 = 0$, there is no way to distinguish the paths, $\beta(\epsilon)$ is unchanged, and the coincidence correlation is therefore also unchanged.

To second order in g_0, the coincidence rate at $\tau = 0$ is reduced by the amount

$$C(0) - C_g(0) = 16(g_0/\gamma_0)^2 (x_1 + x_2)^2 (A - B), \tag{10.54}$$

where

$$A = \left| \int_{-\infty}^{\infty} d\epsilon \frac{\beta(\epsilon)}{(1 + 4(\epsilon/\gamma_0)^2)} \right|^2 \tag{10.55}$$

$$B = \lambda \int_{-\infty}^{\infty} d\epsilon \frac{\beta(\epsilon)}{[1 + 4(\epsilon/\gamma_0)^2]^2}. \tag{10.56}$$

Finally, we need to average over the state of the mechanics. In that case

$$\int_{-\infty}^{\infty} dX_1 dX_2 P(X_1, X_2)(X_1 + X_2)^2 = 2(2\bar{n} + 1). \tag{10.57}$$

The drop in coincidence rate at zero delay is thus an effective measurement of the temperature of the mechanical system. As we have seen, this will have two components, one due to gravitational decoherence that is independent of temperature and the other due to real thermal fluctuations. The key signature of gravitational decoherence will be to resolve the constant background reaction on the coincidence rate as the thermal component is reduced by reducing the temperature of the system.

10.3 TESTS OF NONSTANDARD GRAVITATIONAL EFFECTS US-ING GEOMETRIC PHASE

The uncertainty principle is the cornerstone of the quantum theory, yet it is easily seen to be incompatible with general relativity. The uncertainty principle implies that any attempt to localise a particle to smaller and smaller regions of space leads to an increase in the momentum variance. As the kinetic energy of a particle is bounded below by the variance in momentum divided by twice the mass, confining the particle to small regions of space necessarily increases it kinetic energy. Eventually this will lead to the formation of a gravitational horizon. Once that occurs, the particle cannot be further localised. This leads to a minimum physical length scale [3]. String theory also leads to a modified uncertainty principle and a minimum length scale [294, 319]. In both cases the general form of the modified uncertainty principle can be written

$$\Delta q \Delta p \geq \frac{\hbar}{2} \left(1 + \beta_0 \frac{\Delta p}{M_P c^2} \right), \tag{10.58}$$

where $M_P = 2$ μg is the Planck mass and β_0 is a dimensionless parameter that depends on the specific theoretical model.

There are two ways to modify standard quantum mechanics to reach an uncertainty principle of this form. As the uncertainty principle depends on both the operator algebra and the state space, we can modify either to reach the generalised form. One version of the modified commutation relation is [152]

$$[\hat{q}, \hat{p}] = i\hbar \left(1 + \beta_0 \left(\frac{\hat{p}}{M_P c^2} \right)^2 \right).$$

Another version is given by Ali et al. [9],

$$[\hat{q}, \hat{p}] = i\hbar \left(1 - \gamma_0 \frac{\hat{p}}{M_P c^2} + \gamma_0^2 \left(\frac{\hat{p}}{M_P c^2} \right)^2 \right).$$

The commutation relations refer to the local unitary representing of canonical displacements in position and momentum. A mathematical equivalent formulation can be given in terms of Heisenberg–Weyl form for consecutive displacements in position and momentum,

$$\hat{V}^\dagger \hat{U}^\dagger \hat{V} \hat{U} = e^{i\phi}, \tag{10.59}$$

where

$$\hat{U} = e^{iA\hat{q}/\hbar} \tag{10.60}$$
$$\hat{V} = e^{-iB\hat{p}/\hbar}, \tag{10.61}$$

and $\phi = AB/\hbar$ represents an area in phase space in units of Planck's constant. The form given in Eq. (10.59) has a simple geometric interpretation. This sequence of transformations corresponds to a closed loop in the classical phase space with area AB, known as the geometric phase. Geometric phase gates are used to implement quantum gates in ion trapping [200, 176]. The geometric phase has also been proposed a a path to new forms of quantum control in optomechanics [156] For the purposes of testing the canonical commutation relations, experimentally implementing a geometric phase is equivalent to testing the canonical commutation relations.

Pikovski et al. [224] proposed controlling the optical field in a pulsed optomechanical system [292] to implement a geometric phase. In this case the displacement parameters A, B are proportional to the cavity photon number operator $\hat{n} = a^\dagger a$. In this case the area becomes an operator on the optical degrees of freedom. Setting $A = \lambda \hat{n}/x_{zp}$ and $B = \lambda \hat{n}/p_{zp}$ with x_{zp}, p_{zp} the ground state uncertainties for position and momentum of the mechanical resonator, and using the standard canonical commutation relations, we find

$$\hat{V}^\dagger \hat{U}^\dagger \hat{V} \hat{U} = e^{-i\lambda^2 \hat{n}^2}. \tag{10.62}$$

In this case the sequence of unitary transformations is equivalent to a Kerr non linear medium acting on the optical cavity mode. If however the modified canonical commutation relation in Eq. (10.3) is used we find that

$$\hat{V}^\dagger \hat{U}^\dagger \hat{V} \hat{U} = e^{-i\lambda^2 \hat{n}^2} e^{-i\beta(\lambda^2 \hat{n}^2 \hat{P}^2 + \lambda^3 \hat{n}^3 \hat{P} + (1/3)\lambda^4 \hat{n}^4)} \tag{10.63}$$

with

$$\beta = \beta_0 \frac{\hbar m \Omega}{(M_{PC})^2}, \tag{10.64}$$

where Ω is the frequency of the mechanical resonator, and, as usual, the dimensionless momentum and position operators for the mechanics are $\hat{P} = -i(a - a^\dagger)/\sqrt{2}$, $\hat{Q} = (a + a^\dagger)/\sqrt{2}$. This is no longer a Kerr-type nonlinear transformation but an entangling operation between the mechanics and the optical field that scales with β.

Suppose the optical field is in a coherent state $|\alpha\rangle$ with $|\alpha|^2 >> 1$ prior to the implementation of the sequence of unitary transformations. Assuming that the mechanics is initially in a thermal state, the resulting mean amplitude of the optical field at the end of the sequence is

$$\langle a \rangle = \alpha_0 e^{-i\Phi_\beta} \tag{10.65}$$

with the standard ($\beta = 0$) quantum transformation given by

$$\alpha_0 = \alpha e^{-i(\lambda^2 - |\alpha|^2(1 - e^{-2i\lambda^2}))} \qquad (10.66)$$

and

$$\Phi_\beta \approx \frac{4}{3}\beta|\alpha|^3\lambda^4 e^{-i6\lambda^2}. \qquad (10.67)$$

The objective now is to find an experimental signature of this result by making a measurement of the mean cavity field. A scheme using pulsed optomechanics is shown in Fig. 10.7. The scheme uses pulsed coherent input

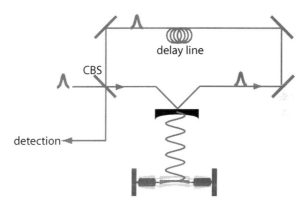

FIGURE 10.7 A scheme for implementing a geometric phase by reinjecting optical pulses into an optomechanical cavity. The controlled beam splitter (CBS) enables a pulse to be reinjected four times with a delay between each re-entry equal to one quarter period of the mechanical period. At the end of four rounds, the CBS switches the pulse out of the device and directs it towards a homodyne detection system.

pulses that are much shorter than the mechanical frequency. The cavity field is then excited to a coherent state with a time-dependent amplitude that is only non-zero over a very short time compared to the mechanical frequency. As the optomechanical interaction is only turned on when the cavity field is not in the vacuum state, this enables us to describe the dynamics in terms of a unitary map acting over the lifetime of the pulse in the cavity. This unitary map is defined by

$$\hat{U} = e^{i\lambda a^\dagger a \hat{Q}}, \qquad (10.68)$$

where $\lambda = g_0/\kappa$ with κ the cavity decay rate. This map implements a displacement in the mechanical momentum.

In order to implement the displacement in position, we use the free evolution of the mechanical resonator between pulses to rotate the phase space by $\pi/4$. This has the effect of transforming a displacement in momentum into a displacement in position. To achieve this, Pikovsky et al. propose that the coherent pulse leaving the cavity after the first unitary operation is delayed in an optical delay line for one quarter of the mechanical period before being injected back into the optomechanical cavity (all indicated by the controlled beam splitter, CBS, in Fig. 10.7). This reinjection can be controlled using a polarising beam splitter together with an electro-optic modulator and quarter wave plates. It is essential that the same pulse be reinjected four times to ensure that the sequence of four displacements act on the same state of the mechanical and the optics. Finally, the optical control is switched to let the final optic pulse exit the system and directed towards a phase detection step.

Pikovsky et al. estimate that a test of the commutator in Eq. (10.3) can be achieved with current technology, while the test of the commutator in Eq. (10.3) will be more challenging. Using a mechanical resonator with a mass of $m = 10^{-11}$ kg and a resonant frequency of $\Omega/2\pi = 10^5$ Hz a sensitivity of $\delta\gamma_0 = 1$ could be reached which would improve the existing bunds on this parameter by ten orders of magnitude. For the commutator in Eq. (10.3) much higher (but achievable) powers are needed to achieve unit sensitivity, but the improvement over existing bounds would be a staggering 33 orders of magnitude and bring the Planck scale within reach. Of course optical loss, thermal noise, and imperfect optics will limit these sensitivities. The mechanical resonator in particular will need to be operated at around 100 mK and with a quality factor greater than 10^6. Despite these challenges, it is remarkable that an optomechanical system might achieve the sensitivity to test a key prediction of string theory and quantum gravity.

Linear detection of optical fields

CONTENTS

A.1   Effect of inefficiencies 319
A.2   Linear detection of optical fields 322
    A.2.1   Homodyne detection 322
    A.2.2   Heterodyne detection 323
A.3   Power spectral density obtained by heterodyne detection .. 323
A.4   Characterising the optomechanical cooperativity 324
A.5   Characterising the temperature of a mechanical oscillator . 326
    A.5.1   Characterisation using the shot noise level 326
    A.5.2   Characterisation using sideband asymmetry 327

In this appendix we briefly introduce some technical details associated with modelling the linear detection of optical fields, including how to include detection inefficiencies and the machinery of homodyne and heterodyne detection. We then provide some useful techniques to accurately calibrate optomechanical parameters from the observed power spectral densities.

A.1   EFFECT OF INEFFICIENCIES

In a real experiment, not all of the photons leaving the cavity optomechanical system are collected in the detection apparatus. Photons may be lost from the cavity, for example, by leaking out from one of the cavity mirrors other than the input-output coupler, or by intracavity absorption, or, after leaving the cavity through the input-output coupler, they might be lost as a result of inefficiencies in the detection process, or by scattering or absorption prior to the detector (see Fig. A.1). While each of these loss mechanisms is physically quite different, they are all linear. As a result, they all have a similar qualitative effect on the optical field.

Losses that occur within the optical cavity can be modelled by introducing an second optical port on the cavity, the output of which is not accessible to

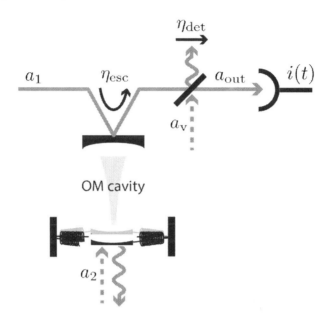

FIGURE A.1 Schematic diagram of a cavity optomechanical system including optical loss channels both within the system (quantified by the escape efficiency η_{esc}) and after the optical field leaves the system prior to and including detection (quantified by the detection efficiency η_{det}). The dashed arrows indicate vacuum fluctuations that are introduced to the field, while the wavy arrows show the dissipation.

the experimenter. In this case, the total cavity decay is apportioned partly to decay through the input-output coupler and partly to decay through the loss port, with respective rates of κ_1 and κ_2, such that the total cavity decay rate $\kappa = \kappa_1 + \kappa_2$. It is natural to define a cavity escape efficiency $\eta_{\text{esc}} = \kappa_1/\kappa$, which quantifies the probability that an intracavity photon will decay through the input-output coupler. The new cavity loss port provides a second path for optical vacuum fluctuations[1] to enter the cavity (or in other words couples the optical cavity to a second optical bath). The additional optical amplitude $\hat{X}_2$ and phase $\hat{Y}_2$ quadrature fluctuations can be included, for instance, via the replacement

$$\hat{X}_{\text{in}} = \sqrt{\eta_{\text{esc}}}\hat{X}_1 + \sqrt{1 - \eta_{\text{esc}}}\hat{X}_2 \tag{A.1a}$$

$$\hat{Y}_{\text{in}} = \sqrt{\eta_{\text{esc}}}\hat{Y}_1 + \sqrt{1 - \eta_{\text{esc}}}\hat{Y}_2, \tag{A.1b}$$

where $\hat{X}_1$ and phase $\hat{Y}_1$ now describe the fluctuations entering the cavity

---

[1]More precisely, thermal fluctuations, but, as we have discussed earlier, at optical frequencies these may be well approximated as vacuum.

through the input port. Since not all of the cavity decay occurs in the reverse direction through the input port, the input-output relations (Eqs. (1.126)) quantifying the accessible output field from the cavity optomechanical system must be modified to

$$\hat{X}_{\text{out}} = \hat{X}_1 - \sqrt{\kappa_1}\hat{X} \qquad (A.2a)$$
$$\hat{Y}_{\text{out}} = \hat{Y}_1 - \sqrt{\kappa_1}\hat{Y}. \qquad (A.2b)$$

The overall effect of the total linear loss that occurs on the optical field between exiting the cavity and detection can be modelled by a beam splitter placed in the output field which reflects part of the incident optical field and introduces further vacuum fluctuations to the field. If the total fractional detection efficiency after the light leaves the cavity (i.e., not including the cavity escape efficiency η_{esc}) is η_{det}, the effect of such inefficiencies translates the quadratures of the output field in the Heisenberg picture to

$$\hat{X}_{\text{out}} \to \sqrt{\eta_{\text{det}}}\hat{X}_{\text{out}} + \sqrt{1 - \eta_{\text{det}}}\hat{X}_{\text{v}} \qquad (A.3a)$$
$$\hat{Y}_{\text{out}} \to \sqrt{\eta_{\text{det}}}\hat{Y}_{\text{out}} + \sqrt{1 - \eta_{\text{det}}}\hat{Y}_{\text{v}}, \qquad (A.3b)$$

where $\hat{X}_{\text{v}}$ and $\hat{Y}_{\text{v}}$ are the introduced vacuum fluctuations.

To take the example of a linearised cavity optomechanical system with on-resonance optical driving treated in Chapter 3, making the above modifications, the output quadrature operators and the mechanical position operator in the presence of both intracavity and detection inefficiencies are found to be

$$\hat{X}_{\text{out}}(\omega) = \hat{X}'_{\text{in}} \qquad (A.4a)$$
$$\hat{Y}_{\text{out}}(\omega) = \hat{Y}'_{\text{in}} + 2\sqrt{\eta\Gamma C_{\text{eff}}}\hat{Q}, \qquad (A.4b)$$
$$= \hat{Y}'_{\text{in}}(\omega) + 2\Gamma\sqrt{2\eta C_{\text{eff}}}\chi(\omega)\left(\hat{P}_{\text{in}}(\omega) - \sqrt{2C_{\text{eff}}}\hat{X}_{\text{in}}(\omega)\right) \qquad (A.4c)$$
$$\hat{Q}(\omega) = \sqrt{2\Gamma}\chi(\omega)\left[\hat{P}_{\text{in}} - \sqrt{2C_{\text{eff}}}\left(\sqrt{\eta_{\text{esc}}}\hat{X}_1 + \sqrt{1 - \eta_{\text{esc}}}\hat{X}_2\right)\right], \qquad (A.4d)$$

where the total quantum efficiency $\eta = \eta_{\text{esc}}\eta_{\text{det}}$, and it is convenient to define the new quadrature operators $\hat{X}'_{\text{in}}$ and $\hat{Y}'_{\text{in}}$ which incorporate the contributions to the output field both from the cavity input field and from the vacuum fields introduced due to inefficiency. $\hat{X}'_{\text{in}}$ is given explicitly by

$$\hat{X}'_{\text{in}} = \eta_{\text{det}}^{1/2}\left[\left(1 - \frac{2\eta_{\text{esc}}}{1 - 2i\omega/\kappa}\right)\hat{X}_1 - \frac{2\sqrt{\eta_{\text{esc}}(1 - \eta_{\text{esc}})}}{1 - 2i\omega/\kappa}\hat{X}_2\right] + \sqrt{1 - \eta_{\text{det}}}\hat{X}_{\text{v}}, \quad (A.5)$$

with an equivalent expression for $\hat{Y}'_{\text{in}}$, but with $\hat{X} \to \hat{Y}$ throughout.

Exercise A.1 *Show that*

$$S_{X'_{\text{in}}X'_{\text{in}}}(\omega) = S_{Y'_{\text{in}}Y'_{\text{in}}}(\omega) = \frac{1}{2}, \qquad (A.6)$$

consistent with Eq. (1.118a) for the power spectral density of a zero temperature bath in the rotating frame.

Note that, in principle, it is possible for the optical field to also experience nonlinear losses due, for example, to two-photon scattering or absorption. Nonlinear losses cannot be modelled via the simple approach given here. However, in typical experiments they are negligible compared to linear losses.

A.2   LINEAR DETECTION OF OPTICAL FIELDS

In this section we provide some details of the homodyne and heterodyne measurement schemes illustrated in Fig. 3.2.

A.2.1   Homodyne detection

In a homodyne detector, the optical field to be detected $a_{\det}$ is interfered on a 50/50 beam splitter with a bright local oscillator field a_{LO} of the same frequency. The outputs of the beam splitter, which we denote respectively as a_- and a_+, are

$$a_\pm = \frac{1}{\sqrt{2}}\left(a_{\mathrm{LO}} \pm a_{\det}\right).\tag{A.7}$$

These outputs are then each independently detected on a photodiode, generating photocurrents described by the detected field operators $\hat{\imath}_\pm = a_\pm^\dagger a_\pm$, which are then subtracted.

Exercise A.2 *Show that, prior to linearisation, the detected field operator for the difference photocurrent is given by*

$$\hat{\imath}(t) = \hat{\imath}_+(t) - \hat{\imath}_-(t) = a_{\mathrm{LO}}^\dagger a_{\det} + a_{\mathrm{LO}} a_{\det}^\dagger.\tag{A.8}$$

In the limit that the local oscillator field is much brighter than the field to be detected ($|\alpha_{\mathrm{LO}}| = |\langle a_{\mathrm{LO}}\rangle| \gg |\alpha_{\det}|$), the local oscillator field can be treated classically, with the homodyne difference photocurrent operator then given by

$$\hat{\imath}(t) = \alpha_{\mathrm{LO}}^* a_{\det} + \alpha_{\mathrm{LO}} a_{\det}^\dagger \tag{A.9}$$

$$= |\alpha_{\mathrm{LO}}| \hat{X}_{\det}^\theta, \tag{A.10}$$

where, similar to the case for direct detection, here $e^{i\theta} = \alpha_{\mathrm{LO}}/|\alpha_{\mathrm{LO}}|$.

It can be seen by comparison of Eqs. (3.32) and (A.10) that direct detection is a subset of homodyne detection with the local oscillator phase angle set to $\theta = \theta_{\det}$.

Homodyne detection has two major advantages over direct detection. Firstly, the detected quadrature can be controlled by controlling the local oscillator phase θ. Secondly, in the limit $|\alpha_{\mathrm{LO}}| \gg |\alpha_{\det}|$, the proportionality constant scaling the detected quadrature into the photocurrent is independent of $\alpha_{\det}$. This enables the measurement to be directly and accurately calibrated to the level of the optical vacuum fluctuations by simply blocking the detected field and thereby replacing it with a vacuum field.

A.2.2   Heterodyne detection

Heterodyne detection is essentially equivalent to homodyne detection, except that the frequency of the local oscillator field is offset from the frequency of the field to be detected by some detuning $\Delta_{\text{LO}} = \Omega_{\text{LO}} - \Omega_L$. In this case, it is straightforward to show that the detected photocurrent operator of Eq. (A.9) becomes

$$
\begin{aligned}
\hat{i}(t) &= \alpha_{\text{LO}}^* a_{\text{det}} e^{i\Delta_{\text{LO}}t} + \alpha_{\text{LO}} a_{\text{det}}^\dagger e^{-i\Delta_{\text{LO}}t} \tag{A.11}\\
&= |\alpha_{\text{LO}}| \hat{X}_{\text{det}}^{\theta - \Delta_{\text{LO}}t}. \tag{A.12}
\end{aligned}
$$

We see, therefore, that the quadrature of the detected field probed by heterodyne detection oscillates in time at a frequency given by the detuning of the local oscillator field. As we will see in the following section and in Section 3.3.6.2, while this has the advantage of providing information about both the phase and amplitude quadratures of the field, it also results in an unavoidably half-quanta noise penalty.

A.3   POWER SPECTRAL DENSITY OBTAINED BY HETERODYNE DETECTION

The power spectral density of a heterodyne measurement of the output from a cavity optomechanical system can be determined relatively straightforwardly. From Eq. (1.42) we have

$$
\begin{aligned}
S_{\hat{i}\hat{i}}(\omega) &= \lim_{\tau \to \infty} \frac{1}{\tau} \left\langle \hat{i}_\tau^\dagger(\omega)\hat{i}_\tau(\omega) \right\rangle \tag{A.13}\\
&= \lim_{\tau \to \infty} \frac{1}{\tau} \iint_{-\tau/2}^{\tau/2} dt\,dt'\, e^{-i\omega t} e^{i\omega t'} \left\langle \hat{i}^\dagger(t)\hat{i}(t') \right\rangle. \tag{A.14}
\end{aligned}
$$

Using Eq. (A.11), the time domain expectation value can be expressed in terms of the detected field annihilation and creation operators as

$$
\left\langle \hat{i}^\dagger(t)\hat{i}(t') \right\rangle = |\alpha_{\text{LO}}|^2 \left[e^{i\Delta_{\text{LO}}(t'-t)} \left\langle a_{\text{det}}^\dagger(t)a_{\text{det}}(t') \right\rangle + e^{i\Delta_{\text{LO}}(t-t')} \left\langle a_{\text{det}}(t)a_{\text{det}}^\dagger(t') \right\rangle \right],
$$

where we have used the rotating wave approximation, neglecting the fast rotating terms. Inserting this expression into Eq. (A.14), after some work we find

$$
\begin{aligned}
S_{\hat{i}\hat{i}}^{\text{het}}(\omega) &= |\alpha_{\text{LO}}|^2 \lim_{\tau \to \infty} \frac{1}{\tau} \left[\left\langle a_{\text{det}}^\dagger(\Delta_{\text{LO}} + \omega)a_{\text{det}}(\Delta_{\text{LO}} + \omega) \right\rangle \right.\\
&\qquad \left. + \left\langle a_{\text{det}}(\Delta_{\text{LO}} - \omega)a_{\text{det}}^\dagger(\Delta_{\text{LO}} - \omega) \right\rangle \right] \tag{A.15}\\
&= |\alpha_{\text{LO}}|^2 \left[S_{a_{\text{det}}a_{\text{det}}}(\Delta_{\text{LO}} + \omega) + S_{a_{\text{det}}^\dagger a_{\text{det}}^\dagger}(\Delta_{\text{LO}} - \omega) \right]. \tag{A.16}
\end{aligned}
$$

As can be seen, the beating between the heterodyne signal and local oscillator fields has the effect of introducing frequency components at $\Delta_{\text{LO}} \pm \omega$ into the power spectral density at frequency ω, and results in a power spectral density of photon number rather than optical quadrature. Expressed in terms of the amplitude and phase quadratures of the detected field and normalised, similar to the case for homodyne detection, such that a value of $1/2$ corresponds to the vacuum noise level, we arrive at the heterodyne power spectral density given in Eq. (3.37) of Chapter 3, which we have equated to the observed classical power spectral density since here, unlike homodyne detection, $S_{\hat{i}\hat{i}}^{\text{het}}(\omega) = S_{\hat{i}\hat{i}}^{\text{het}}(-\omega) = \bar{S}_{\hat{i}\hat{i}}^{\text{het}}(\omega)$.

A.4 CHARACTERISING THE OPTOMECHANICAL COOPERATIVITY

The optomechanical cooperativity of a cavity optomechanical system can be experimentally characterised via several different methods. For example, a common method used for Fabry–Perot-based cavity optomechanical systems is to apply static deflections to the mechanical oscillator and characterise the frequency shift this causes on the optical cavity. As long as the motion of the mechanical oscillator matches the static displacement (i.e., the mechanical eigenmode causes a simple deflection of the mechanical oscillator along the axis of the Fabry–Perot cavity), this, combined with knowledge of the zero-point motion, frequency, and decay rate of the mechanical oscillator, allows the optomechanical cooperativity to be determined through relations (2.17), (3.14), and (3.13). A more generally applicable and direct method is to apply a known force to the specific eigenmode of the mechanical oscillator and calibrate the optomechanical cooperativity via the phase shift that results on the output optical field. Two natural ways to apply a known force are to (a) allow the mechanical oscillator to interact with a thermal bath at a known temperature, or (b) use radiation pressure shot noise as the force. In both cases, the optomechanical cooperativity can then be determined directly from either of the homodyne or heterodyne power spectral densities.

Considering the more common homodyne measurement, in the limit that radiation pressure heating is negligible ($|C_{\text{eff}}(\Omega)| \ll \bar{n} + 1/2$) and that the coherent optical drive is on-resonance ($\Delta = 0$), Eq. (3.43) can be rearranged to give the optomechanical cooperativity as a function of the peak of the power spectral density, the optical quantum efficiency, and the bath temperature

$$|C_{\text{eff}}(\Omega)| = \frac{1}{8\eta} \frac{S_{ii}^{\text{homo}}(\Omega) - 1/2}{\bar{n} + 1/2}. \tag{A.17}$$

Similarly, in the limit that heating from the bath is negligible compared to radiation pressure heating ($|C_{\text{eff}}(\Omega)| \gg \bar{n} + 1/2$), the optomechanical cooperativity is given by

$$|C_{\text{eff}}(\Omega)| = \sqrt{\frac{S_{ii}^{\text{homo}}(\Omega) - 1/2}{8\eta}}. \tag{A.18}$$

In this case, we see that the only calibrations required to determine the optomechanical cooperativity are the calibration of the optical vacuum noise level (the normalisation of $S_{ii}(\Omega)$) and the optical quantum efficiency including both the cavity escape efficiency and detection efficiency. However, the requirement to reach the radiation pressure heating dominated regime is stringent, and while the experimental calibration of escape efficiency is generally straightforward, calibration of detection efficiency is more challenging and prone to errors.

An alternative approach to calibrate the optomechanical cooperativity is to drive the mechanical resonator with a coherent optical force and measure the response to this force. This approach can remove the requirement for both reaching the radiation pressure heating dominated regime and for calibrating the detection efficiency. For instance, consider amplitude modulating the incident optical field upon the optomechanical cavity at the mechanical resonance frequency Ω so that

$$\hat{X}_{\text{in}}(t) \rightarrow \hat{X}_{\text{in}}(t) + A \cos \Omega t, \tag{A.19}$$

where A is the amplitude of modulation. Through radiation pressure, this modulation drives a coherent oscillation of the mechanical resonator which is subsequently imprinted on the phase quadrature of the output optical field. From Eqs. (A.4) and (A.5) we then find that

$$\hat{X}_{\text{out}}(\omega) \rightarrow \hat{X}_{\text{out}}(\omega) + \sqrt{\eta_{\text{det}}} A \left(\frac{1}{2} - \frac{\eta_{\text{esc}}}{1 - 2i\omega/\kappa} \right) (\delta(\Omega) + \delta(-\Omega)) \tag{A.20a}$$

$$\hat{Y}_{\text{out}}(\omega) \rightarrow \hat{Y}_{\text{out}}(\omega) - 2\eta_{\text{esc}}\Gamma \sqrt{\eta_{\text{det}}} A\chi(\omega) C_{\text{eff}} (\delta(\Omega) + \delta(-\Omega)) \tag{A.20b}$$

$$\hat{Q}(\omega) \rightarrow \hat{Q}(\omega) - A\chi(\omega)\sqrt{\Gamma \eta_{\text{esc}}} C_{\text{eff}} (\delta(\Omega) + \delta(-\Omega)). \tag{A.20c}$$

A key thing to note here is that the terms introduced by the modulation scale in the same way on both output optical quadratures as a function of the modulation amplitude A and the detection efficiency η_{det}. Consequently, these factors cancel out if consecutive homodyne measurements are made of the optical amplitude and phase quadratures, and the ratio $\mathcal{R}$ is taken of the power introduced by the modulation. Calculating this ratio at the mechanical resonance frequency, we find

$$\mathcal{R} = \frac{\bar{S}_{Y_{\text{out}}Y_{\text{out}}}(\Omega) - \bar{S}_{Y_{\text{out}}Y_{\text{out}}}^{A=0}(\Omega)}{\bar{S}_{X_{\text{out}}X_{\text{out}}}(\Omega) - \bar{S}_{X_{\text{out}}X_{\text{out}}}^{A=0}(\Omega)} \tag{A.21}$$

$$= 16\eta_{\text{esc}}^2 |C_{\text{eff}}(\Omega)|^2 \left[\frac{1 + 4(\Omega/\kappa)^2}{(1 - 2\eta_{\text{esc}})^2 + 4(\Omega/\kappa)^2} \right]. \tag{A.22}$$

The optomechanical cooperativity can then be experimentally determined, without calibration of the detection efficiency or reaching the radiation pressure heating dominated regime, as

$$|C_{\text{eff}}(\Omega)| = \frac{\mathcal{R}^{1/2}}{4\eta_{\text{esc}}} \left[\frac{(1 - 2\eta_{\text{esc}})^2 + 4(\Omega/\kappa)^2}{1 + 4(\Omega/\kappa)^2} \right]^{1/2}. \tag{A.23}$$

A.5 CHARACTERISING THE TEMPERATURE OF A MECHANICAL OSCILLATOR

A.5.1 Characterisation using the shot noise level

Once the optomechanical cooperativity is known, it is relatively straightforward to experimentally determine the temperature of a mechanical oscillator from the detected power spectral density of the phase quadrature of the output field (which for the case of on-resonance optical driving is given by Eq. (A.4c)). We know from Section 1.3.4 that the integral of the symmetrised power spectral density of an oscillator over all frequencies is equal to $\bar{n} + 1/2$, where here $\bar{n} = \bar{n}_b$ is the oscillator's thermal occupancy, including the effect of radiation pressure heating (Eq. (3.19)). Integrating the detected homodyne power spectral density in Eq. (3.44) and rearranging, we find immediately that $\bar{n}_b$ can be experimentally determined as

$$\bar{n}_b = \frac{1}{4\eta\Gamma\,|C_{\text{eff}}(\Omega)|} \int_0^\infty d\omega \left(S_{ii}^{\text{homo}}(\omega) - \frac{1}{2} \right) - \frac{1}{2}, \qquad (A.24)$$

where we have assumed that the mechanical power spectrum is sufficiently strongly peaked around the mechanical resonance frequency that the effective optomechanical cooperativity is essentially constant across the full spectral region that contains appreciable mechanical power. It should be noted that, to apply this expression to determine the occupancy (or Eq. (A.26) that follows), the homodyne power spectral density $S_{ii}^{\text{homo}}(\omega)$ must be calibrated and normalised such that the optical shot noise level is equal to $1/2$. It is often the case that the lasers in optomechanics experiments are not shot noise limited due to classical laser phase noise or other classical noise sources. In such circumstances, rather than subtracting $1/2$ from the power spectrum within the integral in Eq. (A.24), the actual shot-noise-normalised optical noise spectral density should be used.

Alternatively, the thermal occupancy can be determined without integration using the ratio of the mechanical and shot noise contributions to the observed power spectral density at the mechanical resonance frequency. This is a form of signal-to-noise ratio, and has the advantage that it can be directly read off the observed power spectral density in experiments. The noise component is equal, in general, to the optical shot noise (normalised to $1/2$ here), while the "signal" component is equal to $S_{ii}^{\text{homo}}(\Omega)$ minus the shot-noise-normalised optical noise power spectral density at Ω, which we take here to equal the shot noise level of $1/2$. Using Eq. (3.43), this shot-noise-limited signal-to-noise ratio (SNR) can be expressed as

$$\text{SNR} = 2S_{ii}^{\text{homo}}(\Omega) - 1 = 8\eta\Gamma\,|C_{\text{eff}}(\Omega)| \left(\bar{n}_b + \frac{1}{2} \right). \qquad (A.25)$$

The mechanical occupation is then

$$\bar{n}_b = \frac{\text{SNR}}{8\eta\Gamma\,|C_{\text{eff}}(\Omega)|} - \frac{1}{2}. \qquad (A.26)$$

A.5.2 Characterisation using sideband asymmetry

As we have already discussed in Chapters 1 and 3, the mechanical occupancy and temperature can also be directly determined from sideband asymmetry observed in a heterodyne measurement. In Section 1.3.3 we found that the thermal occupancy of a high-quality oscillator can be determined uniquely using only the ratio of the power spectral densities of the oscillator at $\pm\Omega$ (Eq. (1.104b)). Inspection of Eqs. (3.46) and (3.48) reveals that this ratio can be determined from the heterodyne power spectral density as

$$\frac{S_{QQ}(\Omega)}{S_{QQ}(-\Omega)} = \frac{S_{ii}^{\text{het}}(\Delta_{\text{LO}} + \Omega) - 1/2}{S_{ii}^{\text{het}}(\Delta_{\text{LO}} - \Omega) - 1/2}. \tag{A.27}$$

Determining the mechanical occupancy in this way has the major advantage that, apart from determining the level of the quantum noise, it is calibration free. In contrast, the methods in the previous section using the homodyne power spectral density required knowledge of the optical efficiency, mechanical decay rate, and effective optomechanical cooperativity. However, since the difference in the positive and negative frequency power spectral densities is at the level of a single phonon, sideband asymmetry can only be used for thermometry if differences at this level are resolvable. In contrast, the previous homodyne-based measurements require only that the total thermal phonon number is resolvable over the shot noise of the measurement.

References

[1] J Aasi, J Abadie, BP Abbott, R Abbott, TD Abbott, MR Abernathy, C Adams, T Adams, P Addesso, RX Adhikari, et al. Enhanced sensitivity of the LIGO gravitational wave detector by using squeezed states of light. *Nature Photonics*, 7(8):613–619, 2013.

[2] B Abbott, R Abbott, R Adhikari, P Ajith, Bruce Allen, G Allen, R Amin, SB Anderson, WG Anderson, MA Arain, et al. Observation of a kilogram-scale oscillator near its quantum ground state. *New Journal of Physics*, 11(7):073032, 2009.

[3] RJ Adler and DI Santiago. On gravity and the uncertainty principle. *Modern Physics Letters A*, 14:1371, 1999.

[4] GS Agarwal and S Huang. Electromagnetically induced transparency in mechanical effects of light. *Physical Review A*, 81(4):041803, 2010.

[5] GS Agarwal and SS Jha. Theory of optomechanical interactions in superfluid He. *Physical Review A*, 90(2):023812, 2014.

[6] U Akram, WP Bowen, and GJ Milburn. Entangled mechanical cat states via conditional single photon optomechanics. *New Journal of Physics*, 15(9):093007, 2013.

[7] U Akram, N Kiesel, M Aspelmeyer, and GJ Milburn. Single-photon opto-mechanics in the strong coupling regime. *New Journal of Physics*, 12(8):083030, 2010.

[8] U Akram, W Munro, K Nemoto, and GJ Milburn. Photon-phonon entanglement in coupled optomechanical arrays. *Physical Review A*, 86(4):042306, 2012.

[9] AF Ali, S Das, and EC Vagenas. Discreteness of space from the generalized uncertainty principle. *Physics Letters B*, 678:497, 2009.

[10] KT Alligood, TD Sauer, and JA Yorke. *Chaos*. Springer, Berlin Heidelberg, 1997.

[11] R Almog, S Zaitsev, O Shtempluck, and E Buks. Noise squeezing in a nanomechanical Duffing resonator. *Physical Review Letters*, 98(7):078103, 2007.

[12] G Amelino-Camelia. Gravity in quantum mechanics. *Nature Physics*, 10(4):254–255, 2014.

[13] RW Andrews, RW Peterson, TP Purdy, K Cicak, RW Simmonds, CA Regal, and KW Lehnert. Bidirectional and efficient conversion between microwave and optical light. *Nature Physics*, 10(4):321–326, 2014.

[14] RW Andrews, RW Peterson, TP Purdy, K Cicak, RW Simmonds, CA Regal, and KW Lehnert. Connecting microwave and optical frequencies with a vibrational degree of freedom. In *Proceedings of the SPIE 9343, Laser Resonators, Microresonators, and Beam Control XVII*, page 934309. International Society for Optics and Photonics, 2015.

[15] T Antoni, K Makles, R Braive, T Briant, P-F Cohadon, I Sagnes, I Robert-Philip, and A Heidmann. Nonlinear mechanics with suspended nanomembranes. *Europhysics Letters*, 100(6):68005, 2012.

[16] O Arcizet, P-F Cohadon, T Briant, M Pinard, and A Heidmann. Radiation-pressure cooling and optomechanical instability of a micromirror. *Nature*, 444(7115):71–74, 2006.

[17] O Arcizet, V Jacques, A Siria, P Poncharal, P Vincent, and S Seidelin. A single nitrogen-vacancy defect coupled to a nanomechanical oscillator. *Nature Physics*, 7(11):879–883, 2011.

[18] DG Aronson, GB Ermentrout, and N Kopell. Amplitude response of coupled oscillators. *Physica D: Nonlinear Phenomena*, 41(3):403–449, 1990.

[19] A Ashkin. Acceleration and trapping of particles by radiation pressure. *Physical Review Letters*, 24(4):156, 1970.

[20] E Babourina-Brooks, A Doherty, and GJ Milburn. Quantum noise in a nanomechanical Duffing resonator. *New Journal of Physics*, 10(10):105020, 2008.

[21] T Bagci, A Simonsen, S Schmid, LG Villanueva, E Zeuthen, J Appel, JM Taylor, A Sørensen, K Usami, A Schliesser, et al. Optical detection of radio waves through a nanomechanical transducer. *Nature*, 507(7490):81–85, 2014.

[22] M Bagheri, M Poot, L Fan, F Marquardt, and HX Tang. Photonic cavity synchronization of nanomechanical oscillators. *Physical Review Letters*, 111(21):213902, 2013.

[23] BQ Baragiola, RL Cook, AM Brańczyk, and J Combes. N-photon wave packets interacting with an arbitrary quantum system. *Physical Review A*, 86(1):013811, 2012.

[24] Sh Barzanjeh, M Abdi, GJ Milburn, P Tombesi, and D Vitali. Reversible optical-to-microwave quantum interface. *Physical Review Letters*, 109(13):130503, 2012.

[25] Sh Barzanjeh, MH Naderi, and M Soltanolkotabi. Steady-state entanglement and normal-mode splitting in an atom-assisted optomechanical system with intensity-dependent coupling. *Physical Review A*, 84(6):063850, 2011.

[26] S Basiri-Esfahani, U Akram, and GJ Milburn. Phonon number measurements using single photon opto-mechanics. *New Journal of Physics*, 14(8):085017, 2012.

[27] S Basiri-Esfanhi, J Myers, CR Combes, and GJ Milburn. Quantum and classical control of single photon states via a mechanical resonator. *New Journal of Physics*, 2015.

[28] G Baym and T Ozawa. Two-slit diffraction with highly charged particles: Niels Bohr's consistency argument that the electromagnetic field must be quantized. *Proceedings of the National Academy of Sciences*, 106(9):3035–3040, 2009.

[29] V Belavkin. Non-demolition measurement and control in quantum dynamical systems. In *Information Complexity and Control in Quantum Physics*, pages 311–329. Springer, New York, 1987.

[30] JS Bennett, LS Madsen, M Baker, H Rubinsztein-Dunlop, and WP Bowen. Coherent control and feedback cooling in a remotely coupled hybrid atom–optomechanical system. *New Journal of Physics*, 16(8):083036, 2014.

[31] DG Blair, EN Ivanov, ME Tobar, PJ Turner, F Van Kann, and IS Heng. High sensitivity gravitational wave antenna with parametric transducer readout. *Physical Review Letters*, 74(11):1908, 1995.

[32] A Blais, R-S Huang, A Wallraff, SM Girvin, and RJ Schoelkopf. Cavity quantum electrodynamics for superconducting electrical circuits: An architecture for quantum computation. *Physical Review A*, 69(6):062320, 2004.

[33] J Bochmann, A Vainsencher, DD Awschalom, and AN Cleland. Nanomechanical coupling between microwave and optical photons. *Nature Physics*, 9(11):712–716, 2013.

[34] N Bogolubov. On the theory of superfluidity. *Journal of Physics*, 11:23–29, 1966.

[35] S Bose, K Jacobs, and PL Knight. Preparation of nonclassical states in cavities with a moving mirror. *Physical Review A*, 56(5):4175, 1997.

[36] WP Bowen, R Schnabel, PK Lam, and TC Ralph. Experimental characterization of continuous-variable entanglement. *Physical Review A*, 69(1):012304, 2004.

[37] VB Braginski and AB Manukin. Ponderomotive effects of electromagnetic radiation. *Soviet Journal of Experimental and Theoretical Physics*, 25:653, 1967.

[38] VB Braginskii, AB Manukin, and MY Tikhonov. Investigation of dissipative ponderomotive effects of electromagnetic radiation. *Soviet Journal of Experimental and Theoretical Physics*, 31:829, 1970.

[39] VB Braginskii and YI Vorontsov. Quantum-mechanical limitations in macroscopic experiments and modern experimental technique. *Physics-Uspekhi*, 17(5):644–650, 1975.

[40] VB Braginskii, Yu I Vorontsov, and FY Khalili. Quantum singularities of a ponderomotive meter of electromagnetic energy. *Soviet Journal of Experimental and Theoretical Physics*, 46:705, 1977.

[41] VB Braginskii, Yu I Vorontsov, and FY Khalili. Optimal quantum measurements in detectors of gravitation radiation. *Journal of Experimental and Theoretical Physics Letters*, 27(5), 1978.

[42] VB Braginsky, ML Gorodetsky, and F Ya Khalili. Optical bars in gravitational wave antennas. *Physics Letters A*, 232(5):340–348, 1997.

[43] VB Braginsky and F Ya Khalili. Low noise rigidity in quantum measurements. *Physics Letters A*, 257(5):241–246, 1999.

[44] VB Braginsky and FY Khalili. *Quantum Measurement.* Cambridge University Press, 1995.

[45] VB Braginsky, SE Strigin, and SP Vyatchanin. Parametric oscillatory instability in Fabry–Perot interferometer. *Physics Letters A*, 287(5):331–338, 2001.

[46] VB Braginsky and YI Vorontsov. Quantum-mechanical restrictions in macroscopic measurements and modern experimental devices. *Uspekhi Fiz. Nauk*, 114:41–53, 1974.

[47] VB Braginsky, YI Vorontsov, and KS Thorne. Quantum nondemolition measurements. *Science*, 209(4456):547–557, 1980.

[48] N Brahms, T Botter, S Schreppler, DWC Brooks, and DM Stamper-Kurn. Optical detection of the quantization of collective atomic motion. *Physical Review Letters*, 108(13):133601, 2012.

[49] GA Brawley and WP Bowen. Quantum nanomechanics: Feeling the squeeze. *Nature Photonics*, 7(11):854–855, 2013.

[50] GA Brawley, MR Vanner, PE Larsen, S Schmid, A Boisen, and WP Bowen. Non-linear optomechanical measurement of mechanical motion. *arXiv preprint arXiv:1404.5746*, 2014.

[51] F Brennecke, S Ritter, T Donner, and T Esslinger. Cavity optomechanics with a Bose-Einstein condensate. *Science*, 322(5899):235–238, 2008.

[52] DWC Brooks, T Botter, S Schreppler, TP Purdy, N Brahms, and DM Stamper-Kurn. Non-classical light generated by quantum-noise-driven cavity optomechanics. *Nature*, 488(7412):476–480, 2012.

[53] RG Brown, PYC Hwang, et al. *Introduction to Random Signals and Applied Kalman Filtering*, volume 3. Wiley, New York, 1992.

[54] AO Caldeira and AJ Leggett. Influence of dissipation on quantum tunneling in macroscopic systems. *Physical Review Letters*, 46(4):211, 1981.

[55] AO Caldeira and AJ Leggett. Quantum tunnelling in a dissipative system. *Annals of Physics*, 149(2):374–456, 1983.

[56] S Camerer, M Korppi, A Jöckel, D Hunger, TW Hänsch, and P Treutlein. Realization of an optomechanical interface between ultracold atoms and a membrane. *Physical Review Letters*, 107(22):223001, 2011.

[57] S Carlip. Is quantum gravity necessary? *Classical and Quantum Gravity*, 25(15):154010, 2008.

[58] HJ Carmichael. Quantum trajectory theory for cascaded open systems. *Physical Review Letters*, 70(15):2273, 1993.

[59] HJ Carmichael, GJ Milburn, and DF Walls. Squeezing in a detuned parametric amplifier. *Journal of Physics A: Mathematical and General*, 17(2):469, 1984.

[60] T Carmon, MC Cross, and KJ Vahala. Chaotic quivering of micron-scaled on-chip resonators excited by centrifugal optical pressure. *Physical Review Letters*, 98(16):167203, 2007.

[61] T Carmon, H Rokhsari, L Yang, TJ Kippenberg, and KJ Vahala. Temporal behavior of radiation-pressure-induced vibrations of an optical microcavity phonon mode. *Physical Review Letters*, 94(22):223902, 2005.

[62] SM Carr, WE Lawrence, and MN Wybourne. Accessibility of quantum effects in mesomechanical systems. *Physical Review B*, 64(22):220101, 2001.

[63] CM Caves. Defense of the standard quantum limit for free-mass position. *Physical Review Letters*, 54(23):2465, 1985.

[64] CM Caves and GJ Milburn. Quantum-mechanical model for continuous position measurements. *Physical Review A*, 36(12):5543, 1987.

[65] CM Caves, KS Thorne, RWP Drever, VD Sandberg, and M Zimmermann. On the measurement of a weak classical force coupled to a quantum-mechanical oscillator. I. Issues of principle. *Reviews of Modern Physics*, 52(2):341, 1980.

[66] J Chan, TP Mayer Alegre, AH Safavi-Naeini, JT Hill, A Krause, S Gröblacher, M Aspelmeyer, and O Painter. Laser cooling of a nanomechanical oscillator into its quantum ground state. *Nature*, 478(7367):89–92, 2011.

[67] DE Chang, AH Safavi-Naeini, M Hafezi, and O Painter. Slowing and stopping light using an optomechanical crystal array. *New Journal of Physics*, 13(2):023003, 2011.

[68] C Chatfield. *The Analysis of Time Series: An Introduction*. CRC Press, 2013.

[69] S Chelkowski, H Vahlbruch, B Hage, A Franzen, N Lastzka, K Danzmann, and R Schnabel. Experimental characterization of frequency-dependent squeezed light. *Physical Review A*, 71(1):013806, 2005.

[70] Y Chen, SL Danilishin, FY Khalili, and H Müller-Ebhardt. QND measurements for future gravitational-wave detectors. *General Relativity and Gravitation*, 43(2):671–694, 2011.

[71] JI Cirac, AS Parkins, R Blatt, and P Zoller. "Dark" squeezed states of the motion of a trapped ion. *Physical Review Letters*, 70(5):556, 1993.

[72] J Clarke and FK Wilhelm. Superconducting quantum bits. *Nature*, 453(7198):1031–1042, 2008.

[73] AA Clerk, MH Devoret, SM Girvin, Florian Marquardt, and RJ Schoelkopf. Introduction to quantum noise, measurement, and amplification. *Reviews of Modern Physics*, 82(2):1155, 2010.

[74] AA Clerk, Florian Marquardt, and K Jacobs. Back-action evasion and squeezing of a mechanical resonator using a cavity detector. *New Journal of Physics*, 10(9):095010, 2008.

[75] P-F Cohadon, A Heidmann, and M Pinard. Cooling of a mirror by radiation pressure. *Physical Review Letters*, 83(16):3174, 1999.

[76] JD Cohen, SM Meenehan, GS MacCabe, S Gröblacher, AH Safavi-Naeini, F Marsili, MD Shaw, and O Painter. Phonon counting and intensity interferometry of a nanomechanical resonator. *Nature*, 520(7548):522–525, 2015.

[77] R Colella, AW Overhauser, and SA Werner. Observation of gravitationally induced quantum interference. *Physical Review Letters*, 34(23):1472, 1975.

[78] LIGO Scientific Collaboration et al. A gravitational wave observatory operating beyond the quantum shot-noise limit. *Nature Physics*, 7(12):962–965, 2011.

[79] MJ Collett and DF Walls. Squeezing spectra for nonlinear optical systems. *Physical Review A*, 32(5):2887, 1985.

[80] T Corbitt, C Wipf, T Bodiya, D Ottaway, D Sigg, N Smith, S Whitcomb, and N Mavalvala. Optical dilution and feedback cooling of a gram-scale oscillator to 6.9 mK. *Physical Review Letters*, 99(16):160801, 2007.

[81] J-M Courty, A Heidmann, and M Pinard. Quantum limits of cold damping with optomechanical coupling. *The European Physical Journal D– Atomic, Molecular, Optical and Plasma Physics*, 17(3):399–408, 2001.

[82] P Cudmore and CA Holmes. Phase and amplitude dynamics of nonlinearly coupled oscillators. *Chaos: An Interdisciplinary Journal of Nonlinear Science*, 25(2):023110, 2015.

[83] BD Cuthbertson, ME Tobar, EN Ivanov, and DG Blair. Parametric back-action effects in a high-Q cyrogenic sapphire transducer. *Review of Scientific Instruments*, 67(7):2435–2442, 1996.

[84] LA De Lorenzo and KC Schwab. Superfluid optomechanics: coupling of a superfluid to a superconducting condensate. *New Journal of Physics*, 16(11):113020, 2014.

[85] CM DeWitt and D Rickles. *The role of gravitation in physics: report from the 1957 Chapel Hill Conference*, volume 5. epubli, Berlin, 2011. See comments by Feynman in Chapter 23.

[86] F Diedrich, JC Bergquist, WM Itano, and DJ Wineland. Laser cooling to the zero-point energy of motion. *Physical Review Letters*, 62(4):403, 1989.

[87] L Diosi. Models for universal reduction of macroscopic quantum fluctuations. *Physical Review A*, 40(3):1165, 1989.

[88] L Diósi. The gravity-related decoherence master equation from hybrid dynamics. *Journal of Physics: Conference Series*, 306(1):012006, 2011.

[89] JM Dobrindt and TJ Kippenberg. Theoretical analysis of mechanical displacement measurement using a multiple cavity mode transducer. *Physical Review Letters*, 104(3):033901, 2010.

[90] AC Doherty, S Habib, K Jacobs, H Mabuchi, and SM Tan. Quantum feedback control and classical control theory. *Physical Review A*, 62(1):012105, 2000.

[91] AC Doherty and K Jacobs. Feedback control of quantum systems using continuous state estimation. *Physical Review A*, 60(4):2700, 1999.

[92] AC Doherty, A Szorkovszky, GI Harris, and WP Bowen. The quantum trajectory approach to quantum feedback control of an oscillator revisited. *Philosophical Transactions of the Royal Society A: Mathematical, Physical and Engineering Sciences*, 370(1979):5338–5353, 2012.

[93] A Dorsel, JD McCullen, P Meystre, E Vignes, and H Walther. Optical bistability and mirror confinement induced by radiation pressure. *Physical Review Letters*, 51(17):1550, 1983.

[94] A Dorsel, JD McCullen, P Meystre, E Vignes, and H Walther. Optical bistability and mirror confinement induced by radiation pressure. *Physical Review Letters*, 51(17):1550, 1983.

[95] PD Drummond, K McNeil, and DF Walls. Quantum theory of optical bistability. I. Nonlinear polarisability model. *Optica Acta*, 28:211, 1981.

[96] PD Drummond and DF Walls. Quantum theory of optical bistability. I. Nonlinear polarisability model. *Journal of Physics A: Mathematical and General*, 13(2):725, 1980.

[97] L-M Duan, G Giedke, JI Cirac, and P Zoller. Inseparability criterion for continuous variable systems. *Physical Review Letters*, 84(12):2722, 2000.

[98] MI Dykman. Critical exponents in metastable decay via quantum activation. *Physical Review E*, 75(1):011101, 2007.

[99] M Eichenfield, J Chan, RM Camacho, KJ Vahala, and O Painter. Optomechanical crystals. *Nature*, 462(7269):78–82, 2009.

[100] A Einstein, B Podolsky, and N Rosen. Can quantum-mechanical description of physical reality be considered complete? *Physical Review*, 47(10):777, 1935.

[101] KL Ekinci and ML Roukes. Nanoelectromechanical systems. *Review of scientific instruments*, 76(6):061101, 2005.

[102] Florian Elste, SM Girvin, and AA Clerk. Quantum noise interference and backaction cooling in cavity nanomechanics. *Physical Review Letters*, 102(20):207209, 2009.

[103] C Fabre, M Pinard, S Bourzeix, A Heidmann, E Giacobino, and S Reynaud. Quantum-noise reduction using a cavity with a movable mirror. *Physical Review A*, 49(2):1337, 1994.

[104] M Fleischhauer, A Imamoglu, and JP Marangos. Electromagnetically induced transparency: Optics in coherent media. *Reviews of Modern Physics*, 77(2):633, 2005.

[105] GW Ford, JT Lewis, and RF OÕConnell. Quantum Langevin equation. *Physical Review A*, 37(11):4419, 1988.

[106] S Forstner, S Prams, J Knittel, ED van Ooijen, JD Swaim, GI Harris, A Szorkovszky, WP Bowen, and H Rubinsztein-Dunlop. Cavity optomechanical magnetometer. *Physical Review Letters*, 108(12):120801, 2012.

[107] S Forstner, E Sheridan, J Knittel, CL Humphreys, GA Brawley, H Rubinsztein-Dunlop, and WP Bowen. Ultrasensitive optomechanical magnetometry. *Advanced Materials*, 26(36):6348–6353, 2014.

[108] C Galland, N Sangouard, N Piro, N Gisin, and TJ Kippenberg. Heralded single-phonon preparation, storage, and readout in cavity optomechanics. *Physical Review Letters*, 112(14):143602, 2014.

[109] AA Gangat, TM Stace, and GJ Milburn. Phonon number quantum jumps in an optomechanical system. *New Journal of Physics*, 13(4):043024, 2011.

[110] C Gardiner. *Stochastic Methods: A Handbook for the Natural and Social Sciences Springer Series in Synergetics*. Springer–Verlag, Berlin Heidelberg, 2009.

[111] C Gardiner and P Zoller. *Quantum Noise: A Handbook of Markovian and Non-Markovian Quantum Stochastic Methods with Applications to Quantum Optics*. Springer–Verlag, Berlin Heidelberg, 2010.

[112] CW Gardiner. Driving a quantum system with the output field from another driven quantum system. *Physical Review Letters*, 70(15):2269, 1993.

[113] CW Gardiner and MJ Collett. Input and output in damped quantum systems: Quantum stochastic differential equations and the master equation. *Physical Review A*, 31(6):3761, 1985.

[114] CW Gardiner and A Eschmann. Master-equation theory of semiconductor lasers. *Physical Review A*, 51(6):4982, 1995.

[115] E Gavartin, P Verlot, and TJ Kippenberg. A hybrid on-chip optomechanical transducer for ultrasensitive force measurements. *Nature Nanotechnology*, 7(8):509–514, 2012.

[116] C Genes, A Mari, P Tombesi, and D Vitali. Robust entanglement of a micromechanical resonator with output optical fields. *Physical Review A*, 78(3):032316, 2008.

[117] S Ghaffari, SA Chandorkar, S Wang, EJ Ng, CH Ahn, V Hong, Y Yang, and TW Kenny. Quantum limit of quality factor in silicon micro and nano mechanical resonators. *Scientific Reports*, 3, 2013.

[118] A Ghesquiere. *Entanglement in a Bipartite Gaussian State*. PhD thesis, National University of Ireland, Maynooth, 2009.

[119] J Gieseler, B Deutsch, R Quidant, and L Novotny. Subkelvin parametric feedback cooling of a laser-trapped nanoparticle. *Physical Review Letters*, 109(10):103603, 2012.

[120] S Gigan, HR Böhm, M Paternostro, F Blaser, G Langer, JB Hertzberg, KC Schwab, D Bäuerle, M Aspelmeyer, and A Zeilinger. Self-cooling of a micromirror by radiation pressure. *Nature*, 444(7115):67–70, 2006.

[121] SM Girvin, MH Devoret, and RJ Schoelkopf. Circuit QED and engineering charge-based superconducting qubits. *Physica Scripta*, 2009(T137):014012, 2009.

[122] RJ Glauber. The quantum theory of optical coherence. *Physical Review*, 130(6):2529, 1963.

[123] P Glendinning. *Stability, Instability, and Chaos: an Introduction to the Theory of Nonlinear Differential Equations*. Cambridge University Press, Cambridge, 1994.

[124] GE Gorelick. *In: Studies in the History of General Relativity*. Boston, Birkhaueser, 1992.

[125] M Goryachev, EN Ivanov, F van Kann, S Galliou, and ME Tobar. Observation of the fundamental Nyquist noise limit in an ultra-high Q-factor cryogenic bulk acoustic wave cavity. *Applied Physics Letters*, 105(15):153505, 2014.

[126] J Gough. Optimal quantum feedback for canonical observables. *Quantum Stochastics and Information: Statistics, Filtering and Control*, pages 262–279, 2008.

[127] JE Gough, MR James, HI Nurdin, and J Combes. Quantum filtering for systems driven by fields in single-photon states or superposition of coherent states. *Physical Review A*, 86(4):043819, 2012.

[128] Ph Grangier, RE Slusher, B Yurke, and A LaPorta. Squeezed-light–enhanced polarization interferometer. *Physical Review Letters*, 59(19):2153, 1987.

[129] DJ Griffiths. *Introduction to Quantum Mechanics, second edition*. Pearson Education Limited, 2014.

[130] S Gröblacher, K Hammerer, MR Vanner, and M Aspelmeyer. Observation of strong coupling between a micromechanical resonator and an optical cavity field. *Nature*, 460(7256):724–727, 2009.

[131] S Gröblacher, JB Hertzberg, MR Vanner, GD Cole, S Gigan, KC Schwab, and M Aspelmeyer. Demonstration of an ultracold micro-optomechanical oscillator in a cryogenic cavity. *Nature Physics*, 5(7):485–488, 2009.

[132] TL Gustavson, P Bouyer, and MA Kasevich. Precision rotation measurements with an atom interferometer gyroscope. *Physical Review Letters*, 78(11):2046, 1997.

[133] S Haroche and J-M Raimond. *Exploring the Quantum: Atoms, Cavities, and Photons*. Oxford University Press, 2006.

[134] GI Harris, UL Andersen, J Knittel, and WP Bowen. Feedback-enhanced sensitivity in optomechanics: Surpassing the parametric instability barrier. *Physical Review A*, 85(6):061802, 2012.

[135] GI Harris, DL McAuslan, E Sheridan, Y Sachkou, C Baker, and WP Bowen. Laser cooling and control of excitations in superfluid helium. *arXiv preprint arXiv:1506.04542*, 2015.

[136] GI Harris, DL McAuslan, TM Stace, AC Doherty, and WP Bowen. Minimum requirements for feedback enhanced force sensing. *Physical Review Letters*, 111(10):103603, 2013.

[137] TT Heikkilä, F Massel, J Tuorila, R Khan, and MA Sillanpää. Enhancing optomechanical coupling via the Josephson effect. *Physical Review Letters*, 112(20):203603, 2014.

[138] G Heinrich, M Ludwig, J Qian, B Kubala, and F Marquardt. Collective dynamics in optomechanical arrays. *Physical Review Letters*, 107(4):043603, 2011.

[139] CA Holmes, CP Meaney, and GJ Milburn. Synchronization of many nanomechanical resonators coupled via a common cavity field. *Physical Review E*, 85(6):066203, 2012.

[140] CA Holmes and GJ Milburn. Parametric self pulsing in a quantum opto-mechanical system. *Fortschritte der Physik*, 57(11-12):1052–1063, 2009.

[141] CJ Isham. *Integrable Systems, Quantum Groups, and Quantum Field Theories*. Springer, 1992.

[142] K Jacobs and PL Knight. Linear quantum trajectories: Applications to continuous projection measurements. *Physical Review A*, 57(4):2301, 1998.

[143] K Jacobs and DA Steck. A straightforward introduction to continuous quantum measurement. *Contemporary Physics*, 47(5):279–303, 2006.

[144] K Jaehne, K Hammerer, and M Wallquist. Ground-state cooling of a nanomechanical resonator via a Cooper-pair box qubit. *New Journal of Physics*, 10(9):095019, 2008.

[145] PS Jessen, C Gerz, PD Lett, WD Phillips, SL Rolston, RJC Spreeuw, and CI Westbrook. Observation of quantized motion of Rb atoms in an optical field. *Physical Review Letters*, 69(1):49–52, 1992.

[146] A Jöckel, A Faber, T Kampschulte, M Korppi, MT Rakher, and P Treutlein. Sympathetic cooling of a membrane oscillator in a hybrid mechanical–atomic system. *Nature Nanotechnology*, 10:55–59, 2015.

[147] C Joshi, U Akram, and GJ Milburn. An all-optical feedback assisted steady state of an optomechanical array. *New Journal of Physics*, 16:023009, 2014.

[148] D Kafri, GJ Milburn, and JM Taylor. Bounds on quantum communication via Newtonian gravity. *New Journal of Physics*, 17(1):015006, 2015.

[149] D Kafri, JM Taylor, and GJ Milburn. A classical channel model for gravitational decoherence. *New Journal of Physics*, 16(6):065020, 2014.

[150] RB Karabalin, XL Feng, and ML Roukes. Parametric nanomechanical amplification at very high frequency. *Nano Letters*, 9(9):3116–3123, 2009.

[151] I Katz, A Retzker, R Straub, and R Lifshitz. Signatures for a classical to quantum transition of a driven nonlinear nanomechanical resonator. *Physical Review Letters*, 99(4):040404, 2007.

[152] G Kempf, A Mangano and RB Mann. Hilbert space representation of the minimal length uncertainty relation. *Physical Review D*, 52:1108, 1995.

[153] KV Kepesidis, SD Bennett, S Portolan, Mikhail D Lukin, and P Rabl. Phonon cooling and lasing with nitrogen-vacancy centers in diamond. *Physical Review B*, 88(6):064105, 2013.

[154] W Ketterle. Nobel lecture: When atoms behave as waves: Bose-Einstein condensation and the atom laser. *Reviews of Modern Physics*, 74(4):1131–1151, 2002.

[155] FY Khalili. Frequency-dependent rigidity in large-scale interferometric gravitational-wave detectors. *Physics Letters A*, 288(5):251–256, 2001.

[156] KE Khosla, MR Vanner, WP Bowen, and GJ Milburn. Quantum state preparation of a mechanical resonator using an optomechanical geometric phase. *New Journal of Physics*, 15(4):043025, 2013.

[157] HJ Kimble, Y Levin, AB Matsko, KS Thorne, and SP Vyatchanin. Conversion of conventional gravitational-wave interferometers into quantum nondemolition interferometers by modifying their input and/or output optics. *Physical Review D*, 65(2):022002, 2001.

[158] P Kinsler and PD Drummond. Quantum dynamics of the parametric oscillator. *Physical Review A*, 43(11):6194, 1991.

[159] TJ Kippenberg, H Rokhsari, T Carmon, A Scherer, and KJ Vahala. Analysis of radiation-pressure induced mechanical oscillation of an optical microcavity. *Physical Review Letters*, 95(3):033901, 2005.

[160] TJ Kippenberg and KJ Vahala. Cavity opto-mechanics. *Optics Express*, 15(25):17172–17205, 2007.

[161] D Kleckner and D Bouwmeester. Sub-kelvin optical cooling of a micromechanical resonator. *Nature*, 444(7115):75–78, 2006.

[162] D Kleckner, B Pepper, E Jeffrey, P Sonin, SM Thon, and D Bouwmeester. Optomechanical trampoline resonators. *Optics Express*, 19(20):19708–19716, 2011.

[163] I Kozinsky, HWC Postma, O Kogan, A Husain, and ML Roukes. Basins of attraction of a nonlinear nanomechanical resonator. *Physical Review Letters*, 99(20):207201, 2007.

[164] AG Krause, M Winger, TD Blasius, Q Lin, and O Painter. A high-resolution microchip optomechanical accelerometer. *Nature Photonics*, 6(11):768–772, 2012.

[165] A Kronwald, F Marquardt, and AA Clerk. Arbitrarily large steady-state bosonic squeezing via dissipation. *Physical Review A*, 88(6):063833, 2013.

[166] R Kubo. The fluctuation-dissipation theorem. *Reports on Progress in Physics*, 29(1):255, 1966.

[167] Y Kuramoto. In H Araki, editor, *International Symposium on Mathematical Problems in Theoretical Physics*, volume 39 of *Lecture Notes in Physics*, page 420. Springer-Verlag, 1975.

[168] A Kuzmich, WP Bowen, AD Boozer, A Boca, CW Chou, L-M Duan, and HJ Kimble. Generation of nonclassical photon pairs for scalable quantum communication with atomic ensembles. *Nature*, 423(6941):731–734, 2003.

[169] MD LaHaye, J Suh, PM Echternach, KC Schwab, and ML Roukes. Nanomechanical measurements of a superconducting qubit. *Nature*, 459(7249):960–964, 2009.

[170] P Langevin. Sur la théorie du mouvement Brownien [on the theory of Brownian motion]. *Comptes Rendus de l'Académie des Sciences (Paris)*, 146:530–533, 1908.

[171] F Lecocq, JB Clark, RW Simmonds, J Aumentado, and JD Teufel. Quantum nondemolition measurement of a nonclassical state of a massive object. *arXiv preprint arXiv:1509.01629*, 2015.

[172] F Lecocq, JD Teufel, J Aumentado, and RW Simmonds. Resolving the vacuum fluctuations of an optomechanical system using an artificial atom. *Nature Physics*, 2015.

[173] KC Lee, MR Sprague, BJ Sussman, J Nunn, NK Langford, X-M Jin, T Champion, P Michelberger, KF Reim, D England, et al. Entangling macroscopic diamonds at room temperature. *Science*, 334(6060):1253–1256, 2011.

[174] KH Lee, TG McRae, GI Harris, J Knittel, and WP Bowen. Cooling and control of a cavity optoelectromechanical system. *Physical Review Letters*, 104(12):123604, 2010.

[175] TE Lee and MC Cross. Quantum synchronization of two coupled cavities with second harmonic generation. *Physical Review A*, 88:013834, 2012.

[176] D Leibfried, R Blatt, C Monroe, and D Wineland. Quantum dynamics of single trapped ions. *Reviews of Modern Physics*, 75(1):281, 2003.

[177] PD Lett, RN Watts, CI Westbrook, WD Phillips, PL Gould, and HJ Metcalf. Observation of atoms laser cooled below the doppler limit. *Physical Review Letters*, 61(2):169, 1988.

[178] T Li, S Kheifets, and MG Raizen. Millikelvin cooling of an optically trapped microsphere in vacuum. *Nature Physics*, 7(7):527–530, 2011.

[179] R Lifshitz and MC Cross. In H. G. Schuster, editor, *Review of Nonlinear Dynamics and Complexity*, volume 1, chapter 1. Weinheim: Wiley-VCH Verlag GmbH & Co. KGaA, 2008.

[180] Q Lin, J Rosenberg, X Jiang, KJ Vahala, and O Painter. Mechanical oscillation and cooling actuated by the optical gradient force. *Physical Review Letters*, 103(10):103601, 2009.

[181] DC Lindberg. The genesis of Kepler's theory of light: Light metaphysics from Plotinus to Kepler. *Osiris*, 2:5–42, 1986.

[182] N Lörch and K Hammerer. Sub-Poissonian phonon lasing in three-mode optomechanics. *Physical Review A*, 91:061803, Jun 2015.

[183] N Lörch, J Qian, AA Clerk, F Marquardt, and K Hammerer. Laser theory for optomechanics: Limit cycles in the quantum regime. *Physical Review X*, 4(1):011015, 2014.

[184] M Ludwig, AH Safavi-Naeini, O Painter, and F Marquardt. Enhanced quantum nonlinearities in a two-mode optomechanical system. *Physical Review Letters*, 109(6):063601, 2012.

[185] Y Ma, SL Danilishin, C Zhao, H Miao, WZ Korth, Y Chen, RL Ward, and DG Blair. Narrowing the filter-cavity bandwidth in gravitational-wave detectors via optomechanical interaction. *Physical Review Letters*, 113(15):151102, 2014.

[186] S Mahajan, N Aggarwal, AB Bhattacherjee, et al. Achieving the quantum ground state of a mechanical oscillator using a Bose–Einstein condensate with back-action and cold damping feedback schemes. *Journal of Physics B: Atomic, Molecular and Optical Physics*, 46(8):085301, 2013.

[187] GD Mahan. *Many-Particle Physics*. Springer Science & Business Media, 2000.

[188] S Mancini, VI Man'ko, and P Tombesi. Ponderomotive control of quantum macroscopic coherence. *Physical Review A*, 55(4):3042, 1997.

[189] S Mancini and P Tombesi. Quantum noise reduction by radiation pressure. *Physical Review A*, 49(5):4055, 1994.

[190] S Mancini, D Vitali, and P Tombesi. Optomechanical cooling of a macroscopic oscillator by homodyne feedback. *Physical Review Letters*, 80(4):688, 1998.

[191] F Marquardt, JP Chen, AA Clerk, and SM Girvin. Quantum theory of cavity-assisted sideband cooling of mechanical motion. *Physical Review Letters*, 99(9):093902, 2007.

[192] F Marquardt, JGE Harris, and SM Girvin. Dynamical multistability induced by radiation pressure in high-finesse micromechanical optical cavities. *Physical Review Letters*, 96(10):103901, 2006.

[193] W Marshall, C Simon, R Penrose, and D Bouwmeester. Towards quantum superpositions of a mirror. *Physical Review Letters*, 91(13):130401, 2003.

[194] JM Martinis, MH Devoret, and J Clarke. Experimental tests for the quantum behavior of a macroscopic degree of freedom: The phase difference across a Josephson junction. *Physical Review B*, 35(10):4682, 1987.

[195] F Massel, TT Heikkilä, J-M Pirkkalainen, SU Cho, H Saloniemi, PJ Hakonen, and MA Sillanpää. Microwave amplification with nanomechanical resonators. *Nature*, 480(7377):351–354, 2011.

[196] SA McGee, D Meiser, CA Regal, KW Lehnert, and MJ Holland. Mechanical resonators for storage and transfer of electrical and optical quantum states. *Physical Review A*, 87(5):053818, 2013.

[197] TG McRae and WP Bowen. Near threshold all-optical backaction amplifier. *Applied Physics Letters*, 100(20):201101, 2012.

[198] C Metzger, M Ludwig, C Neuenhahn, A Ortlieb, I Favero, K Karrai, and F Marquardt. Self-induced oscillations in an optomechanical system driven by bolometric backaction. *Physical Review Letters*, 101(13):133903, 2008.

[199] CH Metzger and K Karrai. Cavity cooling of a microlever. *Nature*, 432(7020):1002–1005, 2004.

[200] GJ Milburn, S Schneider, and DFV James. Ion trap quantum computing with warm ions. *Fortschritte der Physik*, 48:801, 2000.

[201] GJ Milburn and MJ Woolley. An introduction to quantum optomechanics. *Acta Physica Slovaca*, 61(5):483–601, 2011.

[202] Ch Monroe, DM Meekhof, BE King, SR Jefferts, WM Itano, DJ Wineland, and P Gould. Resolved-sideband Raman cooling of a bound atom to the 3D zero-point energy. *Physical Review Letters*, 75(22):4011, 1995.

[203] Ch Monroe, DM Meekhof, BE King, and DJ Wineland. A "Schrödinger cat" superposition state of an atom. *Science*, 272(5265):1131–1136, 1996.

[204] CM Mow-Lowry, AJ Mullavey, S Goßler, MB Gray, and DE McClelland. Cooling of a gram-scale cantilever flexure to 70 mK with a servo-modified optical spring. *Physical Review Letters*, 100(1):010801, 2008.

[205] F Mueller, S Heugel, and LJ Wang. Observation of optomechanical multistability in a high-Q torsion balance oscillator. *Physical Review A*, 77(3):031802, 2008.

[206] WJ Munro and CW Gardiner. Non-rotating-wave master equation. *Physical Review A*, 53(4):2633, 1996.

[207] KW Murch, KL Moore, S Gupta, and DM Stamper-Kurn. Observation of quantum-measurement backaction with an ultracold atomic gas. *Nature Physics*, 4(7):561–564, 2008.

[208] A Naik, O Buu, MD LaHaye, AD Armour, AA Clerk, MP Blencowe, and KC Schwab. Cooling a nanomechanical resonator with quantum back-action. *Nature*, 443(7108):193–196, 2006.

[209] AH Nayfeh and DT Mook. *Nonlinear Oscillations*. Wiley, New York, 1979.

[210] M Nielsen and I Chuang. *Quantum Computation and Quantum Information*. Cambridge University Press, Cambridge, 2000.

[211] PBR Nisbet-Jones, J Dilley, D Ljunggren, and A Kuhn. Highly efficient source for indistinguishable single photons of controlled shape. *New Journal of Physics*, 13(10):103036, 2011.

[212] J Nunn, K Reim, KC Lee, VO Lorenz, BJ Sussman, IA Walmsley, and D Jaksch. Multimode memories in atomic ensembles. *Physical Review Letters*, 101(26):260502, 2008.

[213] AD O'Connell, M Hofheinz, M Ansmann, RC Bialczak, M Lenander, E Lucero, M Neeley, D Sank, H Wang, M Weides, et al. Quantum ground state and single-phonon control of a mechanical resonator. *Nature*, 464(7289):697–703, 2010.

[214] M Orszag. *Quantum Optics: Including Noise Reduction, Trapped Ions, Quantum Trajectories, and Decoherence*. Springer–Verlag, Berlin Heidelberg, 2007.

[215] M Ozawa. Measurement breaking the standard quantum limit for free-mass position. *Physical Review Letters*, 60(5):385, 1988.

[216] TA Palomaki, JW Harlow, JD Teufel, RW Simmonds, and KW Lehnert. Coherent state transfer between itinerant microwave fields and a mechanical oscillator. *Nature*, 495(7440):210–214, 2013.

[217] TA Palomaki, JD Teufel, RW Simmonds, and KW Lehnert. Entangling mechanical motion with microwave fields. *Science*, 342(6159):710–713, 2013.

[218] Y-S Park and H Wang. Resolved-sideband and cryogenic cooling of an optomechanical resonator. *Nature Physics*, 5(7):489–493, 2009.

[219] AS Parkins and CW Gardiner. Effect of finite-bandwidth squeezing on inhibition of atomic-phase decays. *Physical Review A*, 37(10):3867, 1988.

[220] R Penrose. On gravity's role in quantum state reduction. *General Relativity and Gravitation*, 28(5):581–600, 1996.

[221] B Pepper, R Ghobadi, E Jeffrey, C Simon, and D Bouwmeester. Optomechanical superpositions via nested interferometry. *Physical Review Letters*, 109(2):023601, 2012.

[222] A Peres and DR Terno. Hybrid classical-quantum dynamics. *Physical Review A*, 63(2):022101, 2001.

[223] A Peters, KY Chung, and S Chu. Measurement of gravitational acceleration by dropping atoms. *Nature*, 400(6747):849–852, 1999.

[224] I Pikovski, MR Vanner, M Aspelmeyer, MS Kim, and C Brukner. Probing Planck-scale physics with quantum optics. *Nature Physics*, 8(5):393–397, 2012.

[225] A Pikovsky, M Rosenblum, and J Kurths. *Synchronization: A Universal Concept in Nonlinear Sciences*, volume 12. Cambridge University Press, Cambridge, 2003.

[226] M Pinard, P-F Cohadon, T Briant, and A Heidmann. Full mechanical characterization of a cold damped mirror. *Physical Review A*, 63(1):013808, 2000.

[227] J-M Pirkkalainen, SU Cho, Francesco Massel, J Tuorila, TT Heikkilä, PJ Hakonen, and MA Sillanpää. Cavity optomechanics mediated by a quantum two-level system. *Nature Communications*, 6:6981, 2015.

[228] J-M Pirkkalainen, E Damskägg, M Brandt, F Massel, and MA Sillanpää. Squeezing of quantum noise of motion in a micromechanical resonator. *arXiv preprint arXiv:1507.04209*, 2015.

[229] MB Plenio and S Virmani. An introduction to entanglement measures. *Quantum Information and Computation*, 7(1):001–051, 2007.

[230] M Poggio, CL Degen, HJ Mamin, and D Rugar. Feedback cooling of a cantilever's fundamental mode below 5 mK. *Physical Review Letters*, 99(1):017201, 2007.

[231] TP Purdy, RW Peterson, and CA Regal. Observation of radiation pressure shot noise on a macroscopic object. *Science*, 339(6121):801–804, 2013.

[232] TP Purdy, P-L Yu, RW Peterson, NS Kampel, and CA Regal. Strong optomechanical squeezing of light. *Physical Review X*, 3(3):031012, 2013.

[233] J Qin, C Zhao, Y Ma, X Chen, L Ju, and DG Blair. Classical demonstration of frequency-dependent noise ellipse rotation using optomechanically induced transparency. *Physical Review A*, 89(4):041802, 2014.

[234] P Rabl. Photon blockade effect in optomechanical systems. *Physical Review Letters*, 107(6):063601, 2011.

[235] P Rabl, P Cappellaro, MVG Dutt, L Jiang, JR Maze, and MD Lukin. Strong magnetic coupling between an electronic spin qubit and a mechanical resonator. *Physical Review B*, 79:041302(R), 2009.

[236] CA Regal and KW Lehnert. From cavity electromechanics to cavity optomechanics. *Journal of Physics: Conference Series*, 264(1):012025, 2011.

[237] MD Reid, PD Drummond, WP Bowen, Eric Gama Cavalcanti, PK Lam, HA Bachor, UL Andersen, and G Leuchs. Colloquium: the Einstein-Podolsky-Rosen paradox: from concepts to applications. *Reviews of Modern Physics*, 81(4):1727, 2009.

[238] J Restrepo, J Gabelli, C Ciuti, and I Favero. Classical and quantum theory of photothermal cavity cooling of a mechanical oscillator. *Comptes Rendus Physique*, 12(9):860–870, 2011.

[239] S Rips, I Wilson-Rae, and MJ Hartmann. Nonlinear nanomechanical resonators for quantum optoelectromechanics. *Physical Review A*, 89(1):013854, 2014.

[240] H Risken. Distribution- and correlation-functions for a laser amplitude. *Zeitschrift für Physik*, 186(1):85–98, 1965.

[241] PP Rohde and TC Ralph. Frequency and temporal effects in linear optical quantum computing. *Physical Review A*, 71(3):032320, 2005.

[242] D Rugar, R Budakian, HJ Mamin, and BW Chui. Single spin detection by magnetic resonance force microscopy. *Nature*, 430(6997):329–332, 2004.

[243] D Rugar and P Grütter. Mechanical parametric amplification and thermomechanical noise squeezing. *Physical Review Letters*, 67(6):699, 1991.

[244] AH Safavi-Naeini, J Chan, JT Hill, TPM Alegre, A Krause, and O Painter. Observation of quantum motion of a nanomechanical resonator. *Physical Review Letters*, 108(3):033602, 2012.

[245] AH Safavi-Naeini, S Gröblacher, JT Hill, J Chan, M Aspelmeyer, and O Painter. Squeezed light from a silicon micromechanical resonator. *Nature*, 500(7461):185–189, 2013.

[246] H Safavi-Naeini, TPM Alegre, J Chan, M Eichenfield, M Winger, Q Lin, JT Hill, DE Chang, and O Painter. Electromagnetically induced transparency and slow light with optomechanics. *Nature*, 472(7341):69–73, 2011.

[247] G Sallen, A Tribu, T Aichele, R André, L Besombes, C Bougerol, S Tatarenko, K Kheng, and JP Poizat. Exciton dynamics of a single quantum dot embedded in a nanowire. *Physical Review B*, 80(8):085310, 2009.

[248] JC Sankey, C Yang, BM Zwickl, AM Jayich, and JGE Harris. Strong and tunable nonlinear optomechanical coupling in a low-loss system. *Nature Physics*, 6(9):707–712, 2010.

[249] DH Santamore, AC Doherty, and MC Cross. Quantum nondemolition measurement of Fock states of mesoscopic mechanical oscillators. *Physical Review B*, 70(14):144301, 2004.

[250] A Schliesser, P Del'Haye, N Nooshi, KJ Vahala, and TJ Kippenberg. Radiation pressure cooling of a micromechanical oscillator using dynamical backaction. *Physical Review Letters*, 97(24):243905, 2006.

[251] A Schliesser, R Riviere, G Anetsberger, O Arcizet, and TJ Kippenberg. Resolved-sideband cooling of a micromechanical oscillator. *Nature Physics*, 4(5):415–419, 2008.

[252] M Schmidt, V Peano, and F Marquardt. Optomechanical Dirac physics. *New Journal of Physics*, 17(2):023025, 2015.

[253] S Schreppler, N Spethmann, N Brahms, T Botter, M Barrios, and DM Stamper-Kurn. Optically measuring force near the standard quantum limit. *Science*, 344(6191):1486–1489, 2014.

[254] J Schwinger. Brownian motion of a quantum oscillator. *Journal of Mathematical Physics*, 2(3):407–432, 1961.

[255] AJ Scott and GJ Milburn. Quantum nonlinear dynamics of continuously measured systems. *Physical Review A*, 63(4):042101, 2001.

[256] A Serafini, F Illuminati, and S De Siena. Symplectic invariants, entropic measures and correlations of Gaussian states. *Journal of Physics B: Atomic, Molecular and Optical Physics*, 37(2):L21, 2004.

[257] SY Shah, M Zhang, R Rand, and M Lipson. Master-slave locking of optomechanical oscillators over a long distance. *Physical Review Letters*, 114(11):113602, 2015.

[258] JH Shapiro, P Kumar, BEA Saleh, MC Teich, G Saplakoglu, and S-T Ho. Theory of light detection in the presence of feedback. *Journal of the Optical Society of America B*, 4(10):1604–1620, 1987.

[259] BS Sheard, MB Gray, CM Mow-Lowry, DE McClelland, and SE Whitcomb. Observation and characterization of an optical spring. *Physical Review A*, 69(5):051801, 2004.

[260] RM Shelby, MD Levenson, SH Perlmutter, RG DeVoe, and DF Walls. Broad-band parametric deamplification of quantum noise in an optical fiber. *Physical Review Letters*, 57(6):691, 1986.

[261] R Simon. Peres-Horodecki separability criterion for continuous variable systems. *Physical Review Letters*, 84(12):2726, 2000.

[262] W Singer. Dynamic formation of functional networks by synchronization. *Neuron*, 69(2):191–193, 2011.

[263] RE Slusher, LW Hollberg, Bernard Yurke, JC Mertz, and JF Valley. Observation of squeezed states generated by four-wave mixing in an optical cavity. *Physical Review Letters*, 55(22):2409, 1985.

[264] MJ Snadden, JM McGuirk, P Bouyer, KG Haritos, and MA Kasevich. Measurement of the earth's gravity gradient with an atom interferometer-based gravity gradiometer. *Physical Review Letters*, 81(5):971, 1998.

[265] MD Srinivas and EB Davies. Photon counting probabilities in quantum optics. *Journal of Modern Optics*, 28(7):981–996, 1981.

[266] DM Stamper-Kurn. *Cavity Optomechanics*. Springer, 2014. See chapter on cavity optomechanics with cold atoms.

[267] K Stannigel, P Rabl, AS Sørensen, P Zoller, and MD Lukin. Optomechanical transducers for long-distance quantum communication. *Physical Review Letters*, 105(22):220501, 2010.

[268] SH Strogatz. *Nonlinear Dynamics and Chaos*. Addison Wesley, 1994.

[269] O Suchoi, K Shlomi, L Ella, and E Buks. Time-resolved phase-space tomography of an optomechanical cavity. *Physical Review X*, 4:011015, 2014.

[270] Junho Suh, Alan J Weinstein, CU Lei, EE Wollman, SK Steinke, Pierre Meystre, AA Clerk, and KC Schwab. Mechanically detecting and avoiding the quantum fluctuations of a microwave field. *Science*, 344(6189):1262–1265, 2014.

[271] J Sulkko, MA Sillanpää, P Häkkinen, L Lechner, M Helle, A Fefferman, J Parpia, and PJ Hakonen. Strong gate coupling of high-Q nanomechanical resonators. *Nano Letters*, 10(12):4884–4889, 2010.

[272] A Szorkovszky, GA Brawley, AC Doherty, and WP Bowen. Strong thermomechanical squeezing via weak measurement. *Physical Review Letters*, 110(18):184301, 2013.

[273] A Szorkovszky, AA Clerk, AC Doherty, and WP Bowen. Detuned mechanical parametric amplification as a quantum non-demolition measurement. *New Journal of Physics*, 16(4):043023, 2014.

[274] A Szorkovszky, AC Doherty, GI Harris, and WP Bowen. Mechanical squeezing via parametric amplification and weak measurement. *Physical Review Letters*, 107(21):213603, 2011.

[275] A Szorkovszky, AC Doherty, GI Harris, and WP Bowen. Position estimation of a parametrically driven optomechanical system. *New Journal of Physics*, 14(9):095026, 2012.

[276] JM Taylor, AS Sørensen, CM Marcus, and ES Polzik. Laser cooling and optical detection of excitations in a LC electrical circuit. *Physical Review Letters*, 107(27):273601, 2011.

[277] MA Taylor, J Janousek, V Daria, J Knittel, B Hage, H-A Bachor, and WP Bowen. Biological measurement beyond the quantum limit. *Nature Photonics*, 7(3):229–233, 2013.

[278] MA Taylor, J Janousek, V Daria, J Knittel, B Hage, H-A Bachor, and WP Bowen. Subdiffraction-limited quantum imaging within a living cell. *Physical Review X*, 4(1):011017, 2014.

[279] J Teissier, A Barfuss, P Appel, E Neu, and P Maletinsky. Strain coupling of a nitrogen-vacancy center spin to a diamond mechanical oscillator. *Physical Review Letters*, 113(2):020503, 2014.

[280] JD Teufel, T Donner, D Li, JW Harlow, MS Allman, K Cicak, AJ Sirois, JD Whittaker, KW Lehnert, and RW Simmonds. Sideband cooling of micromechanical motion to the quantum ground state. *Nature*, 475(7356):359–363, 2011.

[281] JD Teufel, JW Harlow, CA Regal, and KW Lehnert. Dynamical backaction of microwave fields on a nanomechanical oscillator. *Physical Review Letters*, 101(19):197203, 2008.

[282] JD Teufel, D Li, MS Allman, K Cicak, AJ Sirois, JD Whittaker, and RW Simmonds. Circuit cavity electromechanics in the strong-coupling regime. *Nature*, 471(7337):204–208, 2011.

[283] JD Thompson, BM Zwickl, AM Jayich, Florian Marquardt, SM Girvin, and JGE Harris. Strong dispersive coupling of a high-finesse cavity to a micromechanical membrane. *Nature*, 452(7183):72–75, 2008.

[284] KS Thorne, RWP Drever, CM Caves, M Zimmermann, and VD Sandberg. Quantum nondemolition measurements of harmonic oscillators. *Physical Review Letters*, 40(11):667, 1978.

[285] P Treutlein, D Hunger, S Camerer, TW Hänsch, and J Reichel. Bose-Einstein condensate coupled to a nanomechanical resonator on an atom chip. *Physical Review Letters*, 99(14):140403, 2007.

[286] M Tsang and CM Caves. Evading quantum mechanics: Engineering a classical subsystem within a quantum environment. *Physical Review X*, 2(3):031016, 2012.

[287] J Tucker and DF Walls. Quantum theory of the parametric frequency converter. *Annals of Physics*, 52:1, 1969.

[288] AB U'Ren, K Banaszek, and IA Walmsley. Photon engineering for quantum information processing. *Quantum Information & Computation*, 3(7):480–502, 2003.

[289] CH van der Wal, MD Eisaman, A André, RL Walsworth, DF Phillips, AS Zibrov, and MD Lukin. Atomic memory for correlated photon states. *Science*, 301(5630):196–200, 2003.

[290] NG Van Kampen. *Stochastic Processes in Physics and Chemistry*, volume 1. Elsevier, Amsterdam, 1992.

[291] MR Vanner, M Aspelmeyer, and MS Kim. Quantum state orthogonalization and a toolset for quantum optomechanical phonon control. *Physical Review Letters*, 110(1):010504, 2013.

[292] MR Vanner, I Pikovski, GD Cole, MS Kim, C Brukner, K Hammerer, GJ Milburn, and M Aspelmeyer. Pulsed quantum optomechanics. *Proceedings of the National Academy of Sciences*, 108(39):16182–16187, 2011.

[293] G Vasilakis, H Shen, K Jensen, M Balabas, D Salart, B Chen, and ES Polzik. Generation of a squeezed state of an oscillator by stroboscopic back-action-evading measurement. *Nature Physics*, 11(5):389–392, 05 2015.

[294] G Veneziano. A stringy nature needs just two constants. *Europhysics Letters*, 2:199, 1986.

[295] E Verhagen, S Deléglise, S Weis, A Schliesser, and TJ Kippenberg. Quantum-coherent coupling of a mechanical oscillator to an optical cavity mode. *Nature*, 482(7383):63–67, 2012.

[296] P Verlot, A Tavernarakis, T Briant, P-F Cohadon, and A Heidmann. Backaction amplification and quantum limits in optomechanical measurements. *Physical Review Letters*, 104(13):133602, 2010.

[297] G Vidal and RF Werner. Computable measure of entanglement. *Physical Review A*, 65(3):032314, 2002.

[298] R Vijay, MH Devoret, and I Siddiqi. Invited review article: The Josephson bifurcation amplifier. *Review of Scientific Instruments*, 80(11):111101, 2009.

[299] D Vitali, S Gigan, A Ferreira, HR Böhm, P Tombesi, A Guerreiro, V Vedral, A Zeilinger, and M Aspelmeyer. Optomechanical entanglement between a movable mirror and a cavity field. *Physical Review Letters*, 98(3):030405, 2007.

[300] D Vitali, S Mancini, L Ribichini, and P Tombesi. Macroscopic mechanical oscillators at the quantum limit through optomechanical cooling. *Journal of the Optical Society of America B*, 20(5):1054–1065, 2003.

[301] D Vitali, P Tombesi, MJ Woolley, AC Doherty, and GJ Milburn. Entangling a nanomechanical resonator and a superconducting microwave cavity. *Physical Review A*, 76(4):042336, 2007.

[302] K Vogel and H Risken. Quantum-tunneling rates and stationary solutions in dispersive optical bistability. *Physical Review A*, 38(5):2409, 1988.

[303] SP Vyatchanin and AB Matsko. Quantum limit on force measurements. *Soviet Journal of Experimental and Theoretical Physics*, 77:218–221, 1993.

[304] SP Vyatchanin and EA Zubova. Quantum variation measurement of a force. *Physics Letters A*, 201(4):269–274, 1995.

[305] DF Walls and GJ Milburn. *Quantum Optics, 2nd Edition*. Springer, 2008.

[306] D Wang, T Itoh, T Ikehara, and R Maeda. Doubling flexural frequency response using synchronised oscillation in a micromechanically coupled oscillator system. *Micro & Nano Letters*, 7(8):717–720, 2012.

[307] H Wang, Z Wang, J Zhang, KK Ozdemir, L Yang, and YX Liu. Phonon amplification in two coupled cavities containing one mechanical resonator. *Physical Review A*, 90:053814, 2014.

[308] AJ Weinstein, CU Lei, EE Wollman, J Suh, A Metelmann, AA Clerk, and KC Schwab. Observation and interpretation of motional sideband asymmetry in a quantum electromechanical device. *Physical Review X*, 4(4):041003, 2014.

[309] S Weis, R Riviere, A Deléglise, E Gavartin, O Arcizet, A Schliesser, and TJ Kippenberg. Optomechanically induced transparency. *Science*, 330(6010):1520–1523, 2010.

[310] U Weiss. Quantum dissipative systems, series in modern condensed matter physics, 1999.

[311] N Wiener. *Extrapolation, Interpolation, and Smoothing of Stationary Time Series*, volume 2. MIT Press, Cambridge, MA, 1949.

[312] DJ Wilson, V Sudhir, N Piro, R Schilling, A Ghadimi, and TJ Kippenberg. Measurement-based control of a mechanical oscillator at its thermal decoherence rate. *Nature*, 524:325–329, 2015.

[313] I Wilson-Rae. Intrinsic dissipation in nanomechanical resonators due to phonon tunneling. *Physical Review B*, 77(24):245418, 2008.

[314] I Wilson-Rae, N Nooshi, W Zwerger, and TJ Kippenberg. Theory of ground state cooling of a mechanical oscillator using dynamical backaction. *Physical Review Letters*, 99(9):093901, 2007.

[315] I Wilson-Rae, P Zoller, and A Imamoglu. Laser cooling of a nanomechanical resonator mode to its quantum ground state. *Physical Review Letters*, 92(7):075507, 2004.

[316] M Winger, TD Blasius, TP Mayer Alegre, AH Safavi-Naeini, S Meenehan, J Cohen, S Stobbe, and O Painter. A chip-scale integrated cavity-electro-optomechanics platform. *Optics Express*, 19(25):24905–24921, 2011.

[317] HM Wiseman and L Diósi. Complete parameterization, and invariance, of diffusive quantum trajectories for markovian open systems. *Chemical Physics*, 268(1):91–104, 2001.

[318] HM Wiseman and GJ Milburn. *Quantum Measurement and Control*. Cambridge University Press, Cambridge, 2009.

[319] E Witten. Reflections on the fate of spacetime. *Physics Today*, page 24, 1996.

[320] EE Wollman, CU Lei, AJ Weinstein, J Suh, A Kronwald, F Marquardt, AA Clerk, and KC Schwab. Quantum squeezing of motion in a mechanical resonator. *Science*, 349(6251):952–955, 2015.

[321] MJ Woolley and AA Clerk. Two-mode back-action-evading measurements in cavity optomechanics. *Physical Review A*, 87(6):063846, 2013.

[322] M Xiao, L-A Wu, and HJ Kimble. Precision measurement beyond the shot-noise limit. *Physical Review Letters*, 59(3):278, 1987.

[323] C Xiong, L Fan, X Sun, and HX Tang. Cavity piezooptomechanics: Piezoelectrically excited, optically transduced optomechanical resonators. *Applied Physics Letters*, 102(2):021110, 2013.

[324] X Xu, M Gullans, and JM Taylor. Quantum nonlinear optics near optomechanical instabilities. *Physical Review A*, 91(1):013818, 2015.

[325] A Xuereb, R Schnabel, and K Hammerer. Dissipative optomechanics in a Michelson-Sagnac interferometer. *Physical Review Letters*, 107(21):213604, 2011.

[326] I Yeo, P-L De Assis, A Gloppe, E Dupont-Ferrier, P Verlot, NS Malik, E Dupuy, J Claudon, J-M Gérard, A Auffeves, et al. Strain-mediated coupling in a quantum dot-mechanical oscillator hybrid system. *Nature Nanotechnology*, 9(2):106–110, 2014.

[327] HP Yuen. Contractive states and the standard quantum limit for monitoring free-mass positions. *Physical Review Letters*, 51(9):719, 1983.

[328] HP Yuen and JH Shapiro. Generation and detection of two-photon coherent states in degenerate four-wave mixing. *Optics Letters*, 4(10):334–336, 1979.

[329] B Yurke and D Stoler. Generating quantum mechanical superpositions of macroscopically distinguishable states via amplitude dispersion. *Physical Review Letters*, 57(1):13, 1986.

[330] S Zaitsev, O Gottlieb, and E Buks. Nonlinear dynamics of a microelectromechanical mirror in an optical resonance cavity. *Nonlinear Dynamics*, 69(4):1589–1610, 2012.

[331] S Zaitsev, O Shtempluck, E Buks, and O Gottlieb. Nonlinear damping in a micromechanical oscillator. *Nonlinear Dynamics*, 67(1):859–883, 2012.

[332] M Zalalutdinov, KL Aubin, M Pandey, AT Zehnder, RH Rand, HG Craighead, JM Parpia, and BH Houston. Frequency entrainment for micromechanical oscillator. *Applied Physics Letters*, 83(16):3281–3283, 2003.

[333] J Zhang, K Peng, and SL Braunstein. Quantum-state transfer from light to macroscopic oscillators. *Physical Review A*, 68(1):013808, 2003.

[334] K Zhang, P Meystre, and W Zhang. Back-action-free quantum optomechanics with negative-mass Bose-Einstein condensates. *Physical Review A*, 88(4):043632, 2013.

[335] M Zhang, GS Wiederhecker, S Manipatruni, A Barnard, P McEuen, and M Lipson. Synchronization of micromechanical oscillators using light. *Physical Review Letters*, 109(23):233906, 2012.

[336] C Zhao, Q Fang, S Susmithan, H Miao, L Ju, Y Fan, D Blair, DJ Hosken, J Munch, PJ Veitch, et al. High-sensitivity three-mode optomechanical transducer. *Physical Review A*, 84(6):063836, 2011.

[337] M Zych, F Costa, I Pikovski, and Č Brukner. Quantum interferometric visibility as a witness of general relativistic proper time. *Nature Communications*, 2:505, 2011.

Index

back-action evading measurement, 166
  general, 167
  parametric, 176
  two-tone driving, 169
bath commutation and correlation relations, 29
bistability, 238
  optical, 234
  optomechanical, 50

cascaded systems, 200, 220
Casimir invariant, 213
cat state, 227
cavity detuning, $\Delta \equiv \Omega_c - \Omega_L$, 43
centre manifold, 244
circuit QED, 256
classical channel, 304
conditional number distribution, 209
conditional state, 208, 209
conditional variance
  back-action evading measurement, 169
  continuous measurement, 163
  parametric back-action evasion, 178
  two-tone back-action evasion, 173
Cooper pair box, 258
coupled cavity, 206
covariance matrix, 128

diffusion
  phase, 248
dimensionless position and momentum operators, 6
dispersive optomechanics, 40
dissipative optomechanics, 55
Duffing, 231, 240

quantum, 233
dynamical back-action, 62, 97

effective quantisation, 39

fluctuation dissipation theorem
  classical, 9
  quantum, 15
Fock state master equation, 201, 217
Fokker–Planck, 235, 237, 290
force sensing
  bandwidth, 88
  classical sensitivity limit, 91
  quantum sensitivity limit, 88

geometric phase, 316
gravitational decoherence, 219, 300, 311
gravitational heating, 306

Heisenberg equation of motion, 8
HOM interferometer, 202, 216
HOM visibility, 217
Hong–Ou–Mandle, 202
Hopf bifurcation, 241, 242, 244, 247

input-output relations, 31

Jaynes–Cummings, 207, 262, 269

Kerr nonlinearity, 140, 234, 316
Krauss operator, 208

Langevin equations
  quantum, 18, 215
  quantum Markovian, 23, 47
  quantum oscillator, 22
  quantum, rotating wave approximation, 28
  single-photon, 197, 202

limit cycle, 241, 242
linearisation approximation, 53, 115,
    170, 239
logarithmic negativity, 129

magnetic force microscopy, 255
membrane-in-the-middle, 243

nitrogen vacancy centre, 255
non-Markovian, 241
nonlinear damping, 240, 242

optomechanical cooling
    coupling to two baths, 103
    criteria to approach ground
        state, 112
    effective temperature of the
        optical bath, 97, 102
    feedback, 156
    resolved sideband regime, 105
    thermodynamical perspective,
        112
    via a two-level system, 261
optomechanical cooperativity, 62
    characterisation, 350
    effective, 61
    effective, at standard quantum
        limit, 77
optomechanical coupling rate
    (linearised), g, 54
optomechanical coupling strength, G,
    41
optomechanical entanglement, 122
    Einstein–Podolsky–Rosen, 127
    quantifying, 128
optomechanical Hamiltonian
    array, 288, 295
    beam splitter, 105
    coupling to superconducting
        circuit, 258
    coupling to superconducting
        junction, 260
    coupling to two-level system, 261
    Duffing, 232
    feedback, 164

in a rotating frame for the light,
    43
Jaynes–Cummings, 213, 262
linearised, 51, 54
mechanically mediated optical
    beam splitter, 206
microwave to optical convertor,
    271
nonlinear damping, 241
nonlinearised, 42
optomechanically induced
    transparency, 116
parametric, 124
quadratic, 240, 250
strong coupling, 95
two-tone back-action evasion,
    171
optomechanical measurement rate,
    μ, 68, 114
optomechanically induced
    absorption, 116
optomechanically induced
    transparency, 114

parametric amplification, 240
    detuned, 173
parametric instability, 126, 242
phase modulation, 204
phasor diagrams, 6, 56, 140, 183, 190
photodetection
    heterodyne, 70, 349
    homodyne, 70, 251, 348
    inefficiencies, 345
    linear, 68
    mechanical quadratures, 157
    photon counting, 197
    stochastic, 153
photon blockade, 194, 227, 230
photon counting, 197
polaron transformation, 142, 194,
    221, 231, 234
ponderomotive squeezing, 140
positive P-function, 6, 235, 237, 241,
    246
post-selection, 223

power spectral density
  bath, 24
  bath, rotating wave
    approximation, 30
  classical, 9
  classical force sensing, 91
  heterodyne detection, 73, 349
  homodyne detection, 72
  Hopf bifurcation, 247, 291
  imprecision noise, 67
  mechanical, from ponderomotive
    squeezing, 191
  mechanical, squeezed input
    light, 185
  photocurrent, 69
  ponderomotive squeezing, 146
  quantum, 12
  radiation pressure, 66, 100
  symmetrised, 16

quadratic optomechanical coupling,
    243
quadrature operators, 34
quantum gravity, 300
quantum memory, 214
quantum nondemolition
    measurement, 167, 209, 250
quantum trajectories, 157
quantum tunnelling, 237, 238
qutrit, 213

radiation pressure interaction, 37
radiation pressure shot noise, 60, 64
Raman, 206, 243
red sideband, 244
rotating wave approximation, 28, 32,
    35, 105, 116, 124, 157, 206,
    243, 250, 262

second harmonic generation, 244
semiclassical dynamics, 48
sideband asymmetry, 14, 21, 24, 26,
    63, 73
single-photon state, 196
standard quantum limit
  force measurement, 88

free mass, 58
  gravitational wave
    interferometry, 86
  position measurement, 76
  strain sensing, 87
  surpassing for position
    measurement, 179
  surpassing using squeezed light,
    179
stochastic master equation
  continuous measurement of
    mechanical position, 159
  continuous optical homodyne
    detection, 251
strong coupling, 94, 109
susceptibility, 25, 32

thermometry of an oscillator, 15, 26,
    352, 353
transforming into a rotating frame,
    34
two-photon, 213

uncertainty principle
  dimensionless position and
    momentum operators, 6
  force-imprecision, 66
  Heisenberg, 5
  modified, 315
  power spectral densities, 30

vacuum optomechanical coupling
    rate, g_0, 42
van der Pol, 241

Wiener filter, 134
Wiener increment, 153
Wigner function, 6, 239